Advanced Oil Crop Biorefineries

RSC Green Chemistry

Series Editors:
James H. Clark, *Department of Chemistry, University of York, York, UK*
George A. Kraus, *Department of Chemistry, Iowa State University, Iowa, USA*
Andrzej Stankiewicz, *Delft University of Technology, Delft, The Netherlands*
Peter Siedl, *Federal University of Rio de Janeiro, Brazil*
Yuan Kou, *Peking University, People's Republic of China*

Titles in the Series:
1: The Future of Glycerol: New Uses of a Versatile Raw Material
2: Alternative Solvents for Green Chemistry
3: Eco-Friendly Synthesis of Fine Chemicals
4: Sustainable Solutions for Modern Economies
5: Chemical Reactions and Processes under Flow Conditions
6: Radical Reactions in Aqueous Media
7: Aqueous Microwave Chemistry
8: The Future of Glycerol: 2nd Edition
9: Transportation Biofuels: Novel Pathways for the Production of Ethanol, Biogas and Biodiesel
10: Alternatives to Conventional Food Processing
11: Green Trends in Insect Control
12: A Handbook of Applied Biopolymer Technology: Synthesis, Degradation and Applications
13: Challenges in Green Analytical Chemistry
14: Advanced Oil Crop Biorefineries

How to obtain future titles on publication:
A standing order plan is available for this series. A standing order will bring delivery of each new volume immediately on publication.

For further information please contact:
Book Sales Department, Royal Society of Chemistry, Thomas Graham House, Science Park, Milton Road, Cambridge, CB4 0WF, UK
Telephone: +44 (0)1223 420066, Fax: +44 (0)1223 420247, Email: books@rsc.org
Visit our website at http://www.rsc.org/Shop/Books/

Advanced Oil Crop Biorefineries

Edited by

Abbas Kazmi
Green Chemistry Centre, University of York, UK

RSCPublishing

RSC Green Chemistry No. 14

ISBN: 978-1-84973-135-5
ISSN: 1757-7039

A catalogue record for this book is available from the British Library

© Royal Society of Chemistry 2012

All rights reserved

Apart from fair dealing for the purposes of research for non-commercial purposes or for private study, criticism or review, as permitted under the Copyright, Designs and Patents Act 1988 and the Copyright and Related Rights Regulations 2003, this publication may not be reproduced, stored or transmitted, in any form or by any means, without the prior permission in writing of The Royal Society of Chemistry or the copyright owner, or in the case of reproduction in accordance with the terms of licences issued by the Copyright Licensing Agency in the UK, or in accordance with the terms of the licences issued by the appropriate Reproduction Rights Organization outside the UK. Enquiries concerning reproduction outside the terms stated here should be sent to The Royal Society of Chemistry at the address printed on this page.

The RSC is not responsible for individual opinions expressed in this work.

Published by The Royal Society of Chemistry,
Thomas Graham House, Science Park, Milton Road,
Cambridge CB4 0WF, UK

Registered Charity Number 207890

For further information see our web site at www.rsc.org

Preface

The twenty first century is a period of human history where natural resources are stretched to the limit, unprecedented levels of carbon emissions and a rapidly rising population, predicted to reach nine billion by 2050. It is indeed the opinion of some that we have excavated a hole so deep that there is now no turning back from catastrophe, however I believe that it is in human nature to win against the odds, but this will only be possible if we focus our attention on creating a sustainable relationship with the Earth.

The Earth provides us with an ample source of renewable biomass which can be used to fulfil our basic requirements for chemicals and materials. There are many types of biomass such as wood, cereal, sugar and algae however in this book we have decided to focus on the by-products of oil crops in Europe. Rapeseed, sunflower and olive oils are commonly used for cooking and are found in a multitude of food products. Although primarily used for food, natural oils have been used for decades in the speciality chemicals industry. With simple chemical modification, the fatty acids can be used as lubricants or as thickening agents in cosmetic products. It is clear from this example that agricultural products can be used for food and non-food applications without any conflict. The emergence of high volume products such as biofuels has disturbed this delicate balance. The food versus fuel debate became front page news all over the world and acted as a reminder that that the food supply chain should not be interfered with.

To avoid this very issue we decided to focus on the by-products of industrial oil crop supply chains. The farmers are at the start of the supply chain and consumer products such as vegetable oil or biodiesel at the end. By using novel techniques to extract the maximum potential of the by-products, we can improve the economics of the supply chain. Many by-products of the oil crop supply chain have current uses, for example rapeseed straw is used for animal feed or left on the field to replenish the soil. Clearly a calculated amount needs to be left on the field however the remaining straw could be valorised. We know that the straw contains a number of antioxidants which could be used in cosmetic or food applications. The high concentration of protein in the straw could be extracted to produce a high value animal feed and the remaining straw could be used to generate energy or biomethane. After the extraction of secondary metabolites and proteins, the material still contains mainly cellulose which can be hydrolysed to sugars, the starting point for numerous high value chemicals. An alternative approach would be to use the protein in the straw as a natural adhesive to create various biodegradable materials such as bio-boards.

It is in the nature of scientists to create ideas and prove them at the laboratory scale, however it is important that vigorous economic costings and life cycle analysis of processes be conducted prior to proceeding to commercialisation. In this book we have done exactly this and have compared various processes which aim to valorise oil crop by-products by using modern modelling techniques. When developing new processes and products we must not forget the impact such changes would have on society, business and government policy. We have dedicated an entire chapter to cover these issues and important findings are discussed.

At this stage I must acknowledge the European Commission's 7th framework program funded project SUSTOIL. I personally thank all 23 partners, which were located in 10 European countries for all the hard work and effort that was put into the project and this book. I would also like to thank all the authors who have spent valuable time writing chapters for this book. I hope to work with you again in the future so that we can make an impact in this ever-changing world, full of mighty challenges.

Contents

Chapter 1 Introduction   1
Abbas Kazmi, Birgit Kamm, Sören Henke, Ludwig Theuvsen and Rainer Höfer

1.1	Green Chemistry and the Biorefinery		1
1.2	The Biorefinery Concept		4
	1.2.1	Introduction	4
	1.2.2	Principles of Biorefineries	6
	1.2.3	Building Blocks, Chemicals and Potential Screening	8
	1.2.4	Biorefinery Systems	11
	1.2.5	Two-platform Concept	19
	1.2.6	Advanced Oil Crops Biorefineries	20
1.3	The Potential of Oil Crops in Europe		23
	1.3.1	Introduction	23
	1.3.2	Vegetable Oil Production in the European Community (EU-27)	24
	1.3.3	Commodity Oils	26
	1.3.4	Speciality Vegetable Oils	36
	1.3.5	Crop Growing Potential for Non-European Oils	36
1.4	Summary and Conclusions		39
Acknowledgement			39
References			40

Chapter 2	**Farming and Harvesting**		**48**
	Katerina Stamatelatou, David Turley, Ruth Laybourn,		
	Francis Flénet, Alain Quinsac, Ray Marriott,		
	Georgia Antonopoulou, Gerasimos Lyberatos,		
	Antoine Rouilly and Carlos Vaca-Garcia		

        2.1   Introduction   48
        2.2   Increasing Oil Yield   49
             2.2.1   The Production of Oilseed Rape   49
             2.2.2   Sunflower   56
        2.3   Valorisation of Straw and Leaves through
             Green Technologies   67
             2.3.1   Chemicals from Supercritical CO_2 Extraction   68
             2.3.2   Biomethane   80
             2.3.3   Biomaterials from Thermocoupling   85
        References   97
        Website Sources   101

Chapter 3	**Primary Processing**		**102**
	Wim Mulder, Paulien Harmsen, Johan Sanders,		
	Patrick Carre, Birgit Kamm, Petra Schönicke and		
	Geertje Dautzenberg		

        3.1   Introduction   102
        3.2   Pre-treatment Processes   103
             3.2.1   Dehulling   103
             3.2.2   Thermal Pre-treatment   115
             3.2.3   Microwave and Radio Frequency   117
             3.2.4   Pulsed Electric Field   118
             3.2.5   Enzymatic Pre-treatment   118
        3.3   Novel Oil Recovery Processes and Valorisation
             of Waste Streams   119
             3.3.1   Introduction   119
             3.3.2   Oil Extraction from Olives   121
             3.3.3   Oil Extraction from Rapeseed   124
             3.3.4   Oil Extraction from Sunflower Seeds   125
             3.3.5   Pressing and Pressing-related Processes   127
             3.3.6   Solvent Extraction   130
             3.3.7   Residual Oil Recovery   132
             3.3.8   Conclusions   132
        3.4   Protein and Amino Acid Isolation   133
             3.4.1   Protein Hydrolysis   134
             3.4.2   Extraction Process of Peptides and Amino
                     Acids   135
             3.4.3   Conclusions   141

Contents

	3.5	Production of Levulinic Acid from Straw	141
		3.5.1 Introduction	141
		3.5.2 A Short Survey on the Development of Levulinic Acid Chemistry	142
		3.5.3 Levulinic Acid Production	143
		3.5.4 Levulinic Acid from Hexoses *via* Formation of Fructose and 5-BHF	144
		3.5.5 The Bofine Demonstration Plants and Outlook on Future Industrial Scale Facilities	148
		3.5.6 Technology Draft for a Low-temperature Conversion Process of LCF to Levulinic Acid	149
		3.5.7 Outlook	150
	3.6	Integrated Biorefinery	151
		3.6.1 Dehulling	151
		3.6.2 Cold Pressing	154
		3.6.3 Improvement of Meal Quality by Significant Reduction of Hexane Retention in Marcs	154
		3.6.4 Supercritical CO_2 Extraction	155
		3.6.5 Gas-assisted Oil Pressing	156
		3.6.6 Use of Alcohols as an Alternative for Hexane	156
		3.6.7 Simultaneous Extraction and Transesterification	157
		3.6.8 Isolation of Oil Bodies (Oleosomes) Table 3.19	158
		3.6.9 Water Extraction	159
		3.6.10 Anaerobic Digestion of Residues	159
		3.6.11 Recovery of Gums from Water Degumming	160
		3.6.12 Integrated Scheme Biorefinery	160
	References		161
Chapter 4	**Secondary Processing of Plant Oils**		**166**
	Zsanett Herseczki, Abbas Kazmi, Rafael Luque and Diego Luna		
	4.1	Applications of Glycerol	166
		4.1.1 Existing and Novel Glycerol Purification Technologies	167
		4.1.2 Transformation of Glycerol into High-quality Products through Green Chemistry and Biotechnology	174
	4.2	Novel Routes to Biodiesel Incorporating Glycerol into Their Composition	187
		4.2.1 Novel Biofuels Integrating Glycerol into Their Composition	189
		4.2.2 Processing of Oils and Fats in the Actual Oil Refining Plants	192

		4.2.3	Second-generation Technologies for the Production of Biodiesel-like Fuels	193
	References			197

Chapter 5 Assessment of Economic and Environmental Cost-benefits of Developed Biorefinery Schemes — 203
Michael Binns, Anestis Vlysidis and Constantinos Theodoropoulos

	5.1	Introduction		203
	5.2	Methodology		205
		5.2.1	Simulation Software	205
		5.2.2	Optimisation Methods	205
		5.2.3	Life Cycle Analysis	207
		5.2.4	Multi-objective Optimisation	208
		5.2.5	Biorefinery Schemes	209
	5.3	Results and Discussion		223
		5.3.1	Economic Optimisation	223
		5.3.2	Environmental and Multi-objective Optimisation	240
		5.3.3	Holistic Comparisons of Process Options	263
	5.4	Conclusions		276
	References			277

Chapter 6 Modelling Stakeholders' Interplay and Policy Scenarios for Biorefinery Implementation — 280
Piergiuseppe Morone, Caterina De Lucia, Antonio Lopolito and Maurizio Prosperi

	6.1	Introduction		280
	6.2	The Micro-economic Approach to Policy Modelling for Biorefineries		281
		6.2.1	The Theoretical Framework	281
		6.2.2	A Three-steps Methodology	285
		6.2.3	Wrapping-Up – an Application of the Proposed Protocol of Analysis	297
	6.3	The Macro-economic Approach: a CGE Model with the Inclusion of Biorefineries in the Production Process		299
		6.3.1	Application of CGE Models to Biofuels	299
		6.3.2	A Theoretical CGE Model for a Bio-based Economy	300
		6.3.3	Summing Up	307
	6.4	Conclusions		307
	References			308

Subject Index — 311

CHAPTER 1
Introduction

ABBAS KAZMI,[a] BIRGIT KAMM,[b] SÖREN HENKE,[c] LUDWIG THEUVSEN[d] AND RAINER HÖFER[e]

[a] Green Chemistry Centre of Excellence, Department of Chemistry, University of York, YO10 5DD, UK; [b] Research Institute Bioactive Polymer Systems e.V. and Brandenburg University of Technology Cottbus, Kantstrasse 55, D-14513 Teltow, Germany; [c] Universität Göttingen, Department für Agrarökonomie und Rurale Entwicklung, Betriebswirtschaftslehre des Agribusiness, Platz der Göttinger Sieben 5, D-37073 Göttingen; [d] Universität Göttingen, Department für Agrarökonomie und Rurale Entwicklung, Platz der Göttinger Sieben 5, D-37073 Göttingen; [e] Editorial Ecosiris, Klever Straße 31, D-40477 Düsseldorf

1.1 Green Chemistry and the Biorefinery

The principles of green chemistry are now having a real impact on industry and key players such as P&G are now providing greener alternatives that could have a global impact. One such example targets the alkyd resins which provide robust, high-gloss coatings at relatively low prices for a variety of applications including architectural finishes, industrial metal and construction equipment. However, these coatings require hazardous solvents to solubilise the organic polymers, which has led to novel greener resins developed by P&G in association with Cook Composites & Polymers, USA. The novel resins are produced from the esterification of sucrose with fatty acids, both of which are renewable resources that are readily available. Furthermore the process requires significantly less VOC content and therefore is much greener.

RSC Green Chemistry No. 14
Advanced Oil Crop Biorefineries
Edited by Abbas Kazmi
© Royal Society of Chemistry 2012
Published by the Royal Society of Chemistry, www.rsc.org

Biodiesel is now a well-established industry, although it has had turbulent times, and is based on renewable resources such as plant oils; however, the transesterification process could be made more green. An alternative method for manufacturing biodiesel, called the 'Mcgyan Process', has been developed by SarTec Corporation and is based on a fixed-bed, flow-through reactor. The fixed-bed zirconia catalyst, which is continuously used, results in no catalyst waste, unlike the conventional acid/base catalyst systems. The novel process not only improves efficiency but also has a positive impact on the economics to the extent that a large-scale 3 million gallon per year facility is to be constructed.

The future biorefinery needs to be based on existing supply chains and product streams. Biorefinery processes need to remove inefficiencies and wastes from existing processes. This is the only way biorefinery concepts will penetrate conventional markets. The plant oil industry is a great example of this as the oils are mainly used for human consumption (126 million tons); however, a considerable amount is used for chemical (15 million tons) and fuel applications (8 million tons). European vegetable oils are used in the oleochemical industry; however, the majority of oils are imported such as soya, palm and castor oil.

The oleochemical industry uses the key components of plant oils to produce chemicals for various applications such as cosmetics, paints, lubricants, biofuels, plastics, soaps and pharmaceuticals. Using fatty acids, glycerine or fatty acid methyl esters a number of important derivatives such as esters, sulfates, ethoxylates and other chemical functionalities can be produced.

In the surfactant market the crude oil derived alkyl benzene sulfonate has the largest market share; however, greener alternatives such as alcohol ether sulfates, alcohol sulfates and alcohol ethoxylates are significantly growing in the market. With pressure from governments and NGOs the paint industry is looking to reduce VOC emissions by using greener resins such as those derived from soya and sunflower oils. Long-chain fatty acids can also be used as biolubricants; however, the estimated volume of such products in the EU is 127 000 tons as of 2006, out of a 5 million ton lubricant market, mainly due to the high cost of biolubricants.

The polymer industry is also currently based on crude oil and with stricter regulations the industry is shifting towards greener alternatives. A host of polymers can be made from plant oils, for example alkyd resins can be made from condensation polymerisation of polyols, organic acids and fatty acids. Furthermore, smaller building blocks based on plant oils can be used in conventional polymers to improve properties such as elasticity, flexibility, strength and hydrophobicity. For example, oleic acid can be used as a building block to produce important products such as linoleum, polyamides, polyurethanes, polyamido amines and non-nylon polyamides. However, with many of these products the properties and pricing are inferior to crude-oil derived polymers, therefore further research is required.

Additional opportunities exist when cross-metathesis reactions are employed with fatty acids and a number of polymers can be produced such as polyesters, polyethers and polyolefins. It has been shown by Rybak and Meier that the cross-metathesis of fatty acid methyl esters with methyl acrylate with only 0.5 mol% of catalyst can be successfully achieved. Furthermore it was shown that oleyl alcohol can be cross-metathesised with methyl acrylate successfully to produce 11-hydroxy-2-undecanoic acid methyl ester and 2-undecanoic acid methyl ester. The former ester is commonly used to make polymers and the latter is used for detergent applications.

Although there is a well-established market for plant oils in the speciality chemicals industry, the biodiesel industry has rapidly grown through government subsidies and high prices of mineral diesel. Therefore a well-defined stream of products from plant oils exists currently and the biorefinery concept can add value to these processes by utilising the by-products such as straw, meal and glycerol.

Wheat straw, rapeseed straw and sunflower stalks are commonly left on the field to replenish the soil or are used as low-grade animal feeds. Although there are environmentally friendly uses of these materials, they do not significantly contribute economically to farming operations. In attempting to increase revenue, the green chemistry approach is the best as it ensures that any additional processing will not harm the environment with a low carbon footprint. A good example of this is the use of supercritical CO_2 extraction technology, which uses compressed CO_2 to extract valuable chemicals from straws. Wheat straws have a waxy surface and the key components of this wax can be selectively extracted with very high efficiencies. A suitable marketable product from biomass that is of a value that can cover the capital cost requirements is yet to be identified. A number of secondary metabolites that have significant potential include cetearyl alcohol, benzoic acid and fumaric acid, which have applications in the personal care, food and chemical industries. In oilseed crops significant quantities of phenolics, falavanoids and sinapine are found, which can all be used as natural antioxidants. These components can easily be extracted from the oilseed cake by using conventional solvents such as methanol, acetone, water and ethyl acetate from the oilseed cake. However, after extraction of such chemicals and proteins the value of the remaining material decreases as it is no longer viable for animal feed. On the other hand it is a legal requirement to remove glucosinolates from rape meal due to their toxicity. Glucosinolates are an important group of chemicals which can easily be broken down by enzymes to produce isothiocyanates, which have good pesticidal properties. Furthermore, low concentrations of glucosinolates in human diets can offer anti-inflammatory, anti-microbial and chemo-preventive effects.

After extracting key secondary metabolites from biomass, the bulk material still contains a rich resource, which can be further utilised to add value to the process. The main components of straw are cellulose, hemicellulose and lignin, all of which have important uses. Cellulosic material can be enzymatically converted to sugars, which can be used as building blocks for a number of commodity chemicals such as succinic acid. The presence of lignin in fermentation broths can reduce the efficiency of enzymes, therefore separation of this component *via* the organosolv process is becoming common practice. The sugars can be converted to commodity chemicals such as glycerol, aspartic acid, levulinic acid and citric acid. These chemicals are currently derived from crude oil economically and for bioderived alternatives to penetrate the market they must be cost effective.

Such biochemical processes can be energy intensive and the enzymes tend to be very expensive. Furthermore, the processes developed to date do not take into account the lignin that is produced as a by-product. Indeed the lignin could be used to provide heat and power to the process; however, other value-added products can also be made to improve the economic viability. Pyrolysis of ligno-cellulosic materials has been well known for many years and recent developments in this technology may offer a more efficient method of producing fuels and chemicals. Microwave pyrolysis is an energy-efficient method of converting biomass into various products at low temperature. The three forms of products for any biomass are char, oil and gas with the respective ratios being specific to the feedstock and processing conditions. The pyrolysis oil contains a cocktail of chemicals which can be used directly in a variety of applications or can be derivatised to form high-value speciality chemicals. With the addition of certain metal oxides, salts or acids the final composition of the bio-oil can be controlled, for example when $MgCl_2$ is added with biomass, the resulting bio-oil mainly contains furfural, an important commodity chemical.

1.2 The Biorefinery Concept
Dedicated to Michael Kamm, Founder of biorefinery.de GmbH.

1.2.1 Introduction
Sustainable economical growth requires safe supply of raw materials for industrial production. Today's most frequently used industrial raw material is petroleum, which is neither sustainable nor environmentally friendly. While the economy of energy can be based on various alternative raw materials, such as wind, sun, water, biomass, as well as nuclear fission and fusion, the economy of substances is fundamentally dependent on biomass, in particular the biomass of plants. The development of biorefineries represents the key for access to an integrated production of food, feed, chemicals, materials, goods and fuels of the future.[1]

Nature is a permanently renewing production chain for chemicals, materials, fuels, cosmetics and pharmaceuticals. Many of the current biobased products are results of a direct physical or chemical treatment and processing of biomass, such as cellulose, starch, oil, protein, lignin and terpene. On one hand one has to mention that with the help of biotechnological processes and methods, chemicals can be produced such as ethanol, butanol, aceton, lactic acid and itaconic acid as well as amino acids, *e.g.* glutaminic acid, lysine and tryptophan. On the other hand, currently only 6 billion tons of the yearly produced biomass, $1.7–2.0 \times 10^{11}$ tons, is used, and only 3 to 3.5% of this amount is used in the non-food area, such as chemistry.[2]

The basic reaction of biomass photosynthesis is according to:

$$nCO_2 + nH_2O = (CH_2O)n + nO_2$$

Industrial utilisation of raw materials for the energetic and material-demanding industry from agriculture and forestry is still at an early stage.

The majority of biological raw materials are produced in agriculture, in forestry and by microbial systems. The forest can provide excellent raw materials for the paper and cardboard industry, the construction industry and the chemical industry. Field fruits represent an organically chemical pool, from which fuels, chemicals and chemical products as well as biomaterials are produced (Figure 1.1).[3] Waste biomass and biomass of nature and landscape cultivation are valuable organic reservoirs of raw material and must be used in accordance with their organic composition. During the development of biorefinery systems the term 'waste biomass' will become obsolete in the medium term.[4] Due to low cost, plentiful supply and amenability to biotechnology, carbohydrates appear likely to be the dominant source of feedstocks for

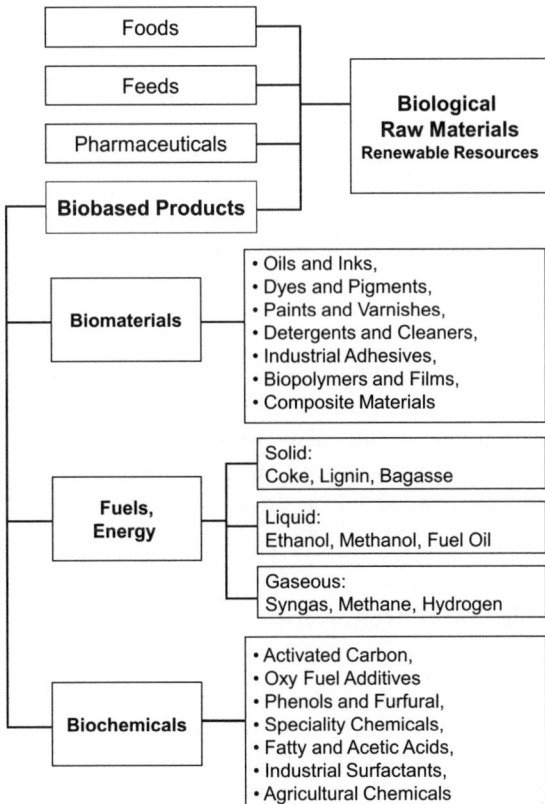

Figure 1.1 Products and product classes based on biological raw materials.[7]

biocommodity processing. Starch-rich and cellulosic materials each have important advantages in this context. Corn is by far the dominant feedstock for biological production of commodity products today. Advantages over cellulosic materials include much larger ultimate supply, lower purchase cost, lower anticipated transfer cost and lower inputs of chemicals and energy required for production.[5] Recently the goal of the US Department of Agriculture and the US Department of Energy is the additional supply of 1 billion tons biomass for a prize of 35 US dollars per ton per year for the industrial chemical and biotechnological utilisation, without restriction of today's applications of biomass from agriculture and forestry.[6] The European Commission and the US Department of Energy have come to an agreement for co-operation in this field (US Department of Energy (DOE), 2005). Based on the European biomass action plan of 2006[47] both strategic EU-projects, (1) BIOPOL, European Biorefineries: Concepts, Status & Policy Implications[48] and (2) Biorefinery Euroview (Current situation and potential of the biorefinery concept in the EU: strategic framework and guidelines for its development)[49] began preparation for the 7th EU framework program.

In order to minimise food-feed-fuel conflicts and in order to use biomass most efficiently, it is therefore necessary to develop strategies and ideas for how one might use biomass fractions, in particular green biomass and agricultural residues, such as straw more efficiently. In future developments, food- and feed-processing residues should therefore also become part of biorefinery strategies since either particular waste fractions may be too small for a cost-efficient specific valorisation treatment *in situ* or the diverse technologies necessary are not available. Fibre-containing food-processing residues may then be pre-treated and processed with other cellulosic material from other sources in order to produce ethanol or other platform chemicals. Foodprocessing residues have, however, a special feature one has to be aware of. Due to their high water content and endogenous enzymatic activity, foodprocessing residues have a comparatively low biological stability and are prone to uncontrolled degradation and spoilage including rapid autoxidation. To avoid extra costs for transportation and conservation, the use of foodprocessing residues should also become part of a regional biomass utilisation network.[50] Thus advanced oil crop biorefineries could produce oil for food, proteins for functional products and straw for platform chemicals and lignin.

1.2.2 Principles of Biorefineries

1.2.2.1 Fundamentals

Biomass is similar to petroleum as it has a complex composition. Therefore its primary separation into main groups of substances is appropriate. Subsequent treatment and processing of those substances leads to a whole palette of products. Petrol-chemistry is based on the principle of separating hydrocarbons simply and in a pure form in refineries. In efficient product lines, a system based

Introduction

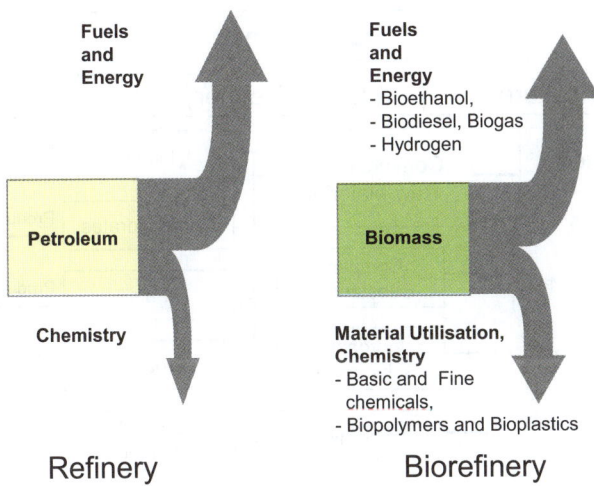

Figure 1.2 Comparison of the basic principles of petroleum-refinery and biorefinery.[8]

on family trees has been built, in which basic chemicals, intermediate products and sophisticated products are produced. This principle of petroleum refineries must be transferred to biorefineries. Biomass contains C:H:O:N, a feature that petroleum does not have and therefore complicates processing. The biotechnological conversion will therefore become, beside the chemical, a big player in the future (Figure 1.2).

Thus biomass can already be modified within the process of genesis in such a way that it is adapted for the purpose of subsequent processing and can particularly target products that have already been formed. For those products the term 'precursors' is used.

Plant biomass always consists of the basic products such as carbohydrates, lignin, proteins and fats, besides various substances such as vitamins, dyes, flavours and aromatic essences of different chemical structures. Biorefineries combine the essential technologies between biological raw materials and the industrial intermediates and final products (Figure 1.3).

A technically feasible separation operation, which would allow a separate use or subsequent processing of all these basic compounds, exists up to now only in the form of an initial attempt. Assuming that out of the estimated annual production of biomass by biosynthesis of 170 billion tons, 75% are carbohydrates, mainly in the form of cellulose, starch and saccharose, 20% lignin and only 5% other natural compounds such as fats (oils), proteins and various substances,[10] the main attention firstly should be focused on an efficient access to carbohydrates, their subsequent conversion to chemical bulk products and corresponding final products. Glucose, accessible by microbial or chemical methods from starch, sugar or cellulose, is among other things predestined for a key position as a basic chemical, because a broad palette of biotechnological or chemical products is accessible from glucose. In the case of starch the advantage of enzymatic compared to chemical hydrolysis is today already realised.[11,12]

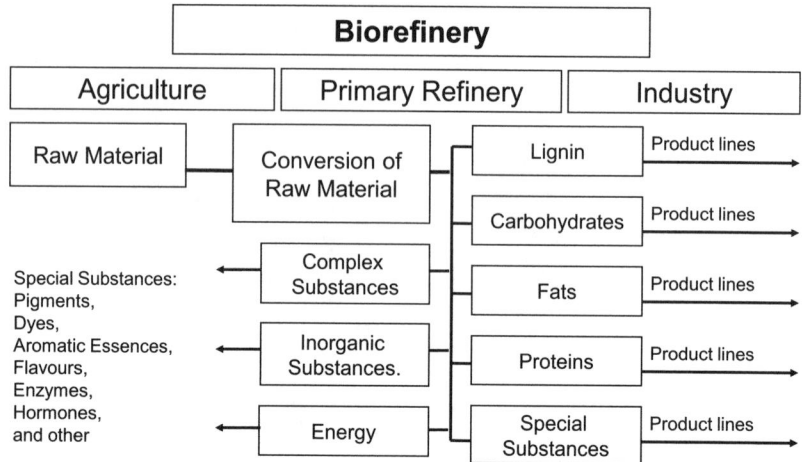

Figure 1.3 Providing code-defined basic substances *via* fractionation for the development of relevant industrial product family trees.[7,9]

In the case of cellulose this is not yet realised. Cellulose-hydrolysing enzymes can only act effectively after pre-treatment to break up the very stable lignin/cellulose/hemicellulose composites.[13] These treatments are still mostly thermal, thermo-mechanical or thermo-chemical and require a considerable input of energy. The arsenal for microbial conversion of substances out of glucose is large, and the reactions are energetically profitable. It is necessary to combine the degradation processes *via* glucose to bulk chemicals with the building processes to their subsequent products and materials (Figure 1.4).

Among the variety of glucose-accessible microbial and chemical products, in particular lactic acid, ethanol, acetic acid and levulinic acid are favourable intermediates for the generation of industrially relevant product family trees. Here, potential strategies are considered: first, the development of new, possibly biologically degradable products (follow-up products of lactic and levulinic acid) or second, the entry as intermediates into conventional product lines (acrylic acid, 2,3-pentandion) of petrochemical refineries.[7]

1.2.3 Building Blocks, Chemicals and Potential Screening

A team from PNNL and NREL submitted a list of 12 potential biobased chemicals.[14] Key areas of the investigation have been biomass-precursors, platforms, building blocks, secondary chemicals, intermediates, products and uses (Figure 1.5).

The final selection of 12 building blocks began with a list of more than 300 candidates. The shorter list of 30 potential candidates was selected using an iterative review process based on the petrochemical model of building blocks, chemical data, known market data, properties, performance of the potential

Introduction

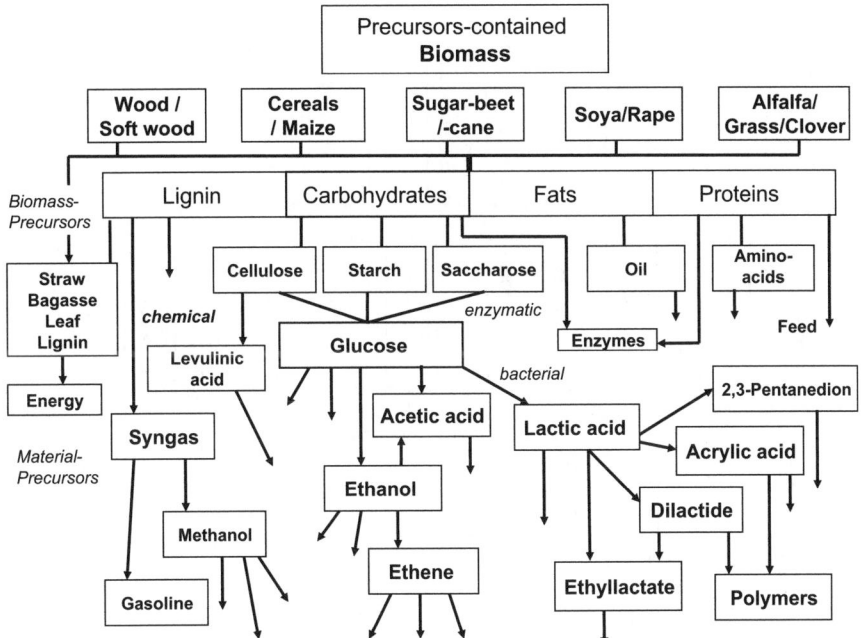

Figure 1.4 A possible biorefinery rough scheme for precursors-containing biomass with preference of carbohydrate line. (From references 7 and 9.)

candidates and the prior industry experience of the team at PNNL and NREL. This list of 30 was ultimately reduced to 12 by examining the potential markets for the building blocks and their derivatives and the technical complexity of the synthesis pathways.

The reported block chemicals can be produced out of sugar *via* biological and chemical conversions. The building blocks can be subsequently converted to a number of high-value biobased chemicals or materials. Building block chemicals, as considered for this analysis, are molecules with multiple functional groups that possess the potential to be transformed into new families of useful molecules. The 12 sugar-based building blocks are 1,4-diacids (succinic, fumaric and malic), 2,5-furan dicarboxylic acid, 3-hydroxy propionic acid, aspartic acid, glucaric acid, glutamic acid, itaconic acid, levulinic acid, 3-hydroxybutyrolactone, glycerol, sorbitol and xylitol/arabinitol.[14]

A second-tier group of building blocks was also identified as viable candidates. These include gluconic acid, lactic acid, malonic acid, propionic acid, the triacids, citric and aconitic, xylonic acid, acetoin, furfural, levuglucosan, lysine, serine and threonine. Recommendations for moving forward include examining top-value products from biomass components such as aromatics, polysaccharides and oils; evaluating technical challenges in more detail related to chemicals and biological conversions; and increasing the suites of potential pathways to these candidates. For the purposes of this study hydrogen and

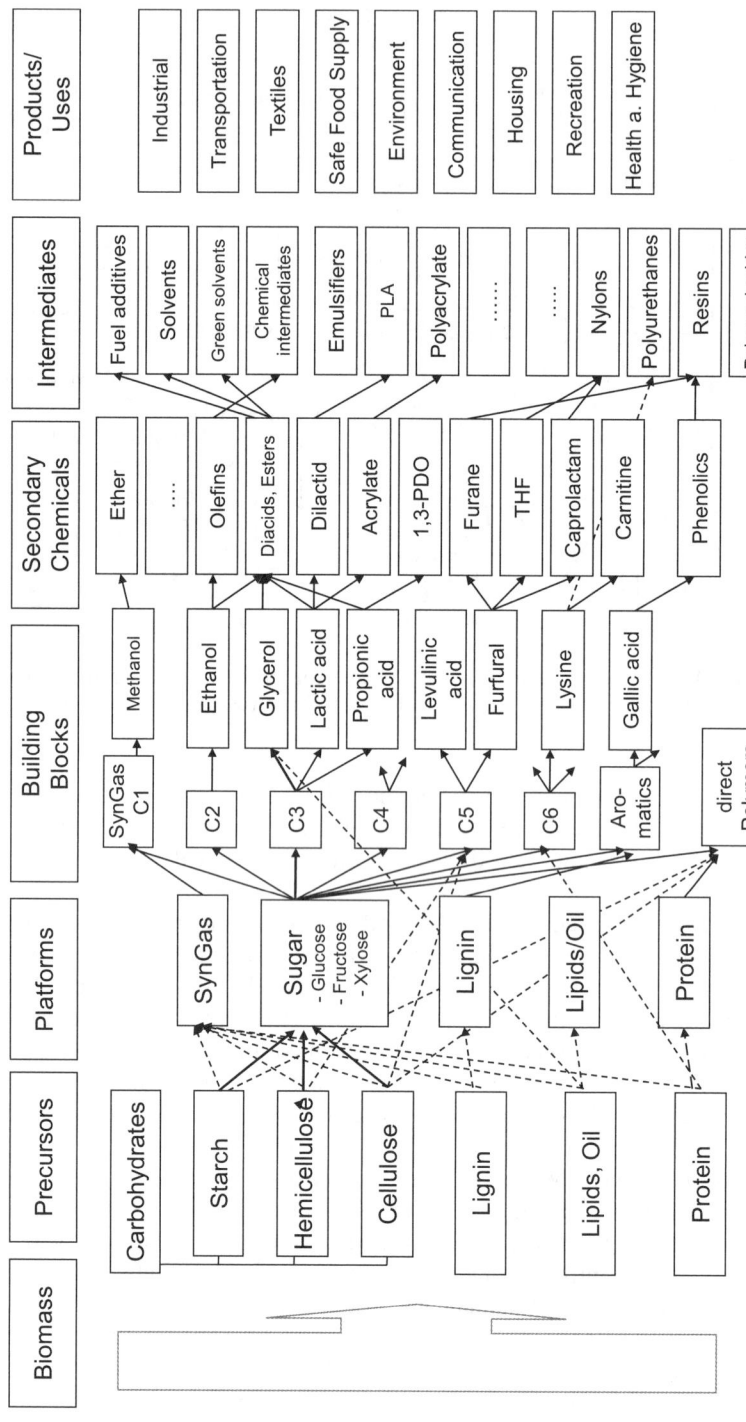

Figure 1.5 Model of a biobased product flow-chart for biomass feedstock. From reference 14.

methanol comprise the best near-term prospects for biobased commodity chemical production because obtaining simple alcohols, aldehydes, mixed alcohols and Fischer–Tropsch liquids from biomass is not economically viable and requires additional development.[14]

1.2.4 Biorefinery Systems

1.2.4.1 Background

Biobased products are prepared for a useable economical use by a meaningful combination of different methods and processes (physical, chemical, biological and thermal). It is therefore necessary that the basic biorefinery technologies have to be developed. For this reason a profound interdisciplinary cooperation of the various compartment disciplines in research and development is inevitable. It appears therefore to be reasonable to refer to the term 'biorefinery design'. Biorefinery design means: Bringing together well-sounded scientific and technological basics, with practical technologies, products and product lines inside biorefineries. The basic conversions of each biorefinery can be summarised as follows:

In the first step, the precursor containing biomass is separated by physical methods. The main products (M1-Mn) and the by-products (B1-Bn) will subsequently be subjected to microbiological or chemical methods. The follow-up products (F1-Fn) of the main and by-products can furthermore be converted or enter the conventional refinery (see Figure 1.4).

Currently four complex biorefinery systems are the focus in research and development:

(1) the 'Lignocellulosic Feedstock Biorefinery' using 'nature-dry' raw materials such as cellulose-containing biomass and wastes. Also straw from oil plants (rape, sunflower) is an excellent starting material for using in the LCF biorefineries.
(2) the 'Whole Crop Biorefinery' uses raw materials such as cereals or maize.
(3) the 'Green Biorefineries' using 'nature-wet' biomasses such as green grass, alfalfa, clover or immature cereal.[7,9]
(4) the 'Biorefinery two-platforms concept' includes the sugar platform and the syngas platform.[14]

1.2.4.2 Lignocellulosic Feedstock Biorefinery

Among the potential large-scale industrial biorefineries the Lignocellulosic Feedstock (LCF) biorefinery will most probably be pushed through with highest success. On the one side the raw material situation is optimal (straw, reed, grass, wood, paper-waste *etc.*); on the other side conversion products

$$\text{Lignocellulose} + H_2O \rightarrow \text{Lignin} + \text{Cellulose} + \text{Hemicellulose}$$
$$\text{Hemicellulose} + H_2O \rightarrow \text{Xylose}$$
$$\text{Xylose } (C_5H_{10}O_5) + \text{acid Catalyst} \rightarrow \text{Furfural } (C_5H_4O_2) + 3H_2O$$
$$\text{Cellulose } (C_6H_{10}O_5) + H_2O \rightarrow \text{Glucose } (C_6H_{12}O_6)$$

Figure 1.6 A possible general equation of conversion at the LCF biorefinery.

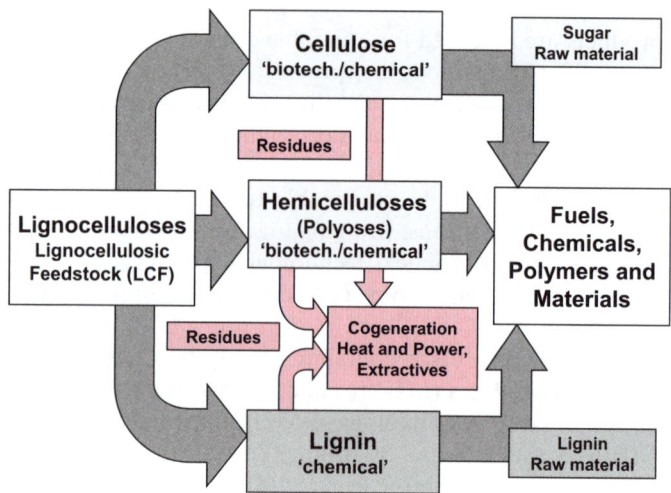

Figure 1.7 Lignocellulosic feedstock biorefinery.[16]

have a good position on the traditional petrochemical as well as on the future biobased product market. An important point for utilisation of biomass as chemical raw material is the cost of raw material. Currently the costs for corn stover or straw are 30 USD/ton; for corn 110 USD/ton (3 USD/bushel).[15]

Lignocellulose materials consist of three primary chemical fractions or precursors: (a) hemicellulose/polyoses, a sugar-polymer of predominantly pentoses; (b) cellulose, a glucose-polymer; and (c) lignin, a polymer of phenols (Figure 1.6).

The lignocellulosic biorefinery regime has a distinct ability for genealogical trees. The main advantage of this method is the fact that the natural structures and structure elements are preserved, the raw materials have a low price and large product varieties are possible (Figure 1.7). Nevertheless there is still development and optimisation demand for these technologies, *e.g.* in the field of separation of cellulose, hemicellulose and lignin as well as lignin utilisation in the chemical industry.

Introduction

In particular furfural and hydroxymethylfurfural are interesting products. Furfural is the starting material for the production of nylon 6,6 and nylon 6. The original process for the production of nylon 6,6 was based on furfural. The last of these production plants was closed in 1961 in the USA for economical reasons (the artificial low price of petroleum). Nevertheless the market for nylon 6 is huge.

However, there are still some unsatisfactory parts within the LCF, such as utilisation of lignin as fuel, adhesive or binder. They are unsatisfactory because the lignin scaffold contains considerable amounts of mono-aromatic hydrocarbons, which, if isolated in an economically efficient way, could add significant value to the primary processes. It should be noticed that there are obviously no natural enzymes to split the naturally formed lignin into basic monomers as easily as this is possible for the also naturally formed polymeric carbohydrates or proteins.[17]

An attractive accompanying process to the biomass-nylon-process is the already mentioned hydrolysis of the cellulose to glucose and the production of ethanol. Certain yeasts give a disproportionation of the glucose-molecule during their generation of ethanol to glucose, which practically shifts its entire reduction ability into the ethanol and makes the last one obtainable in 90% yield (w/w; regarding the formula turnover).

Based on recent technologies a plant was conceived for the production of the main products furfural and ethanol from LC-feedstock for the area of West Central Missouri (USA). Optimal profitability can be reached with a daily consumption of about 4360 tons of feedstock. Annually the plant produces 47.5 million gallons of ethanol and 323 000 tons of furfural.[17]

Ethanol may be used as a fuel additive. Ethanol is also a connecting product for a petrochemical refinery. Ethanol can be converted into ethene by chemical methods. As it is well known from petrochemically produced ethene, it is at the start of a whole series of large-scale technical chemical syntheses for the production of important commodities, such as polyethylene, or polyvinylacetate. Furthermore petrochemically produced substances can similarly be manufactured by microbial conversion of glucose, such as hydrogen, methane, propanol, acetone, butanol, butandiol, itaconic acid and succinic acid.[18–20] DuPont has entered into a six-year alliance with Diversa to produce sugar from husks, straw and stovers and develop a process to co-produce bioethanol and value-added chemicals, such as 1,3-propandiol.[19] Through metabolic engineering an *Escheria coli* K12 microorganism produces 1,3-propandiol (PDO), in a simple glucose fermentation process developed by DuPont and Genencor. In a pilot plant operated by Tate & Lyle, the PDO yield reaches $135\,\text{gl}^{-1}$ at the rate of $4\,\text{gl}^{-1}\text{h}^{-1}$. PDO is used for the production of PTT (polytrimethylenterephthalate), a new polymer that is used for the production of high-quality fibres branded Sorona.[20]

Recently there have been examples of LCF pilot plants in Europe. Abengoa Bioenergy are participating in the FP6 Biosynergy project to construct a pilot plant for 70 t/day of corn stover, wheat straw, grasses and wood residues in

Salamanca, Spain, with the main product being ethanol.[49] Supported by the German government, the Leuna Chemical Site is a pilot plant designed for a capacity of about 1300 kg wood (poplar, beech (50% dry matter)) per week. The central elements of the pilot plant concept are two reactors for the Organosolv pulping for production of cellulose, lignin and xylose. Cellulose is the starting material for conversion to monomeric sugar as a starting material for platform chemicals by using enzymatic hydrolysis.[51]

1.2.4.3 Whole Crop Biorefinery

Raw materials for the 'Whole Crop Biorefinery' are cereals, such as rye, wheat, triticale and maize (Figure 1.8). The first step is the mechanical separation into corn and straw, whereas the portion of corn is approximately 10% (w/w) and the portion of straw is 90% (w/w).[21] Straw means a mixture of chaff, nodes, ears and leaves. The straw represents an LC feedstock and may be further processed in an LCF biorefinery regime.

On one side there is the possibility of separation into cellulose, hemicellulose, lignin and their further conversion within separate product lines, which is shown in the LCF biorefinery. Furthermore the straw is a starting material for the production of syngas *via* pyrolysis technologies. Syngas is the basic material for the synthesis of fuels and methanol (Figure 1.9).

The corn may be either converted into starch or directly used after grinding to meal. Further processing may be carried out in the four directions (a) breaking up, (b) plasticisation, (c) chemical modification or (d) biotechnological conversion *via* glucose. The meal can be treated and finished by extrusion

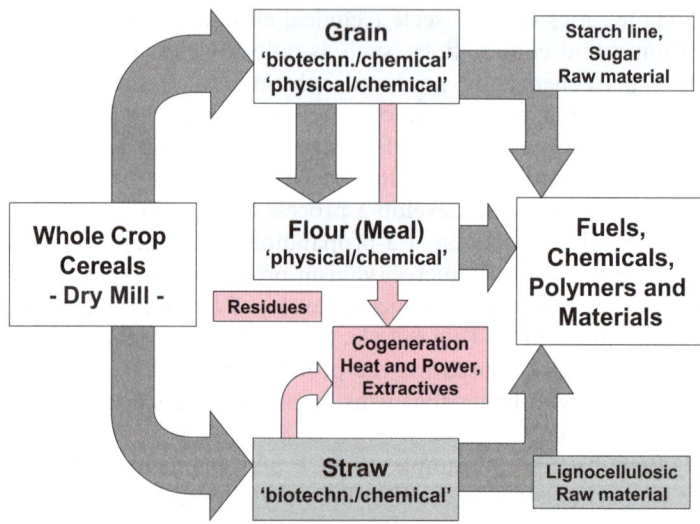

Figure 1.8 Whole crop biorefinery – based on dry milling.[16]

Introduction

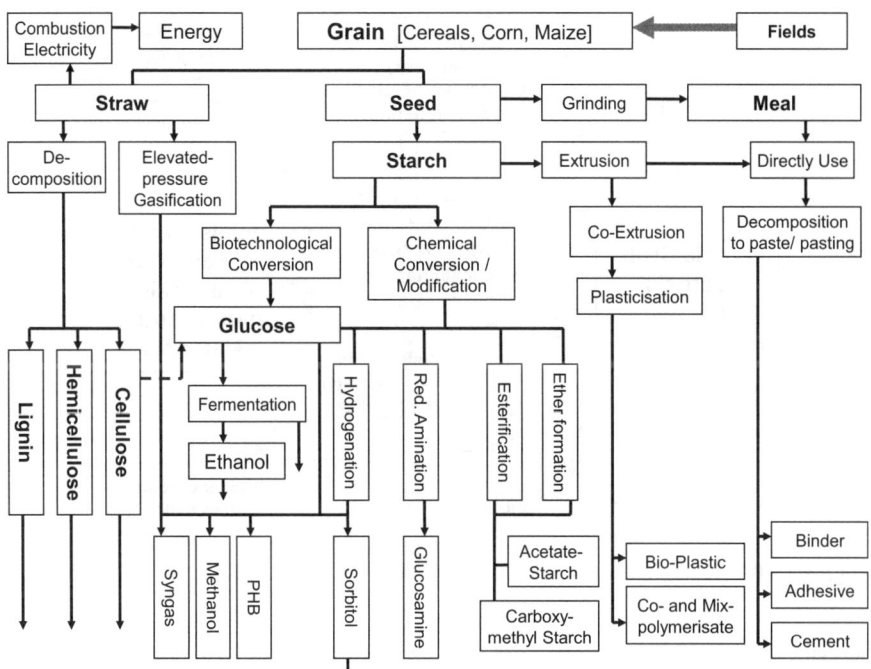

Figure 1.9 Products from whole crop biorefinery.[7,9]

into binder, adhesives and filler. Starch can be finished *via* plasticisation (co- and mix-polymerisation, compounding with other polymers), chemical modification (etherification into carboxy-methyl starch; esterification and re-esterification into fatty acid esters *via* acetic starch; splitting reductive amination into ethylen diamine a. o. and hydrogenative splitting into sorbitol, ethylenglycol, propylenglycol and glycerin.[3,22,23]

Furthermore, starch can converted by biotechnological methods into poly-3-hydroxybutyricacid in combination with production of sugar and ethanol.[24,25]

Biopol, the copolymer poly-3-hydroxybutyrate/3-hydroxyvalerte, developed at ICI is produced from wheat carbohydrates by fermentation using Alcaligenes eutropius.[26]

An alternative to traditional dry fractionation of mature cereals into sole grains and straw has been developed by Kockums Construction Ltd (Sweden), later called Scandinavian Farming Ltd. In this whole crop harvest system, whole immature cereal plants are harvested. The whole harvested biomass is conserved or dried for long-term storage. When convenient, it can be processed and fractionated into kernels, straw chips of internodes and straw meal (leaves, ears, chaff and nodes).

Fractions are suitable as raw materials for the starch polymer industry, feed industry, cellulose industry, particle board producers, gluten for the chemical industry and as a solid fuel. Such a dry fractionation of the whole crop to optimise the utilisation of all botanical components of the biomass has been

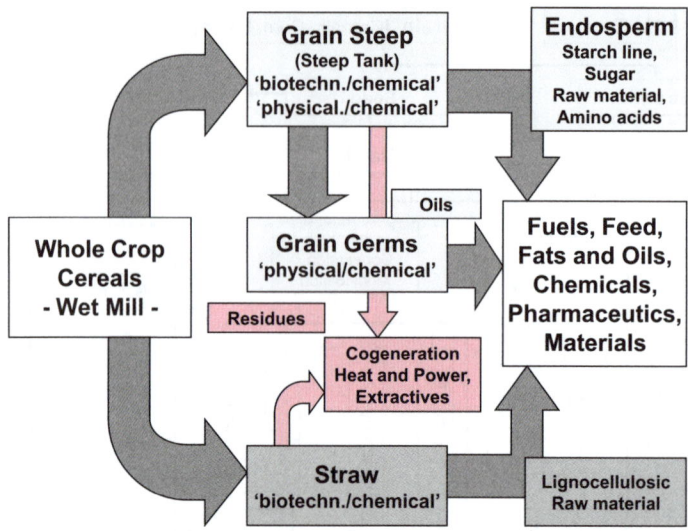

Figure 1.10 Whole crop biorefinery, wet-milling.[16]

described.[27,28] A biorefinery and its profitability has also been described in detail.[29]

An expansion of the product lines to grain processing represents the 'whole crop wet mill based biorefinery'. The grain is swelled and the grain germs are pressed, generating highly valuable oils.

The advantages of the whole crop biorefinery based on wet milling is that the receipt of the natural structures and structure elements like starch, cellulose, oil and amino acids (proteins) are kept intact to a high extent, and well-known basic technologies and processing lines can still be used. High raw material costs and costly technologies are the disadvantages. However, many products could generate high value *e.g.* in the pharmacy and cosmetics industries (Figures 1.10 and 1.11).

However, the basic biorefinery technology, the corn wet mills used in 11% of the US corn harvest in 1992, made $7 billion from products, and employed almost 10 000 people.[1]

Wet milling of corn yields corn oil, corn fibre and corn starch. The starch products of the US corn wet-milling industry are fuel alcohol (31%), high-fructose corn syrup (36%), starch (16%) and dextrose (17%). Corn wet-milling also generates other products (*e.g.* gluten meal, gluten feed, oil).[30] An overview of the product range is shown in Figure 1.11.

The French Biohub Programme aims to develop a portfolio of cereal-based platform chemicals (*e.g.* isosorbide) as intermediates for monomers and polymery, for speciality or commodity markets. A demonstration platform is organised, which will produce 2000 t per annum of L-succinic acid. The product will go to make solvents, coolants, plasticisers, fuel additives, intermediates and fine chemicals, in addition to uses in food, cosmetics and pharmaceuticals.[52]

Introduction

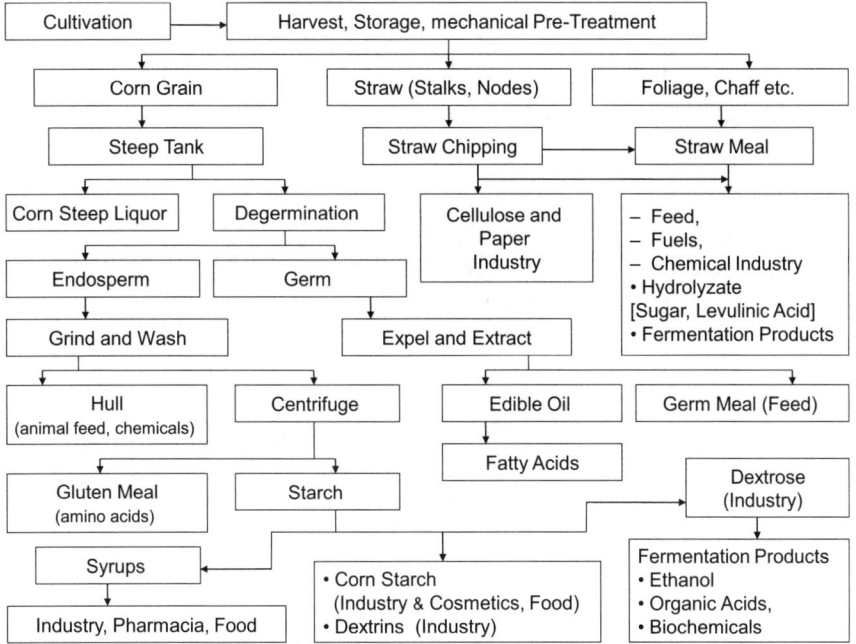

Figure 1.11 Products from whole crop wet-mill based biorefinery.

1.2.4.4 Green Biorefinery

Often the economy of bioprocesses is still the problem because in the case of bulk products the price is affected mainly by raw material costs.[31] The advantages of the Green Biorefinery are a high biomass profit per hectare and a good coupling with agricultural production, whereas the price segment of the raw materials is still low. On one hand simple basic technologies can be used and have a good biotechnical and chemical potential for further conversions (Figure 1.12).

On the other hand a fast primary processing or the use of preservation methods like silage or drying are necessary, for both the raw materials and the primary products. However, each preservation method changes the content materials.

Green biomass comes from green crops, for example grass from cultivation of permanent grass land, closure fields, nature preserves or green crops, such as lucerne, clover and immature cereals from extensive land cultivation. Thus, green crops represent a natural chemical factory and food plant.

Green crops are primarily used today as forage and a source of leafy vegetables. A process called wet-fractionation of green biomass, green crop fractionation, can be used for simultaneous manufacturing of both food and non-food items.[32]

Scientists in several countries have developed green crop fractionation in Europe and elsewhere.[33–35] Green crop fractionation is now studied in about

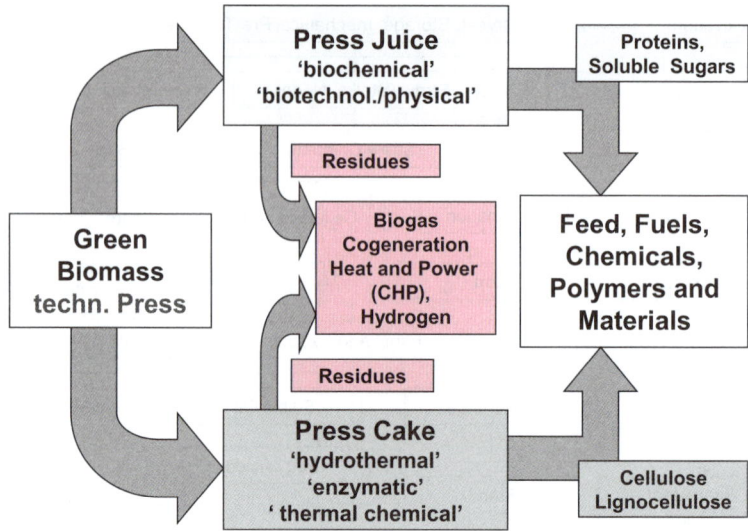

Figure 1.12 A system 'green biorefinery'.[16]

80 countries.[41] Several hundred temperate and tropical plant species have been investigated for green crop fractionation.[35–37] However, more than 300 000 higher plant species are left for investigations.[33,34,38–41,44]

Green biorefineries can, by fractionation of green plants, process a few tonnes of green crops per hour (farm scale process) to more than 100 tonnes per hour (industrial scale commercial process).

The careful wet fractionation technology is used as a first step (primary refinery) to isolate the content-substances in their natural form. Thus, the green crop goods (or humid organic waste goods) are separated into a fibre-rich press cake (PC) and a nutrient-rich green juice (GJ).

Besides cellulose and starch, the press cake contains valuable dyes and pigments, crude drugs and other organics. The green juice contains proteins, free amino acids, organic acids, dyes, enzymes, hormones, other organic substances and minerals. In particular the application of the methods of biotechnology is predestined for conversions, because the plant water can simultaneously be used for further treatments. Starting from green juice the main focus is directed to products such as lactic acid and corresponding derivatives, amino acids, ethanol and proteins. The press cake can be used for production of green feed pellets, as raw material for production of chemicals, such as levulinic acid, as well as for conversion to syngas and hydrocarbons (synthetic biofuels). The residues of substantial conversion are suitable for the production of biogas combined with generation of heat and electricity (Figure 1.13). Reviews of green biorefinery concepts, contents and goals have been published.[16,45,46]

The green biorefinery demonstration plant in Brandenburg, Germany, produces high-value proteins and fermentation media from 20 kt per annum of alfalfa and wild mix grass. The press cake is used for fodder (primary

Introduction

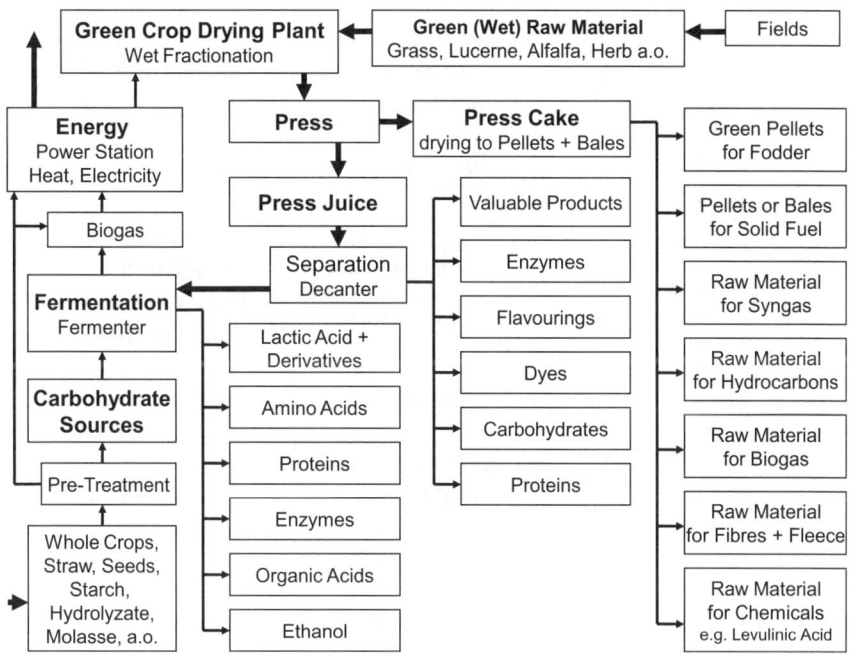

Figure 1.13 Products from green biorefinery. A system green biorefinery combined with a green crop drying plant.[7,9]

refinery).[53] The fermentation medium is an excellent starting material for the production of platform chemicals, such as lactic acid or lysine. Also the direct production of lysine lactate for application in the cosmetic industry is a possible secondary product line.[54]

1.2.5 Two-platform Concept

The 'two-platform concept' is one that uses biomass that consists on average of 75% of carbohydrates, which can be standardised over an 'intermediate sugar platform' as a basis for further conversions. Alternatively, the biomass can be converted thermochemically to synthesis gas which can be converted to additional higher value products.

The 'sugar platform' is based on biochemical conversion processes and focuses on the fermentation of sugars extracted from biomass feedstocks.

The 'syngas platform' is based on thermochemical conversion processes and focuses on the gasification of biomass feedstocks and by-products from conversion processes.[14,33,42] In addition to gasification, other thermal and thermochemical biomass conversion methods are also of interest: hydrothermolysis, pyrolysis, thermolysis, burning. The application is chosen according to the water content of biomass.[44]

The gasification and other thermochemical concepts concentrate on the utilisation of the precursor carbohydrates as well as their carbon and hydrogen

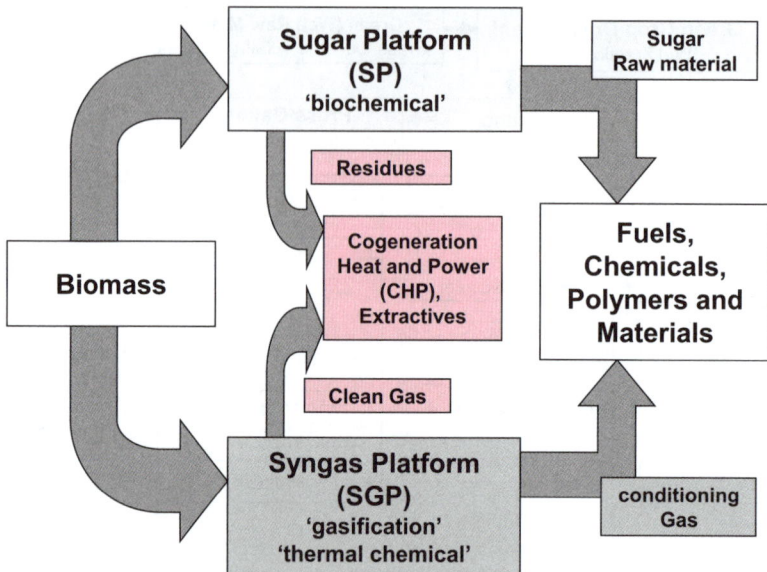

Figure 1.14 Sugar platform and syngas platform.[43]

content. The proteins, lignin, oils and lipids, amino acids and other ingredients including N- and S-compounds are not taken into account in this case (Figure 1.14).

1.2.6 Advanced Oil Crops Biorefineries

Oil crop plants, such as rape and sunflower, are cultivated worldwide, in particular in Europe, North America and South America. Recently rapeseed and rape oil production averages 19 M tons per annum. Sunflower production amounts to 5 M tons per annum.

Today's cultivation and harvesting of rape and sunflower, the rape and sunflower oil recovery by pressing of rape and sunflower seed as well as the remaining rape straw is the object of consideration for development of future advanced oil crop biorefineries.

Rape needs to be harvested when green in order to avoid losses due to germination. At the harvest of rape plants, about 50 cm of the rape straw is left on the field. A combine harvester can recover 60% of the rape straw. Assuming a corn-straw ratio of rapeseed of about 1:2.9 at an average corn yield of about 3.5 tons per ha and year, about 10 tons of crop wastes remain on the field in the form of rape straw. After chaffing, a part of the rape straw is left on the field as humus forming substrate and nutrient supplier. The maximum recovery rate amounts to about 50–80% of the whole crop residues, *i.e.* about 5–8 t/ha rape straw per year. In Germany rapeseed is cultivated on around 1.4 MM ha per year; therefore the outcome of rapeseed straw amounts to about 7 MM t.[55]

Introduction

There are two methods of rapeseed harvest:[56]

(1) Direct thresh: grains are black and are rustling in the pod; straw can be partially green; harvests mostly in the second half of July; use of combine harvesters with additional equipment for rape.
(2) Swath thresh: plants are cut and put on swath, when the grains start to burnish on both sides; after drying on the field, threshed with special swath-reapers by the pick-up method.

After swath thresh, the straw has a dry matter content of about 90% and can be pressed into square bales (80 × 80 × 240 cm) for transportation. One square bale corresponds to nearly 250 kg of rapeseed straw. Figure 1.15 shows the combination of today's oil production in central oil plants and local oil plants[57] as well as the use of straw for production of levulinic acid (see Section 3.3).

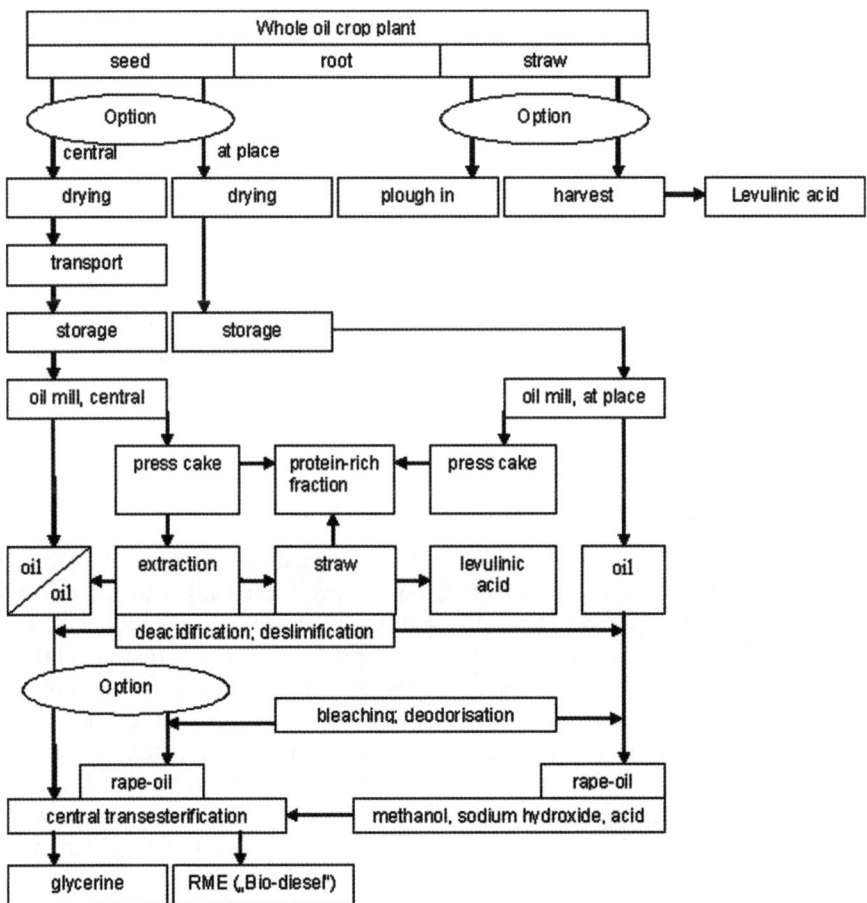

Figure 1.15 Today's rape oil production combined with future straw use for production of levulinic acid.

(1) Central Rapeseed Oil Production

The fractionation of oilseeds is described by the example of rape. Depending on the oil content, the following process steps are implemented:

If the oil content is less than 20%, the fruits are extracted by the use of a solvent after grinding. At oil content higher than 20%, after the grinding process the fruits are pre-pressed and extracted afterwards.

After removal of plant remains and other impurities the rapeseeds are ground in order to create a larger surface area by the destruction of the soft tissue and the seed coats. The pre-disintegration is realised by a corrugated roll, the fine-grinding by plain rolls. Successively the obtained pulp is rendered less viscous by a conditioning process that includes moistening and heat treatment. Finally the oil is pressed out of the pre-treated rapeseeds. On the one hand the desired rapeseed oil is generated, successively purified from small seed particles by filtration and dried. On the other hand, a rapeseed cake is obtained, which afterwards is crushed by the use of a cake breaker and a crimper for the extraction of the remaining oil.

The crushed rapeseeds are extracted with technical *n*-hexane, which, in principle, is operated in a loop. Two fractions are obtained from the extractor, the highly oil enriched hexane, called *Miscella,* and the extraction meal, which is still interspersed with hexane. The hexane is recycled from the *Miscella* by a multi-fold evaporator and is reintroduced into the circular flow. The remaining hexane in the extraction meal is recycled by processing the meal with a steam-heated screw-conveyor drier and subsequent evaporation and is fed again into the circular flow.

The yield of rapeseed oil by pre-pressing and extraction amounts to about 98%. A yield of 394 kg oil from 1 ton rapeseed, which corresponds to 98.5%, is reported in ref. 58. The ratio of extracted to pressed rapeseed oil is 1 : 2.43 according to ref. 58.

(2) Decentralised Rapeseed Oil Production

In decentralised rapeseed oil production in agricultural farms, namely in decentralised oil mills, the transport from the producer's farm to the central mill is no longer required. The mills operate small presses and therewith the process differs significantly from the central process. For logistic and mainly financial reasons, the decentralised rapeseed oil production is principally operated in a single stage, *i.e.* the extraction *via* solvents is omitted. Furthermore, unlike during the central process, the rapeseed is not pre-treated prior to the pressing process. Cold pressed rapeseed oil is obtained as the product.

On average 353 kg rapeseed oil per ton of pressed rapeseed are gained with this process. Assuming an average oil content of 40% in the rapeseed, this amount stands for an average yield of oil of about 88.3%.[59]

In addition Figure 1.15 shows the possibility of extraction of straw residues from press cake of central oil crop plant. Crude fibres extracted from press cake are an additional raw material for platform chemicals, such as levulinic acid production (see Section 3.3). The part of crude fibres from rapeseed amount to 10.3% on dry matter. The part of crude fibres from sunflower seed amount to 17.3% on dry matter.

1.2.6.1 Conclusion

For the development of advanced oil crop biorefineries it is necessary to combine today's oil crop industry and future LCF biorefineries. Straw can be considered as cellulose-containing biomass and wastes for production of platform chemicals, such as levulinic acid. Thus straw from oil crop plants (rape, sunflower) is an excellent starting material for use in the LCF biorefineries. In additional it is necessary to develop improved applications for proteins (see Section 3.2.2) separated from oil crop press cake.

1.3 The Potential of Oil Crops in Europe

1.3.1 Introduction

Because of their high energy content, fats and oils have been of crucial importance for nutrition since the evolution of the human species. Their calorific value is more than double that of carbohydrates and proteins. In addition, fats allow us to consume fat-soluble vitamins and provide the human metabolism with essential fatty acids. Vegetable oilseeds have been known to European peoples since prehistoric times. Environmental archaeologists have found evidence of the use of hazelnuts (*Corylus avellana*) in Mesolithic dwelling places and, more recently, oil crops such as camelina (*Camelina sativa* Crtz.), linseed (*Linum usitatissimum*) and opium poppy (*Papaver somniferum*) were part of the diet among the northern European sedentary food-producing farmers of the Neolithic age.[60–62] In southern Europe the cultivation of the olive tree has been known since biblical times. Olive oil obtained from the fruit of the olive tree is an important component in the wholesome diet of Mediterranean people and in Mediterranean cuisine.[63–65]

However, shortages in basic foodstuffs were frequent in medieval northern Europe, leading to periods of hunger and famine and in turn to upheaval and even civil war. The *Bauernkrieg* (Peasants' War) of 1525 was perhaps the most famous of these conflicts. The invention of fertilisers and improvements in agrarian cultivation since the beginning of the last century led to a great increase in yield on fertilised fields. After the devastating years of World War II, the Common Agricultural Policy[66] (CAP) in the emerging European Community was aimed at encouraging better agricultural productivity, at creating a viable agricultural sector providing a fair standard of living for farmers and at ensuring a stable supply of affordable food for consumers.[7] The underlying concept of subsidies and price guarantees to farmers achieved its goals. The Common Market moved towards self-sufficiency and the price of food started playing less of a role in determining nutrition than it had before.

However, a steady increase in agricultural productivity leading to almost permanent oversupply of the major farmyard commodities prompted the common market of the European Union to reform the CAP. The MacSharry reforms in 1992 can be seen as the trigger for the rise in farmers' interest in finding new sales channels for their products. The results of the CAP negotiations included a 15% set-aside rate of the cultivated area in the European Union to reduce surpluses in the grain sector. This area became available for the potential production of regenerative raw materials.[67] Additionally, a considerable body of legislation and policies has provided incentives to encourage the use of energy crops for the production of renewable energies, in particular biofuels. Oil crops became part of the European landscape.

Biofuel production, meanwhile, is absorbing more and more raw agricultural products since – besides starchy and sugary crops – oleiferous crops are currently the only products that can act as substitutes for fossil fuels. Vegetable oils, mostly in the form of the fatty acid methyl ester (FAME), are well established as transportation fuel (biodiesel)[68] or can be used in combined heat and power (CHP or co-generation) systems, generating electricity and heat at the same time.[69] Crude vegetable oil (CTO) can be submitted directly to catalytic hydrotreatment or can be used as feedstock in the Fluid Catalytic Cracker (FCC) together with fossil feedstock (co-processing).[70] In other words, the generation of different forms of energy is a new outlet for vegetable oils that will fundamentally alter the traditional use pattern of vegetable oils for food and oleochemistry. Recently, the EU27 has become the world's largest producer of rapeseed oil, which is produced mainly in central and northern Europe, and the second-largest producer (behind the Commonwealth of Independent States, CIS) of sunflower seed and oil, which is dominant in the southern European countries.[71]

However, the spectrum of vegetable oils is much broader, also including niche products such as hemp seed oil and specialty oils, which have recently been attracting more interest, such as jatropha oil.[72,73] Vegetable oils can be used for a broad spectrum of purposes, including human consumption; raw materials for cosmetics, oil paints or varnishes and other chemical products; and inputs for biogenic lubricant and fuel production.[74,75]

The demand for vegetable oils has been rising for many years due not only to the growing world population but also to a variety of new uses. The world production of oil crops is estimated at 137 Mio mto in 2010. The most important vegetable oil is palm oil, with an annual production volume of 45 Mio mto. In second place, with 37 Mio mto, is soybean oil. Other important oils are canola oil (21.3 Mio mto) and sunflower oil (11.4 Mio mto).[76]

1.3.2 Vegetable Oil Production in the European Community (EU-27)

Europe is the world leader in olive oil production. It is also a major producer of rapeseed oil but only a minor producer of soybean oil (Figure 1.16).

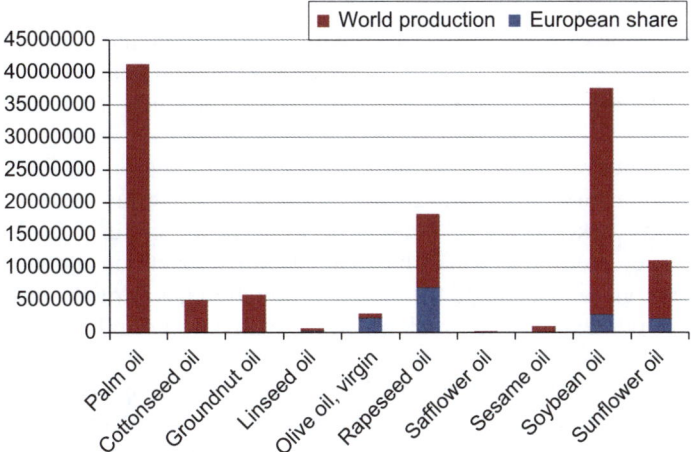

Figure 1.16 Vegetable oils: world production and European share in mto 2008 (FAOSTAT 2010).[77]

In 2010 the area devoted to oil crops in Europe was 10.7 Mio ha, an increase of 7% over 2009. Also, the average yield increased from 2.68 to 2.7 mto per hectare.[78] The major European oil crops are olives, sunflower, soybean and rapeseed; the last of these is the dominant oil crop in the European Union.[79] The cultivation area of oil crops in Europe depends on two factors: the climatic requirements of oil crops and the monetary yield they provide compared to alternative crops, since farmers select the culture that provides the highest monetary return.

Only plants that tolerate European climate conditions can be grown in Europe. However, with regard to climatic requirements, it has to be taken into account that Europe spreads over several climatic zones. Therefore, generalised statements about the cultivation of specific crops in Europe are difficult. Nonetheless, crops that require tropical or sub-tropical climate conditions cannot be grown in Europe or can be grown only in selected regions of southern Europe. Where climatic requirements are not met, plants can be adapted to new climates through breeding or genetic engineering – although the latter is widely rejected by consumers and public opinion in most EU member states.[80]

The second determinant of farmers' crop choice is monetary earnings, which are determined by yields, prices and production costs. In addition to supply and demand, vegetable oil prices are determined by the absolute levels of stocks and changes in world stocks of vegetable oils and fossil fuels. Furthermore, the performance of possible substitutes influences vegetable oil prices. Whether or not farmers cultivate oil crops depends not only on monetary returns from oil crops but also on their competitiveness compared to alternative crops within the crop rotation. Since the prices, yields and production costs of substitutes and competing crops are very difficult to predict, the potential cultivation areas of specific cultures are difficult to estimate from an economic perspective. This is even more the case in this era of growing volatility on agricultural (and oil) markets, which has been triggered by increasing demand from newly emerging markets, melting global stocks of agricultural products and global yield losses.[81]

Another determinant of the economic potential of oil crops in Europe is the political framework. Agricultural production in Europe (as well as in many other regions) takes place in a dense network of agricultural (and environmental) policy measures, which have significant impacts on global competitiveness. The growing demand for biofuel production has strongly increased the influence of political decision-making on the cultivation of oil crops in Europe and abroad.

In summary, the growing potential of oil crops in Europe is determined by the climatic adaptability of crops as well as economic considerations. The latter are influenced by a variety of determinants, including the yields, market prices and costs of oil crops and competing crops as well as crude oil prices. Furthermore, agricultural markets are strongly influenced by various policies, such as agricultural, environmental and bioenergy policies. Against this background, this paper presents a description of the production and use of commodity oils and specialty oils. Key aspects of the following analysis are worldwide and European cultivation areas and the production of and demand for selected vegetable oils. We conclude by presenting a brief outlook on oils that might have a potential use in Europe.

1.3.3 Commodity Oils

Due to its ability to withstand spring frost and to germinate and grow at low temperatures, rapeseed (also called rape or oilseed rape and *colza* in French and Spanish) is one of the few oil crops suitable for farming in the moderate climates of the north (and the far south). Historically, the oil pressed from the rapeseed has been used for cooking, lighting and industrial applications. Traditional rapeseed contains high quantities of eicosenoic and docosenoic (gadoleic and erucic) acids, on the one hand, and glucosinolate (a thioether that acts as defence compound against herbivores) on the other, which make the seed meal unpalatable and possibly dangerous to livestock if fed in large quantities. Nonetheless, this traditional brassica variety (HEAR, High Erucic Acid Rape), continues to offer growers significant opportunities to maximise the use of set-aside land. HEAR oil has preserved an important niche market in industrial applications, and Germany and the UK are the largest producers within Europe. HEAR oil is used to produce erucamide, a high-melting wax, used as a lubricant in PVC processing and as a slip and anti-blocking agent in polyolefin film (Figure 1.17). Micronised erucamide wax is used in flexo and gravure inks. Here, besides high temperature stability, slip and anti-blocking, it provides good recoatability and wet print properties. Erucic acid esters are especially valued in tribology as biobased synthetic lubricants. Erucic acid can be hardened to yield behenic acid and like other fatty acids can be converted

Figure 1.17 Uses of HEAR-based erucamide as a slip and anti-blocking agent in packaging film.

into behenyl alcohol by catalytic high-pressure hydrogenation of the methyl or butyl ester. Behenyl alcohol is used traditionally as an emollient, emulsifier and viscosity regulator in cosmetic O/W-emulsions.[82] Total European demand for erucic acid is actually in the order of 40 000 mto.[16] US production of erucamide in 2008 was 12 800 mto[83] and European production should be in the same order of magnitude.

Canola is an edible variety of rapeseed with a low percentage of erucic acid and low levels of glucosinolates. It was developed through conventional plant breeding at the University of Manitoba in the 1970s. The denomination 'Canola' is used mainly in North America and Australia for Low Erucic Acid Rape (LEAR or 00 rape) varieties of rapeseed, *i.e.*, 'an oil that must contain less than 2% erucic acid and less than 30 micromoles of glucosinolates per gram of air-dried oil-free meal'.[84] Rapeseed is the most important oil crop in the European Union. Much of the rapeseed grown in Europe is of canola quality but retains the name rapeseed. The annual world production of rapeseed totalled 59.7 Mio mto, of which 21.57 Mio mto are produced in Europe. The main growing areas in the European Union are France and Germany.[85]

Rapeseed oil has a unique fatty acid composition compared to other oil crops: the lowest proportion of saturated fats, a high amount of oleic-fatty-acid (average: 62%) and a well-balanced proportion of omega-3 and -6 fatty acids. The main uses of rapeseed oil are as a cooking oil and as a raw material in margarine production. Other applications range from mayonnaise and salad dressings to infant food. As an ingredient in margarines and bakery products, rapeseed oil ensures the right consistency and good nutritional values. Specialty rapeseed oils like high oleic acid rape and oils with higher than average ω-3 levels or high oxidation stability for deep frying applications (HOLLi [high oleic/low linolenic] rape) are grown under contract in northern Europe.[86,87] Furthermore, rapeseed oil can be formulated into total-loss lubricating systems, like chain-saw oils, agrochemical products, printing inks, pharmaceuticals and cosmetics.[88] In such applications, they are considered more environmentally friendly than their mineral oil counterparts.

While soybeans are the major oilseed used for biodiesel production in the USA, Brazil and Argentina, edible rapeseed is the most common oilseed used for biodiesel in Europe.

Biodiesel manufacturing consists of a catalyst-induced transesterification of a vegetable oil to create a fatty acid methyl ester (FAME) (cf. Figure 1.18). The catalyst is usually a strong base, such as NaOH or KOH.[89] Once the ester chains are broken off, the leftover glycerol molecule is a valuable by-product.[90] Typically, the volume of the biodiesel is equivalent to the volume of the original plant oil. In principle, biodiesel can be produced from a wide variety of oilseed crops and animal fats, including soapstock, used cooking oil and waste grease. However, the main challenges to biodiesel production from used oils and greases are colour and the high percentage of free fatty acids (FFAs) in the feedstock, which may necessitate additional process steps, like esterification, prior to transesterification, as well as careful filtration and drying operations.

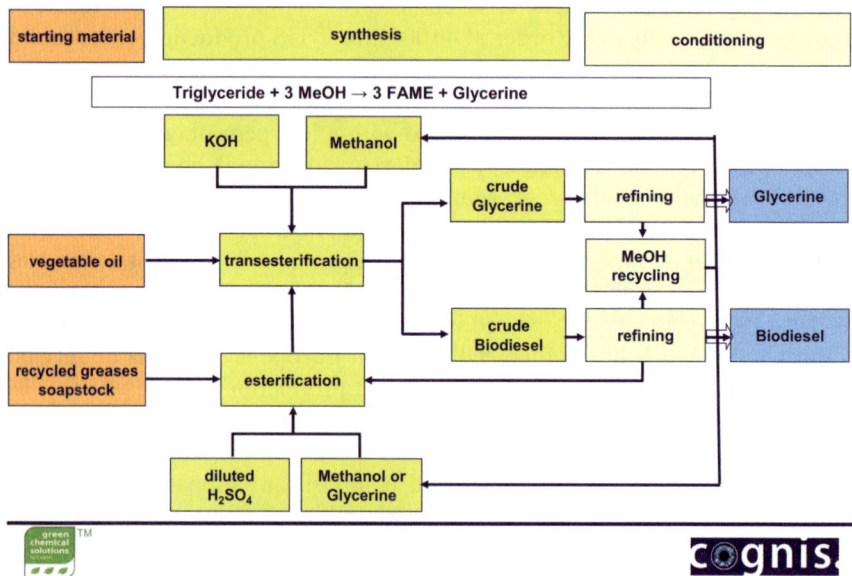

Figure 1.18 Production of biodiesel (reproduced from E. Dinjus, U. Arnold, N. Dahmen, R. Höfer and W. Wach, in R. Höfer (ed.), *Sustainable Solutions for Modern Economies*, RSC Publ., Cambridge, 2009).

Table 1.1 Influence of saturated fatty acid fraction on the low-temperature behaviour of biodiesel.[91]

Oil	$CFPP^*$ (FAME)	Fraction of saturated fatty acids
Rapeseed oil	$-12\,°C^{**}$	7–8%
Soybean oil	$-4\,°C$	12–15%
Palm oil	$10-14\,°C$	Approx. 45%

CFPP = cold filter plugging point, DIN EN 14 214, the temperature at which a fuel will cause a fuel filter to plug due to fuel components, which have begun to crystallise or gel.
Like petroleum diesel fuel, commercial biodiesel qualities are sold as winter quality (CFPP –20 °C), transition quality (CFPP –10 °C) and summer quality (CFPP 0 °C).

As a biodiesel source, rapeseed has a great many advantages over other vegetable oils: oilseed rape, to start with, reaches up to four times the yield per hectare compared to soybeans and more than double the yield compared to sunflower. Biodiesel made from edible rapeseed gels at a significantly lower temperature than biodiesel produced from other feedstock, making rapeseed biodiesel a more suitable fuel for colder regions (cf. Table 1.1).

The EU is home to the world's largest biodiesel industry and markets. Facing the need for up-to-date law to regulate the use of renewable energy sources, the EU has recognised biofuels as a strategic tool to reduce greenhouse gas emissions from the road transport sector and to increase the security of energy supply. The European Union established its first legal framework for biofuels in 2003 with the EU Biofuel Directive (Directive 2003/30/EC), which set indicative targets to promote the use of renewable fuels in the transport sector, followed in December 2008 by the Renewable Energy Directive (Directive 2009/28/EC) and the Fuel Quality Directive (Directive 2009/30/EC) – two major pieces of legislation for the biofuels industry. On the one hand, the Renewable Energy Directive aims at reaching an overall 20% renewable energy target in the EU by 2020, and will drive the future biofuels policies of the EU member states. Its core elements are the 10% binding target for renewables in transport and the introduction of a comprehensive and unparalleled set of sustainability criteria that at first biofuels have to fulfil to be counted towards the target or as a precondition for tax relief. The Fuel Quality Directive, on the other hand, sets technical specifications for fuels, together with a target for the reduction of life cycle greenhouse gas emissions.[92] In response to these stimuli, rapeseed is expected to remain the dominant oilseed crop in the EU for years to come. Provided that no paradigm shift occurs regarding second-generation biofuels,[9] the consumption of rapeseed oil is expected to grow on average by 2.3% annually, driven by biodiesel demand, and to reach 11.8 Mio mto by 2019/2020. Growth will be driven by biodiesel and other industrial uses (6 Mio mto in 2009) while food demand (2.8 Mio mto in 2009) may increase slightly[93] but may also remain static in the future or even decline.[94]

Soy is the most important seed oil crop in the world; it represents 53% of global oilseed production. In 2009/2010 the world production was 258 Mio mto.[18] With a yearly harvest of 1.01 Mio mto, the EU-27 is only a modest producer.[95] European soybean oil production is 2.6 Mio mto (2009/2010) mainly based on imported soybeans; this is only a small share of the world annual production of soybean oil of 37.5 Mio mto. The main areas for growing soybeans are North and South America, India and China. The main growing area in the European Union is Italy, where yields of up to 3.8 mto per hectare (European average 2.7 mto/ha) are possible.[18] The rest of Europe has disadvantages with regard to soy production because these areas are located in unfavourable climate zones.[96] In recent years, soybean cultivation has been expanded on a global basis because the soybean plant has been adapted to new climate conditions through genetic engineering.[97] However, the widespread use of GM soybeans in the Americas has caused problems with exports. GM crops require extensive certification before they can be legally imported into the European Union, where there is considerable supplier and consumer reluctance to use GM products for human or animal consumption.

Soybeans are distinguished from other oil crops in that they only yield approx. 21% oil but 40–50% protein. Together, oil and protein content account for about 60% of dry soybeans by weight. Because soybeans are an important protein source, they play a paramount role in the global food and animal feed industries. However, because of the high protein content, soybean oil can be regarded as a kind of by-product and, historically, soybeans have been a protein rather than an oilseed crop.

Soybean oil is the world's most widely used edible oil. Based on 1998/1999 soybean utilisation in the USA, 95% of the total soybean oil produced was used in food applications and was made into margarine, salad dressings, baking, frying and cooking oil and frozen foods.[98] In 2009 according to FAPRI, this use pattern had significantly changed, with 85% used for food and 15% for biodiesel and other industrial applications.[35] The same source claims a use pattern of 55% for food and 45% for industrial, biodiesel and other applications in Europe. Industrial uses for soybean oil comprise:

- Soy alkyd resins, urethane oils, oil-modified phenolic and urea resins for coatings and inks.
- Epoxidised soybean oil (ESO) as a plasticiser for PVC and nitrocellulose or as a sticker for agrochemical applications.
- Soy polyols as building blocks for polyurethanes and composites and as radiation curing monomers.
- Soy fatty acid and other oleochemicals derived from it, including methyl soyate as a green solvent and biofuel.[9,12]

Although global soybean oil production is expected to grow by more than 3% annually, climatic conditions in most parts of the EU-27 do not favour an increase in European soybean plantations. Growth in European demand is expected to be covered by imports.

European sunflower production is 6.87 Mio mto a year; this represents 21.9% of global production.[26] Sunflower plants grow well in average to rich soils in a moderate climate in full sun with temperatures mainly between 20 and 25 °C.

Table 1.2 Types of sunflower oil and their fatty acid profiles.[102]

	Oleic/ Monounsaturated	Linoleic acid/ Polyunsaturated	Saturated
Linoleic	−20%	69%	11%
High oleic	82%	9%	9%
NuSun	65%	26%	9%

They need their roots to grow deep and wide in order to withstand strong winds. The main growing areas in Europe are Bulgaria, France, Hungary, Italy, Romania and Spain.[17] Sunflower oil contains more than 85% unsaturated fatty acids with linoleic acid accounting for 60% (Table 1.2). The linoleic acid content and the relation between oleic and linoleic acid are greatly influenced by environmental factors during the growing season, especially temperature. Sunflower oil is one of the most important edible oils. It has a light golden colour, a mild flavour and more vitamin E than any other vegetable oil.

In addition to the original linoleic sunflower oil, two new cultivars have been developed since the 1980s and are commercially available.

High oleic sunflower oil (HOSO) is premium sunflower oil with monounsaturated levels of 80% and above. The oil has a neutral taste, provides excellent oxidative stability without hydrogenation and is used extensively for frying. A diet rich in HOSO has favourable outcomes for blood lipids.[99] However, some have claimed that its crop growing properties are inferior to those of linoleic sunflower.[100] Besides its use in food, HOSO fulfils the needs in industrial applications where high monounsaturated levels are required.[101] European HOSO plantations are currently 450 000 ha, or 5% of the total vegetable oil crop area in the EU-27.[16]

NuSun is a mid-oleic sunflower oil actively promoted by the US National Sunflower Association and especially designed to meet the needs of the US food industry. Developed by standard hybrid procedures, it requires no hydrogenation, thus eliminating concerns regarding trans fatty acids.

The main uses of sunflower oils are in salads and salad dressings, for cooking and as a raw material in margarine production. Moreover, they are also used as a raw material in the chemical industry and biofuel production. However, the quality of biodiesel produced from sunflower oil varies, depending on weather conditions and agricultural practices.[103] The largest exporters of sunflower oil are Argentina, Russia, the European Union and the USA. In the marketing year 2009/2010, the USDA expected global sunflower oil production to be 11.5 Mio mto, which cannot satisfy global demand. For this reason, the markets expect prices to rise, which will increase the attractiveness of sunflower cultivation in Europe.[104]

Cotton is the most important fibre plant. One hundred countries plant cotton on 34 Mio ha, but 75% of all cotton is produced in China, Russia, the USA, India, Pakistan, Uzbekistan and Turkey.[105] In 2009/2010, global production was 39.15 Mio mto.[17,26] The main growing areas in Europe are Greece and Spain, with an output of 481 000 mto a year. Cottonseed oil is edible and a source of unsaturated fatty acids, which is why it is used mainly for nutritional purposes. However, it is also used in other ways, for instance, as an input factor for the cosmetic industry. From an economic perspective, the profitability of cotton production has increased compared to soy production because the price of cottonseed oil is 15% higher and the production costs are lower. The demand for cotton seed oil is high and, as a result, its availability is limited. While production meets the commercial demand for cottonseed oil, it is usually not available at the retail level. In 2008 cottonseed oil production was 4.9 Mio mto, of which just 70,574 mto were produced by European countries due to unfavourable climate conditions.[106]

In 2009/2010 worldwide groundnut (also called peanut) production was 22.8 Mio mto and the area harvested was 24 Mio ha in 2008.[26] European production is only 9591 mto a year. Groundnuts are produced in Bulgaria, Cyprus and Greece, which are the only European countries that meet their climatic requirements. Groundnut oil is used especially in Asian cuisine, but it is also an ingredient in some cosmetics and medicines. Because of its high smoke point (229 °C), refined peanut oil is often used in deep-fat frying.[107] Besides food, peanut oil is used in alkyd resins, as carrier oil in crop protection and in mineral oil-free pulp and paint defoamers.[108] The global production of groundnut oil is 5.7 Mio mto a year; European groundnut oil production is about 98 000 mto a year. This exceeds European groundnut production since the European Union imports groundnuts and processes them into oil.[18] The main demand for groundnuts comes from the European Union, Canada and Japan. Although groundnuts have economic potential in Europe, the European climate prevents an increase in production.

Global sesame seed production is 3.6 Mio mto a year. Major producing countries include Mexico, India, China and Sudan. In 2009/2010, the world sesame oil production was 0.9 Mio mto.[26] As a part of the modern lifestyle, sesame is used in food and cuisine, cosmetics and wellness treatment and alternative medicine.[109] Although European demand is high, the crop potential is low. Because the sesame plant requires tropical climate conditions,[18] Italy is the only European sesame seed producer, yielding 1360 mto a year.

The European Union produced 12.6 Mio tons of olives in 2008; the rest of the world produces 5.4 Mio mto a year. In Europe the area harvested is 5 Mio ha. In 2008 olive oil production in Europe totalled 2.3 Mio mto, which is a major share of the global annual production of 3.1 Mio mto. The main growing areas in Europe are Spain, Italy, Greece, Portugal, Slovenia and Croatia. Olive oil is used mainly in cooking.[18,26] It is a major component of the diet in the countries surrounding the Mediterranean Sea. For people living in this region, olive oil is the main source of fat in their cuisine. In recent years, olive oil has also become more popular among consumers in Northern Europe, the US and Canada. Growing enthusiasm for the Mediterranean diet and for olive oil is due largely to studies indicating that this diet plays a positive role in the prevention of certain diseases, especially coronary heart disease.[110] Olive oil contributes flavours that are reflected throughout the whole dish. A good quality olive oil complements green vegetables. Traditional dishes are prepared

Introduction

with seasonal vegetables, various greens, parsley and grains. In vegetarian dishes, olive oils with herbal flavours are usually preferred. In salads or in cooking, olive oil is usually mixed with herbs and spices like oregano, rosemary or thyme, which are also important elements of the Mediterranean diet.

The following internationally recognised quality definitions were promulgated by the International Olive Oil Council (IOOC)[111] and have been adopted in the Joint FAO/WHO Food Standard Program:

> *Olive oil* is the oil obtained solely from the fruit of the olive tree (*Olea europaea* L.), to the exclusion of oils obtained using solvents or re-esterification processes and of any mixture with oils of other kinds. It is marketed in accordance with the following designations and definitions:
>
> *Virgin olive* oils are the oils obtained from the fruit of the olive tree solely by mechanical or other physical means under conditions, particularly thermal conditions, that do not lead to alterations in the oil, and which have not undergone any treatment other than washing, decantation, centrifugation and filtration.
>
> Virgin olive oils fit for consumption as they are include:
>
> (i) *Extra virgin olive oil*: virgin olive oil that has a free acidity, expressed as oleic acid, of not more than 0.8 grams per 100 grams.
> (ii) *Virgin olive oil*: virgin olive oil that has a free acidity, expressed as oleic acid, of not more than 2 grams per 100 grams.
> (iii) *Ordinary virgin olive oil*: virgin olive oil that has a free acidity, expressed as oleic acid, of not more than 3.3 grams per 100 grams.
>
> Virgin olive oil not fit for consumption as it is, designated lampante virgin olive oil, is virgin olive oil that has a free acidity, expressed as oleic acid, of more than 3.3 grams per 100 grams. It is intended for refining or for technical use.
>
> Refined olive oil is the olive oil obtained from virgin olive oils by refining methods that do not lead to alterations in the initial glyceridic structure. It has a free acidity, expressed as oleic acid, of not more than 0.3 grams per 100 grams.
>
> Olive oil is the oil consisting of a blend of refined olive oil and virgin olive oils fit for consumption as they are. It has a free acidity, expressed as oleic acid, of not more than 1 gram per 100 grams.
>
> Olive-pomace oil is the oil obtained by treating olive pomace with solvents or other physical treatments, to the exclusion of oils obtained by re-esterification processes and of any mixture with oils of other kinds.

Throughout human history olives have been hand harvested, a laborious process that represents the major proportion of the costs of production (Figure 1.19). Hand harvest is accomplished by three techniques: (1) collection of fallen fruit from the ground late in the growing season, (2) stripping of fruit directly

Figure 1.19 Traditional olive harvesting (Attic amphora attributed to the unknown Antimenes Painter, 530 BC–510 BC; reproduced with the kind permission of The British Museum, London).[112]

from the tree into picking bags or onto nets below the tree, (3) beating limbs with large sticks to dislodge fruit, which is also collected on nets.

However, mechanical olive gathering is gaining in most of the more intensive orchards around the world.[113]

European olive cultivation is strongly subsidised by the European Union, which guarantees intervention prices above world market prices. 'Approximately 2.5 Mio producers – roughly one-third of all EU farmers – are involved' in the olive sector; this reflects the special significance of European olive production. The maximum growing area has already been reached in the EU-27, and no further increases are expected.[114] However, other states bordering the Mediterranean Sea, particularly Turkey, are making efforts to increase their olive oil production.

In 2008 world production of linseed oil totalled 0.6 Mio mto, of which 200 000 mto are produced in Europe.[17] Producers in Europe are Belgium, Holland, France and Germany. The main growing areas outside the European Union are the USA, Russia, Canada, China, India and South America. Linseed oil is used for cooking, cosmetics, paints and animal feed.[115,116] European flax cultivation is of little importance due to a lack of competitiveness compared to alternative crops, and it is therefore subsidised by the European Union. In summary, flax cultivation in Europe is possible, but the monetary income it yields is currently low.[117]

Table 1.3 summarises the cultivation areas of the commodity oils described.

Another vegetable oil of significant importance for industrial uses is tall oil. Crude tall oil (CTO) is separated from black liquor in the Kraft sulfate pulping of mainly coniferous trees. Coniferous trees store triglycerides, fatty acids, resin

Introduction

Table 1.3 Oil crops: area harvested (ha) in 2008.[18]

Oil crop (commodities)	European Union	World
Soybeans	236 317	96 870 395
Cotton seed	373 306	31 432 045
Canola/rapeseed	6 127 566	30 308 662
Groundnuts	10 625	24 590 075
Sunflower seed	3 742 142	25 023 511
Sesame seed	265	7 534 201
Olives	5 022 512	10 839 026
Linseed	121 870	2 436 657

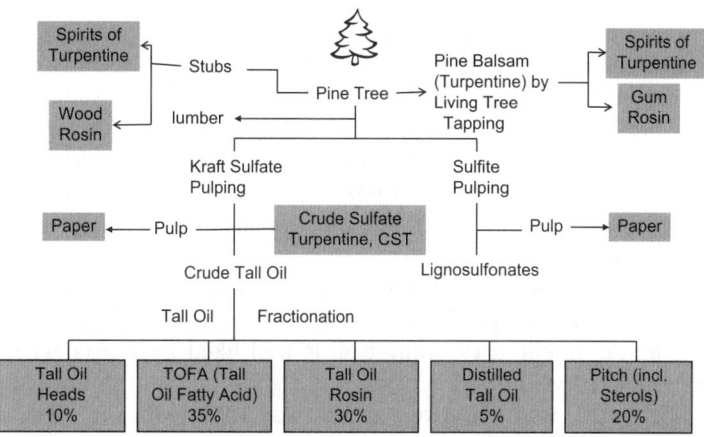

Figure 1.20 The tall value chain.

acids, sterols and sterol esters as nutrients in their parenchyma cells, while the radial resin ducts contain resin acids and turpentine for healing wounds in the bark (Figure 1.20). That is why pine balsam obtained by tapping is a source of rosin and terpenes but not of CTO.

During Kraft pulping, wood chips are cooked in an alkaline liquor to produce pulp. As the light organic phase of the cooker steam condenses, Crude Sulfate Turpentine (CST) is liberated. CST distillation creates α-pinene and β-pinene. Both pinenes are converted to pine oil and polyterpene resins that are used as tackifiers for adhesives and for chewing gums. Alkaline cooking converts resin and fatty acids into soaps. The soap is converted to crude tall oil (CTO) by acidulation with sulfuric acid. CTO is not a triacylglyceride like most vegetable oils, but is actually a mixture of five components with different boiling points, which are split by fractionation into heads (which boil first), then tall oil fatty acids (TOFA), distilled tall oil (DTO, a mixture of fatty and rosin acids), tall oil rosin (TOR) and pitch (the unsaponifiable residue).

TOFA is similar to seed oil oleic acid. Important applications of TOFA are the manufacture of TOFA esters as solvents, lubricants and carrier oils, alkyd resins and dimer acids. In 2008 European TOFA production capacity was 181 000 mto, or roughly 15% of European fatty acid production by fat splitting.[118,119] The supply of TOFA is ultimately constrained by the supply of CTO and the capacity of the Kraft pulping industry that is its source. TOFA is preferred in those applications where a high degree of unsaturation is required. In these applications, it competes with other unsaturated fatty acids, such as soy and rapeseed fatty acid.

1.3.4 Speciality Vegetable Oils

Specialty oils are oils that have only a small market volume. The demand is not high, because specialty oils are niche products. Among the specialty oils are nut oils, such as almond oil, macadamia nut oil, hazelnut oil and walnut oil. Cold or expeller pressed nut oils are the latest darlings of the cooking world. Most nut oils taste like the nut from which they are extracted. Nut oils in general are excellent used in salad dressing, over pasta with some cheese, in baked goods or for dipping French bread.[12] But nut oils are also particularly beneficial in cosmetic applications.

Walnut oil is a case in point. The walnut tree takes six years from planting to first harvest, and it can produce for up to 100 years.[58] In addition to its nutritional and cosmetic uses, walnut oil is well liked by artists. The Renaissance painters liked walnut oil because it is a good oil paint thinner and brush cleaner. The world production of walnuts was 1.7 Mio mto in 2008, and the European share totalled 182 016 mto.[18] The yield from one ton of walnuts is 500 litres of oil. The main suppliers are France, Italy, Turkey and Iran.[58]

The world production of pumpkin was 20.8 Mio mto in 2008.[18] The main producing countries are China and India. The European share of pumpkin production was 1.3 Mio mto in 2008; Austria and Slovenia are the traditional pumpkin seed oil producers in Europe. Pumpkin seed oil is used primarily in cooking, but it is also an active substance in folk medicine.[120] On the one hand, the European potential for producing pumpkin oil is high because the climate conditions are favourable. On the other hand, the production is expensive because the crop requires high labour input. Therefore, low-wage countries like India and China have a better competitive position due to their lower production costs.

1.3.5 Crop Growing Potential for Non-European Oils

Besides those oils already used in Europe, there are other potentially usable oils. Examples are lallemantia oil, safflower oil, common evening primrose oil, cannabis oil, castor oil, crambe oil and jatropha oil. These oils have been of little significance so far, and only few market data are available. In particular, the volume of production is difficult to determine.

Introduction

Lallemantia oil (also called oil of dragonhead) is a seed oil extracted from the seeds of the *Lallemantia iberica*. Its main growing area is Eastern Europe. Lallemantia oil has a fatty acid pattern very close to linseed oil and, like linseed oil, is a drying oil. It is mainly regarded as a vegetable oil source for purposes such as paint, lacquer and linoleum production.[121] Perennial cultivation trials in Thüringen have shown that lallemantia is an undemanding crop with good yield under favourable weather conditions. However, in years with cool and rainy weather the plant is very susceptible to infection by fungi, which reduces yield and oil content of the kernels to nearly half. Only new resistant cultivars would allow lallemantia to be grown successfully as an alternative or complement to flax oil.[35,62]

Safflower oil is an edible oil that has a fatty acid composition similar to sunflower oil. Because of its high linoleic acid content and high oxidative stability, it is used in salad oils and dietetic margarines. World production of the thistle-like safflower is around 0.6 Mio mto. The European production is roughly 70 mto a year.[18] As in the case of lallemantia, safflower is highly susceptible to fungi, and its production is not profitable in most northern European countries.[35]

The most important part of the evening primrose (*Oenothera*) is its seeds, which contain the oil. Evening primrose oil is used mainly for medical applications, such as treatment of allergies, eczema, symptoms of PMS, arthritis or diabetic neuropathy.[58] Evening primrose can be cultivated in Europe, and there are already some German farmers growing it for oil production.

Hemp is a potential agricultural crop for Europe because European climate conditions favour its cultivation. France is the world's largest producer of hemp seed, with a 31% share of world production, or an annual production of 26 213 mto. Hemp seed oil is used in cosmetics and medicines and as lamp oil. Its use as a fuel is insignificant because it is too expensive and not available in high quantities.[58]

The castor oil plant is a tropical plant produced mainly in India, which accounts for 70% of the global production of castor seeds, also called beans, or 1.5 Mio mto. Demand for castor oil comes primarily from Europe (EU-27), China, Japan and the USA. Castor oil is taken in conventional medicine as a laxative and used as an eye drop to treat some eye irritations. It is also an ingredient in some hair conditioners and skin products. Oncologists now use castor oil as a vehicle for delivering some chemotherapy drugs to cancerous tumours. Industrial applications of castor oil include soaps, lubricants and greases, coatings and adhesives, personal care and detergent, and surfactants and oleochemicals.[122] Castor oil is the raw material for sebacic acid, Turkey red oil and the green polyamide PA 11.

The chemical structure of castor oil (Figure 1.21) is unusual in that it is a ricinoleic acid triglyceride; thus, it has a hydroxyl functional group on the twelfth carbon of each of the three ricinoleic acid groups. The functional groups cause castor oil to be unusually polar and also allow chemical derivatisation that is not practical with most other seed oils.

The cultivation of the castor oil plant is poorly mechanised and labour intensive. Due to its climate requirements, it cannot be grown in Europe.[63]

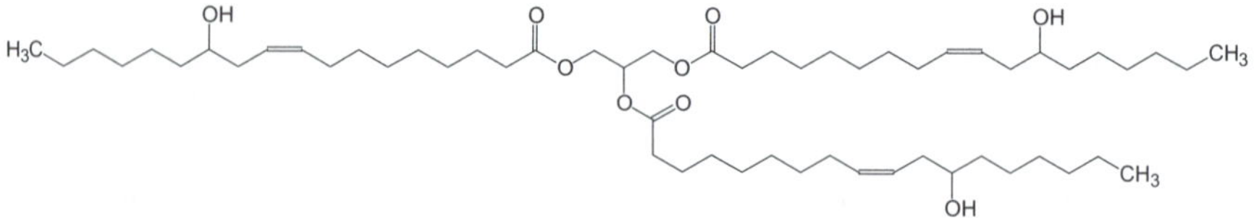

Figure 1.21 Chemical structure of castor oil.

Figure 1.22 *Jatropha curcas* (personal communication Jirawate Chedchant, Kasetsart University, Bangkok, and Laurent Vaysse, CIRAD, Montpellier).

Crambe is a plant in the mustard family and a source of erucic acid, which is produced exclusively from rapeseed and crambe. Moreover, crambe includes 9% more erucic acid than rapeseed. Nevertheless, the production volume is much smaller than that of rapeseed because there is no industrial demand and a lack of specialised industrial facilities. This is the main reason why only government research or private pilot projects have cultivated crambe so far.[123] If the industry changes its production, the demand for crambe will grow, and Europe is in a preferred climate zone for cultivating crambe.[35]

Jatropha is a succulent plant which grows in tropical and sub-tropical areas (Figure 1.22).

The jatropha seed is composed of an outer cover (31% of weight), an inner cover (26% of weight) and a kernel (43% of weight); the kernel contains 22% protein and 46% oil. With approximately 21% of the saturated palmitic and stearic acids, 42% oleic acid and 36% linoleic acid jatropha shows a fatty acid pattern not unlike that of cottonseed and peanut oil.[124] The cultivation volume is 1 Mio ha in several rural development projects in Africa and Asia. However, recent reports based on jatropha cultivation projects in Kenya and Mozambique have concluded 'that the dominant arguments about Jatropha as a food-security safe biofuel crop, a source of additional farm income for rural farmers, and a potential driver of rural development were misinformed at best and dangerous at worst'. The reports fuel internationally emerging evidence that Jatropha failed to meet expected outcomes, that it will not make a major contribution to solve the problems of climate change, energy security or poverty and in fact endangers food sovereignty and rural livelihoods.[125] In any case, climate conditions in the European Union are unfavourable for jatropha cultivation.[126,127]

Another specialty oil crop cultivated in a country adjacent to the Mediterranean Sea is the Argan tree. Argan oil prepared by mechanical extraction (cold pressing) from the almond-like kernels of the fruit satisfies the criteria of 'extra-virgin'. In terms of its composition, Argan oil offers a fatty acid pattern similar to jatropha and is exceptionally rich in polyunsaturated fatty acids (including linoleic acid, ω-6) and natural tocopherols. Argan oil is widely used by Moroccan women, both in their cooking for its taste and for many traditional beauty rituals. Argan oil, also called 'Berber's gold', is regarded as a precious oil in modern cosmetics with many varied applications in skin, hair and nail care.[128]

1.4 Summary and Conclusions

In summary, due to favourable biological and climate factors, the agronomic potential of various oil crops in Europe, such as rapeseed oil, olive oil, sunflower oil, linseed oil and nut oils, is high. The surface area covered by oil crops is expected to grow moderately. However, due to favourable climate conditions, the southern hemisphere of the Earth has become the dominant source for leading commodities, notably soybean oil and palm oil.

The creation of new vegetable oil cultivars that respond to consumer demands, market needs and ecological constraints could open opportunities for the European agro-industry. Vegetable oil species with high ω-3 fatty acid levels as alternatives to marine oils, for example, would offer an acceptable macronutrient level, on the one hand, and a sufficient supply of essential fatty acids, on the other, without further deteriorating the remaining marine fish stocks.

The importance of vegetable oils for energy generation will increase and become a new segment of vegetable oil use creating a noticeable impact on the applications for food and oleochemistry purposes. With prices for fossil crude oil increasing, maintaining the balance between the food, oleochemical and energy uses of vegetable oils will be a formidable challenge – not only on a regional basis in Europe but on a global basis in general.

Acknowledgement

The authors would like to thank Dieter Bockey, ufop, Berlin, for his kind assistance.

References

1. National Research Council (NRC, USA) *Biobased Industrial Products: Priorities for Research and Commercialization*, National Academic Press, Washington DC, 2000.
2. H. Zoebelin (ed.), *Dictionary of Renewable Resources*, Wiley-VCH, Weinheim, 2001, http://www.wiley-vch.de/publish/dt/books/bySubjectLS00/ISBN3-527-30114-3/?sID = e6eff66da518de522eba6d389014cef5
3. D. J. Morris and I. Ahmed, *The Carbohydrate Economy, Making Chemicals and Industrial Materials from Plant Matter*, Institute of Local Self Reliance, Washington DC, 1992.
4. B. Kamm, M. Kamm, K. Richter, B. Linke, I. Starke, M. Narodoslawsky, K. D. Schwenke, S. Kromus, G. Filler, M. Kuhnt, B. Lange, U. Lubahn, A. Segert and S. Zierke, Grüne BioRaffinerie Brandenburg – Beiträge zur Produkt- und Technologieentwicklung sowie Bewertung, Brandenburgische Umwelt Berichte, BUB 8, 260-69, ISSN 1434-2375, 2000.
5. L. R. Lynd, C. E. Wyman and T. U. Gerngross, *Biocommodity Engineering, Biotechnol. Progr.*, **15**, 777–793, 1999, http://dx.doi.org/10.1021/bp990109e
6. US Department of Agriculture (USDA) and US Department of Energy (DOE) (ed.), *Biomass as Feedstock for a Bioenergy and Bioproducts Industry: The Technical Feasibility of a Billion-Ton Annual Supply*, US Department of Energy, Office of Scientific and Technical Information, PO Box 62, Oak Ridge, TN, 2005.
7. B. Kamm and M. Kamm, *Appl. Microbiol. Biotechnol.*, 2004, **64**, 137–145, http://www.springerlink.com/content/r6budmagfule/
8. B. Kamm, M. Kamm and P. Gruber (ed.), *Biorefineries – Industrial Processes and Products*, Wiley-VCH, Weinheim, 2006, ISBN: 3-527-31027-4.
9. B. Kamm and M. Kamm, *Chem. Biochem. Eng. Q.*, 2004, **18**(1), 1–6, http://www.fkit.hr/cabeq/pdf/18_1_2004/Biorefinary%20CABEQ_2004_01.pdf
10. H. Röper, Perspektiven der industriellen Nutzung nachwachsender Rohstoffe, insbesondere von Stärke und Zucker. *Mitteilung der Fachgruppe Umweltchemie und Ökotoxikologie der Gesellschaft Deutscher Chemiker*, 2001, **7**(2), 6–12, http://www.oekochemie.tu-bs.de/ak-umweltchemie/mblatt/2001/b1h201.pdf
11. Y. Y. Linko and P. Javanainen, *Enzym. Microb. Tech.*, 1996, **19**, 118–123, doi:10.1016/0141-0229(95)00189-1.
12. K. J. Zielinska, K. M. Stecka, A. H. Miecznikowski and A. M. Suterska, *Pr. Inst. Lab. Badaw. Przem. Spozyw.*, 2000, **55**, 22–29.
13. B. Kamm, M. Kamm, M. Schmidt, I. Starke and E. Kleinpeter, *Chemosphere*, 2006, **62**, 97–105, doi:10.1016/j.chemosphere.2005.03.073.
14. T. Werpy and G. Petersen (ed.), *Top Value Added Chemicals from Biomass*, US Department of Energy, Office of Scientific and Technical Information, 2004, No. DOE/GO-102004-1992, www.osti.gov/bridge

15. B. Dale, *Encyclopedia of Physical Science and Technology*, 3rd edn, vol. 2, pp. 141–157, 2002.
16. S. Kromus, B. Kamm, M. Kamm, P. Fowler and M. Narodoslawsky, in *Biorefineries – Industrial Processes and Products*, ed. B. Kamm, M. Kamm and P. Gruber, Wiley-VCH, Weinheim, 2006, vol. 1, pp. 253–294, ISBN: 3-527-31027-4, http://www.wiley-vch.de/publish/dt/books/bySubjectCH00/ISBN3-527-31027-4/?sID = 8814a51c10ced4486b38faa8076c735b
17. M. Ringpfeil, *Biobased Industrial Products and Biorefinery Systems – Industrielle Zukunft des 21. Jahrhunderts*, 2001, www.biopract.de
18. K. D. Vorlop, Th. Willke and U. Prüße, in *Biorefineries – Industrial Processes and Products*, ed. B. Kamm, M. Kamm and P. Gruber, Wiley-VCH, Weinheim, 2006, vol. 2, pp. 385–406, ISBN: 3-527-31027-4.
19. J. F. Jenck, F. Agterberg and M. J. Droescher, *Green Chem.*, 2004, **6**, 544–556.
20. DuPont, *US Pat.* 5 686 276, 2004, http://www.dupont.com/sorona/home.html
21. O. Wurz, *Zellstoff- und Papierherstellung aus Einjahrespflanzen*, Eduard Roether Verlag, Darmstadt, 1960.
22. J. J. Bozell, in *Encyclopedia of Plant and Crop Science*, ed. R. M. Goodman, Dekker, New York, 2004, 0-8247-4268-0.
23. C. Webb, A. A. Koutinas and R. Wang, *Adv. Biochem. Eng./Biotechn.*, 2004, **87**, 195–268, http://www.springerlink.com/content/lt7h0rcclaj8/
24. R. V. Nonato, P. E. Mantellato and C. E. V. Rossel, *Appl. Microbiol. Biotechnol.*, 2001, **57**, 1–5, http://www.springerlink.com/content/3eye5-a15bw9ttrcp/?p=7c312bb21cb64b858dd56c457d5c97e0&pi=0
25. C. E. V. Rossel, P. E. Mantellato, A. M. Agnelli and J. Nascimento, in *Biorefineries – Industrial Processes and Products*, ed. B. Kamm, M. Kamm and P. Gruber, Wiley-VCH, Weinheim, 2006, vol. 1, pp. 209–226, ISBN: 3-527-31027-4.
26. A. Fiechter, *Plastics from Bacteria and for Bacteria: Poly(β-hydroxyalkanoates) as Natural, Biocompatible, and Biodegradable Polyesters*, Springer-Verlag, New York, 1990, pp. 77–93.
27. F. Rexen, *New industrial application possibilities for straw*, Documentation of Svebio Phytochemistry Group (Danish), Fytokemi i Norden, Stockholm, Sweden, 1986-03-06, 1986, 12.
28. J. Coombs and K. Hall, in *Cereals – Novel Uses and Processes*, ed. G. M. Campbell, C. Webb and S. L. McKee, Plenum Publ. Corp., New York, 1997, pp. 1–12.
29. E. Audsley and J. E. Sells, in *Cereals – Novel Uses and Processes*, ed. G. M. Campbell, C. Webb and S. L. McKee, Plenum Publ. Corp., New York, 1997, pp. 191–294.
30. A. J. Hacking, in *Economic Aspects of Biotechnology*, Cambridge University Press, New York, 1986, pp. 214–221.
31. Th. Willke and K. D. Vorlop, *Appl. Microbiol. Biotechnol.*, 2004, **66**(2), 131–142.
32. R. Carlsson, in *Handbook of Plant and Crop Physiology*, ed. M. Pessarakli, Marcel Dekker Inc., New York, 1994, pp. 941–963.

33. N. W. Pirie, *Leaf Protein – Its Agronomy, Preparation, Quality, and Use*, Blackwell Scientific Publications, Oxford/Cambridge, UK, 1971.
34. N. W. Pirie, *Leaf Protein and Its By-products in Human and Animal Nutrition*, Cambridge University Press, UK, 1987.
35. R. Carlsson, in *The Green Biorefinery, Proceedings of 11th International Green Biorefinery Conference, Neuruppin, Germany, 1997*, ed. S. Soyez, B. Kamm and M. Kamm, Verlag GÖT, Berlin, 1998, ISBN 3-929672-06-5.
36. R. Carlsson, in *Leaf Protein Concentrates*, ed. L. Telek and H. D. Graham, AVI Publ. Co., Inc., Westport, CT, 1983, pp. 52–80.
37. L. Telek and H. D. Graham (ed.), *Leaf Protein Concentrates*, AVI Publ. Co., Inc., Westport, CT, 1983.
38. R. J. Wilkins (ed.), *Green Crop Fractionation*, The British Grassland Society, c/o Grassland Research Institute, Hurley, Maidenhead, SL6 5LR, UK, 1977.
39. I. Tasaki (ed.), *Recent Advances in Leaf Protein Research, Proc. 2nd Int. Leaf Protein Res. Conf.*, Nagoya, Japan, 1985.
40. P. Fantozzi (ed.), *Proc. 3rd Int. Leaf Protein Res. Conf.*, Pisa-Perugia-Viterbo, Italy, 1989.
41. N. Singh (ed.), *Green Vegetation Fractionation Technology*, Science Publ. Inc., Lebanon, NH 03767, USA, 1996.
42. D. H. White and D. Wolf, *Research in Thermochemical Biomass Conversion*, ed. A. V. Bridgewater and J. L. Kuester, Elsevier Applied Science, New York, 1988.
43. National Renewable Energy Laboratory (NREL), 2005, http://www.nrel.gov/biomass/biorefinery.htm
44. C. Okkerse and H. van Bekkum, *Green Chem.*, 1999, **4**, 107–114.
45. B. Kamm, M. Kamm and K. Soyez (ed.), *Die Grüne Bioraffinerie/The Green Biorefinery, Technologiekonzept, 1st International Symposium Green Biorefinery/Grüne Bioraffinerie*, Oct. 1997, Neuruppin, Germany, Proceedings, Berlin, 1998, ISBN 3-929672-06-5.
46. M. Narodoslawsky (ed.), *Green Biorefinery 2nd International Symposium Green Biorefinery*, October 13–14, 1999, Feldbach, Austria, Proceedings, SUSTAIN, Verein zur Koordination von Forschung über Nachhaltigkeit, Graz TU, Austria, 1999.
47. Biomasse Action Plan, 2005, http://www.euractiv.com/en/energy/biomass-action-plan/article-155362
48. EU-Projekt BIOPOL Specific Support Action, Priority Scientific Support to Policies, 2007, http://www.biorefinery.nl/biopol
49. EU-Projekt Biorefinery-Euroview Specific Support Action, Priority Scientific Support to Policies, 2007, http://www.biorefinery-euroview.eu
50. B. Mahro and M. Timm, *Eng. Life Sci.*, 2007, **7**(5), 457–468, DOI: 10.1002/elsc.200620206.AQ
51. J. Michels and K. Wagemann, *Biofuels Bioprod. Bioref., Special Issue Biorefinery*, 2010, **4**, 263–267.
52. A. Baylis, *Biofuels Bioprod. Bioref.*, 2010, **4**, 115–117.

53. B. Kamm, Ch. Hille, P. Schönicke and G. Dautzenberg, *Biofuels Bioprod. Bioref., Special Issue Biorefinery*, 2010, **4**, 253–262.
54. S. Leiß, J. Venus and B. Kamm, *Chemie Ingenieur Technik*, 2010, **82**(7), 1091–1095.
55. *Halmgutartige Biomasse: Ölsaatenstroh* In: Martin Kaltschmitt, Hans Hartmann, Hermann Hofbauer (Hrsg.): *Energie aus Biomasse. Grundlagen, Techniken und Verfahren*. Springer Verlag, Berlin und Heidelberg 2009; ISBN 978-3-540-85094-6.
56. Private information, Rape farmer, Region Potsdam-Mittelmark, 2008.
57. G. A. Reinhardt, *Energy and CO_2 balance of renewable commodities. Theoretical fundamentals and case study rape seed*. Vieweg, Wiesbaden, Germany, 1993.
58. K. Weber, in *International Congress on Oilseed and Oils*, New Dehli, 1979.
59. Federal Minister for Food, Agriculture and Forestry (ed.), *Report of the Federal and State Governments on Renewable Raw Materials*, BT-Printed Matter, Bonn, 1989.
60. T. Lechler, *Die Ernährung als Einflussfaktor auf die Evolution des Menschen*, Dissertation, Univers. Hannover, 2001.
61. H. Küster in *The Cambridge World History of Food*, K. F. Kiple and K. Conèe Ornelas (eds.), Cambridge Univers. Press: Cambridge, 2000.
62. http://www.leindotter.de/tradition.html (retrieved 25.08.2010).
63. M. Nestle, *Mediterranean diets: historical and research overview*, Am J Clin Nutr, 1995, **61**, 1313S–1320S.
64. A. Trichopoulo and P. Lagiou, *Healthy traditional Mediterranean diet: Expression of culture, history, lifestyle*, Nutrition Reviews, 1997, **55**, 383–389.
65. F. Sofi, M. Cesari, R. Abbate, G. F. Gensini and A. Casini, *Adherence to Mediterranean diet and health status: meta-analysis*, BMJ (Clinical research ed.) **337**, a1344 (2008) http://bmj.com/cgi/pmidlookup?view=long&pmid=18786971.
66. European Commission, Directorate-General for Agriculture and Rural Development, *The Common Agricultural Policy explained*, s.a.; http://ec.europa.eu/agriculture/publi/capexplained/cap_en.pdf (retrieved 21.09.2010).
67. D. Bockey, *Potentials for raw materials for the production of biodiesel*, ufop, Berlin, s.a., http://www.ufop.de/downloads/ufop_brochure_06.pdf (retrieved 22.09.2010).
68. E. Dinjus, U. Arnold, N. Dahmen, R. Höfer and W. Wach, *Green Fuels – Sustainable Solutions for Transportation*, in Sustainable Solutions for Modern Economies, R. Höfer (ed.), RSC Publ.: Cambridge, 2009.
69. M. Kaltschmitt and D. Thrän, *Biomass-based Green Energy Generation*, in Sustainable Solutions for Modern Economies, R. Höfer (ed.), RSC Publ.: Cambridge, 2009.
70. M. Kaltschmitt, U. Andrée and St. Majer, *Co-processing of Vegetable Oils in Crude Oil Refineries – Opportunities and Limitations*, Erdöl – Erdgas – Kohle, 2010, **126**(5), 203–210.

71. K. Hill and R. Höfer, *Natural Fats and Oils*, in Sustainable Solutions for Modern Economies, R. Höfer (ed.), RSC Publ.: Cambridge, 2009.
72. J. Heller, *Physic Nut Jatropha Curcas L.*, Gatersleben, 1996.
73. L. Baldoni and A. Belaj, *Olive*, in Oil Crops (Handbook of Plant Breeding), J. Vollmann and R. Istvan (eds.), Vienna, pp. 397–422, 2009.
74. E. Cahoon, H. Damude and A. Kinney, *Modifying Vegetable Oils for Food and Non-food Purposes*, in Oil Crops (Handbook of Plant Breeding), J. Vollmann and R. Istvan (eds.), Vienna, pp. 31–56, 2009.
75. R. Kühl and V. Hart, *Marktstruktur- und Verwendungsanalyse von Öl- und Eiweißpflanzen*, Ufop Schriften Heft 34, Berlin, 2010.
76. United States Department of Agriculture, *Oilseeds: World Markets and Trade*, http://www.fas.usda.gov/oilseeds/circular/2009/December/oilseedsfull12-09.pdf, 2009, (retrieved 25.08.2010).
77. Faostat, *FAOSTAT*, http://faostat.fao.org/, 2010, (retrieved 25.08.2010).
78. European Union Directorate-General for Agriculture and Rural Development, *Agriculture in the European Union – Statistical and Economic Information*, http://ec.europa.eu/agriculture/agrista/2009/table_en/2009enfinal.pdf, 2009, (retrieved 25.07.2010).
79. Eurostat, *Main Tables for Agriculture*, http://epp.eurostat.ec.europa.eu/portal/page/portal/agriculture/data/main_tables, 2010, (retrieved 18.07.2010).
80. Commission of the European Communities, *Commission Staff Working Document on Risk and Crisis Management in Agriculture*, http://ec.europa.eu/agriculture/publi/communications/risk/workdoc_en.pdf, 2005, (retrieved 18.07.2010).
81. Ista Mielke, *Palm Oil Price Cycles: The Factors Determining Price-making of Palm Oil and of Oils Generally*, Hamburg, 2003.
82. http://www.thefreelibrary.com/Cognis'+ageless+beauty+rejuvenating+shampoo-a0190748603 (retrieved 25.08.2010).
83. M. P. Malveda, M. Blagoev, C. Funada, *Natural Fatty Acids*, ed. SRI Consulting, CEH Marketing Research Report, March 2009.
84. *Canola Varieties*, http://www.canolacouncil.org/chapter2.aspx (retrieved 26.08.2010).
85. Ista Mielke, *Oil World Annual 2010*, Hamburg, 2010.
86. http://www.aak.com/Global/Brochures/Rapeseedoil_from_AAK.pdf (retrieved 26.08.2010).
87. C. A. Barth, *Nutritional value of rapeseed oil and its high oleic/low linolenic variety – A call for differentiation*, European Journ. Lipid Science and Technol., 2009, **111**(10), 953–956.
88. V. Cardoza and C. Stewart, *Transgenic Crops VI: Canola*, Biotechnology in Agriculture and Forestry, 2007, **61**, 29–37.
89. B. Gutsche, *Technologie der Methylesterherstellung — Anwendung für die Biodieselproduktion*, Fett/Lipid, 1997, **99**(12), 418–427.
90. F. Bohmert, *75 Jahre Henkel Glycerin*, Schriften des Werksarchivs, Bd. 18, Henkel KGaA, Düsseldorf, 1985; R. Christoph, B. Schmidt, U. Steinberner and W. Dilla, *Glycerol*, Ullmann's Encyclopedia of Industrial

Chemistry, 6th Ed., Wiley-VCH, Weinheim, 2000; A. Behr, J. Eilting, K. Irawadi, J. Leschinski and F. Lindner, *Improved utilisation of renewable resources: New important derivatives of glycerol, Green Chem.*, 2008, **10**, 13–30; M. Pagliaro and M. Rossi, *The Future of Glycerol*, 2nd Ed., RSC Publ.: Cambridge, 2010.

91. R. Höfer and J. Bigorra, *Biomass-based Green Chemistry – sustainable solutions for modern economies*, 7th Green Chemistry Conference, Barcelona, 2007.
92. EREC – European Renewable Energy Council, *Biofuels in the EU legislation*, Brussels, 2007, http://www.erec.org/renewableenergysources/biofuels.html (retrieved 22.09.2010).
93. UFOP Bericht 2008/2009, Berlin (2009), http://www.ufop.de/downloads/UFOP_GB_2008_09_WEB.pdf.
94. *FAPRI 2010 U.S. and World Agricultural Outlook*, Iowa State Univers.: Ames, 2010, http://www.fapri.iastate.edu/outlook/2010/ (retrieved 26.08.2010).
95. Factsheet Soy 2009, *Product Board MVO*, April 2009, http://www.mvo.nl/Portals/0/statistiek/nieuws/2009/MVO_Factsheet_Soy_2009.pdf (retrieved 26.08.2010).
96. J. Southworth, R. Pfeifer, M. Habeck, J. Randolph, O. Doering, J. Johnston, D. Rao, *Changes in Soybean Yields in the Midwestern United States as a Result of Future Changes in Climate, Climate Variability and CO_2 Fertilization, Climatic Change*, 2002, **53**, 447–475.
97. Ensa, *All about Soyfoods*, http://www.ensa-eu.org/documents/all_about_soyfood_en.pdf, 2010, (retrieved 25.07.2010).
98. P. Golbitz, *Soya & Oilseed Blue Book*, Soyatech Inc.: Bar Harbor (2000); T. Wang, *Soybean oil*, in VEGETABLE OILS IN FOOD TECHNOLOGY: Composition, Properties and Uses, F. D. Gunstone, Ed., Blackwell/CRC: Oxford, Boca Raton, 2002.
99. E. L. Ashton, J. D. Best and M. J. Ball, *J. Am. Coll. Nutr.*, 2001, **20**(4), 320; A. Allman-Farinelli, K. Gomes, E. J. Favaloro and R. A. Petocz, *JADA*, July 2005, **105**(7), 1071–1079.
100. T. Graf, G. Wurl, A. Biertümpfel, *Chancen und Möglichkeiten der Bereitstellung maßgeschneiderter Pflanzenöle*, in Fachagentur Nachwachsende Rohstoffe (Hrsg.), Gülzower Fachgespräche, Lacke und Farben aus nachwachsenden Rohstoffen, 18-32, FNR, Gülzow, 1999.
101. H. Käb, *Marktanalyse – Industrielle Einsatzmöglichkeiten von High Oleic Pflanzenölen*, in Fachagentur Nachwachsende Rohstoffe (Hrsg.), Gülzower Fachgespräche, Bd. 19., FNR, Gülzow, 2001, http://www.fnr-server.de/ftp/pdf/literatur/pdf_32gfg19_ho_sonnenblume.pdf.
102. http://www.sunflowernsa.com/health/default.asp?contentID=45 (retrieved 20.08.2010).
103. G. A. Pereyra-Iruja, N. G. Izquierdo, M. Covi, S. M. Nolasco, F. Quiroz and L. A. N. Aguirrezábal, *Variability in sunflower oil quality for biodiesel production: A simulation study*, *Biomass and Bioenergy*, March 2009, **33**(3) 459–468.

104. The National Sunflower Association, *The National Sunflower Association*, http://www.sunflowernsa.com/, 2010, (retrieved 18.07.2010).
105. T. Nagata, H. Löorz, J. Widholm, M. Khadi, V. Santhy, M. Yadav, *Cotton: An Introduction*, in Cotton, U. Zehr (ed.), Berlin and Heidelberg, pp. 14–65, 2010.
106. National Cottonseed Products Association, *Cottonseed Oil Fact Sheet*, http://www.cottonseedoiltour.com/cso-factsheet/, 2010, (retrieved 28.07.2010).
107. J. G. Woodroof, *Peanut oil*, in Peanuts—Production, Processing, Products, J. G. Woodroof, ed., 293–307, AVI Publ.: Westport, 1983.
108. R. Höfer, F. Jost, M. J. Schwuger, R. Scharf, J. Geke, J. Kresse, H. Lingmann, R. Veitenhansl, W. Erwied, Foams and Foam Control, in *Ullmann's Encyclopedia of Industrial Chemistry*, A11, 465, VCH Weinheim, 1988.
109. W. Collinge, *The American Holistic Health Association Complete Guide to Alternative Medicine*, Warner Books, New York, 1996; S. P. Kochhar, *Sesame, Rice-bran and Flaxseed oils*, in VEGETABLE OILS IN FOOD TECHNOLOGY: Composition, Properties and Uses, F. D. Gunstone, Ed., Blackwell/CRC: Oxford, Boca Raton 2002; G. Manteljan, *The World's Healthiest Foods – Essential Guide for the Healthiest Way of Eating*, gmf pub, Seattle, 2007.
110. D. Boskou, *Olive Oil*, in VEGETABLE OILS IN FOOD TECHNOLOGY: Composition, Properties and Uses, F. D. Gunstone, Ed., Blackwell/CRC: Oxford, Boca Raton, 2002.
111. COUNCIL REGULATION (EC) No 865/2004 of 29.4.2004 on the common organisation of the market in olive oil and table olives and amending Regulation (EEC) No 827/68; International Olive Oil Council, *TRADE STANDARD APPLYING TO OLIVE OILS AND OLIVE-POMACE OILS*, COI/T.15/NC No 3/Rev. 4, November 2009.
112. W. Nicol, *A Catalogue of the Greek and Etruscan Vases in the British Museum*, London, 1851, Old Catalogue 538, Vase B226, CVA British Museum 4 III H e Pl. 55, 4.
113. L. Ferguson, *Trends in Olive Harvesting*, Grasas y Aceites, Enero-Marzo 2006, **57**(1), 1–7.
114. European Commission, *The Olive Oil Sector in the European Union*, ec.europa.eu/agriculture/publi/fact/oliveoil/2003_en.pdf, 2003, (retrieved 28.07.2010).
115. S. Thormahlen, *Sustainable Solutions for Nutrition: A Consumer Expectation*, in Sustainable Solutions for Modern Economies, R. Höfer (ed.), RSC Publ.: Cambridge, 2009.
116. H. Nieder, *Chancen für Leinöl – der Markt für Farben und Lacke*, in Fachagentur Nachwachsende Rohstoffe (Hrsg.), Nachwachsende Rohstoffe – Von der Forschung zum Markt, 85-92, FNR, Gülzow, 1998.
117. S. Kirst, G. Buchbauer, C. Klausberger, *Lexikon der pflanzlichen Fette und Öle*, Vienna, 2007.
118. M. Kjellin, I. Johansson, *Surfactants from Renewable Resources*, Chippenham, 2010.

119. Greno http://greno-forstservice.de/images/Berichte.htm, 2010, (retrieved 28.07.2010).
120. T. Lelley, L. Brent, *Hull-Less Oil Seed Pumpkin*, in *Oil Crops (Handbook of Plant Breeding)* J. Vollmann, R. Istvan (eds.), Vienna, pp. 469–492, 2009.
121. Tlu, *Anbautelegramm Iberischer Drachenkopf*, http://www.tll.de/ainfo/pdf/idra0708.pdf, 2008, (retrieved 28.07.2010); G. Wurl, T. Graf, A. Biertümpfel and A. Vetter, *Ergebnisse mehrjähriger Anbauversuche mit Iberischem Drachenkopf (Lallemantia iberica) als Linolensäurelieferant* http://www.tll.de/ainfo/pdf/aidr0403.pdf (retrieved 29.08.2010).
122. F. C. Naughton, *Production, chemistry, and commercial applications of various chemicals from castor oil*, *JAOCS*, 1974, **51**(3) 65; D. S. Ogunniyi, *Castor oil: A vital industrial raw material*, *Bioresource Technology*, June 2006, **97**(9) 1086–1091; http://castoroil.in/downloads/castor_oil_report_preview_ebook.pdf; H. Mutlu and M. Meier, *Castor Oil as Renewable Resource of the Chemical Industry*, *European Journal of Lipid Science and Technology*, 2010, **112**, 10–30.
123. R. Hansen, *Crambe Profile*, http://www.agmrc.org/commodities__products/grains__oilseeds/crambe_profile.cfm 2009, (retrieved 28.07.2010).
124. J. Chedchant, *Study on quality and quantity of oil extracted from Jatropha curcas L.*, Kasetsart University, Bangkok, s.a.
125. D. Ribeiro and N. Matavel, *Jatropha! A socio-economic pitfall for Mozambique*, Justiça Ambiental & União Naconal de Camponeses, August 2009; Endelevu Energy, World Agroforestry Centre, Kenya Forestry Research Institute, *Jatropha Reality Check: A field assessment of the agronomic and economic viability of Jatropha and other oilseed crops in Kenya*, Deutsche Gesellschaft für Technische Zusammenarbeit (GTZ), Eschborn, December 2009.
126. o.V., *Jatropha curcas – eine genügsame Pflanze für die Biodiesel-Produktion*, Entwicklung und ländlicher Raum, 2006, **6**, 25–28.
127. R. Devappa, R. Maes, H. Makkar, W. De Greyt and K. Becker, *Quality of Biodiesel Prepared from Phorbol Ester Extracted Jatropha Curcas*, 2010.
128. Stussi, F. Henry, Ph. Moser, L. Danoux, C. Jeanmaire, V. Gillon, I. Benoit, Z. Charrouf, G. Pauly, *Argania spinosa – How Ecological Farming, Fair Trade and Sustainability Can Drive the Research for New Cosmetic Active Ingredients*, *SÖFW-Journal*, 10-2005, **131**, 35–46; C. d'Erceville, F. Henry, P. Lago and A. Rathjens, *Plant-based Biologically Active Ingredients for Cosmetics*, in Sustainable Solutions for Modern Economies, R. Höfer (ed.), RSC Publ.: Cambridge, 2009.

CHAPTER 2
Farming and Harvesting

KATERINA STAMATELATOU,*[a] DAVID TURLEY,[b] RUTH LAYBOURN,[c] FRANCIS FLÉNET,[d] ALAIN QUINSAC,[e] RAY MARRIOTT,[f] GEORGIA ANTONOPOULOU,[g] GERASIMOS LYBERATOS,[g] ANTOINE ROUILLY[h] AND CARLOS VACA-GARCIA[h]

[a] Institute of Chemical Engineering and High Temperature Chemical Processes, PO Box 1414, 26504, Patras, Greece; current address: Department of Environmental Engineering, Democritus University of Thrace, GR-67100, Xanthi, Greece; [b] National Non-Food Crops Centre, York, UK; [c] Food and Environment Research Agency, Sand Hutton, York, UK; [d] CETIOM – Centre de Grignon, Avenue L. Brétignières, 78850 Thiverval-Grignon, France; [e] CETIOM, 11, rue Monge, Parc Industriel, 33600 Pessac, France; [f] Green Chemistry Centre of Excellence, Chemistry Department, University of York, York, YO10 5DD, UK; [g] Institute of Chemical Engineering and High Temperature Chemical Processes, PO Box 1414, 26504, Patras, Greece; [h] Laboratoire de Chimie Agro-Industrielle UMR 1010 INRA/INP-ENSIACET, 118, Route de Narbonne, F-31077 Toulouse Cedex 04, France

2.1 Introduction

The competitiveness and sustainability of the biodiesel and vegetable oil market can be achieved through increasing the quantity of total biomass cultivated and the yield of the vegetable oil. Furthermore, pre-treatment technologies aiming to increase the bulk density and decrease the water content of the biomass can be employed to reduce the cost of biomass transportation and storage. The valorisation of the by-products (straw, stalk, leaves) through conversion processes into high value-added chemicals and biomaterials as well as energy also contributes to improving the economics of the whole biorefinery scheme.

2.2 Increasing Oil Yield

Oilseed rape and sunflower are important oil-rich crops grown in Europe. Improving the vegetable oil and biomass yields is possible through the employment of specific strategies as identified and assessed in the following sections.

2.2.1 The Production of Oilseed Rape

2.2.1.1 Introduction

Commercial oilseed rape cultivars are derived from the genus Brassica and the family Cruciferae. Within this family, numerous species have been inter-bred to form a number of sub-species. Most commercial cultivars are derived from hybrids of *B. nigra, B. oleracea* and *B. rapa* (synonymous with *B. campestris*), which through hybridisation produced the new species *B. carinata* (a more drought-tolerant brassica oilseed), *B. juncea* and *B. napus*. The term 'Canola' is also widely used in reference to oilseed rape, particularly in North America. The term was introduced by Canadian breeders as a marketing name for oilseed rape cultivars producing seeds that are both low in glucosinolates and the fatty acid erucic acid, also referred to as double-low (or 'OO') and are used in food applications. These were originally developed to reduce the erucic acid content to less than 1% to improve human digestibility. Seed glucosinolate levels were also reduced to improve ruminant and monogastric palatability and digestion of the oil meal produced as a by-product of oil extraction, which led to widespread use of rapemeal as a protein supplement in animal feed.

B. napus and *B. rapa* (slightly lower yielding than *B. napus*) are the two most important widely grown species. Spring/summer types tend to be grown in North America whilst a mixture of spring/summer and predominantly winter-tolerant varieties are grown in Europe. *B. juncea* and *B. rapa* are mainly grown in India and the Far East. Globally, *B. napus* is the dominant commercial species.

Winter cultivars require a vernalisation period of a few weeks where air temperatures are below 5 °C. Spring cultivars require a relatively short growth period and are therefore cultivated in areas with very severe winters. Oilseed rape is suited to cooler temperate climates and produces the highest oil yield of any crop in such situations, whereas warmer and drier climates favour sunflower and olive production.

2.2.1.1.1 Commercial Importance of Oilseed Rape. The global production area of oilseed rape (OSR) has grown from just 7 million hectares in 1965 to currently around 29.6 million hectares, with China currently growing the largest amount of OSR by area. Current world production is around 52 million tonnes of seed. In Europe, the planting of oilseed rape increased significantly during the 1970s to reduce the food vegetable oil deficit in the EU. Growers realised that, despite its comparatively low yields, it was a valuable and effective break crop in cereal-dominated arable rotations and provided good financial rewards for little input in comparison to other arable crops. Within the UK alone, the area under oilseed rape increased from 5000 hectares in 1971 to 0.4 million hectares in 1999, an 83-fold increase in less than 30 years (National Statistics Online). The EU as a whole is the largest producer of oilseed rape in the world (18 million tonnes) followed by China (11 million tonnes), Canada (9.5 million tonnes) and India (6.5 million tonnes). Rapeseed production accounts for 70% of all oilseed production within the EU-27 and the EU currently meets its internal demand from domestic production, importing only small volumes of seed or rape oil.

2.2.1.1.2 Main Issues for Growers. As a small seed with few water reserves, dry conditions at sowing can compromise crop emergence leading to poor and patchy crop establishment. Oil production is also reliant on a large plant canopy to photosynthetically capture solar energy, which equates to a relatively large nitrogen requirement (similar to that of winter cereals). In recent years sulfur applications have also increased in importance. In the wetter climates that favour oilseed rape production fungal diseases can be a common problem significantly affecting yields. Oilseed rape crops are also threatened by a number of insect pest species. However, in contrast, the dense canopy of oilseed rape means that once well established, it can effectively suppress competing weed species. An additional problem for growers is lodging (stem bending or failure and root dislodging) in oilseed rape, typically caused by heavy rains or wind events close to harvest time. If this occurs in oilseed rape crops it can significantly reduce yield.

2.2.1.2 Current Yields of Oilseed Rape

The average seed yield of oilseed rape crops in Europe is around 3–3.3 t/ha, falling to 1.8 t/ha in China and Canada, and 0.8 t/ha in India (in addition to climatic effects, such differences relate to the type of Brassica species grown and whether spring or winter types). Climatic conditions as well as technical developments make Europe the most productive region for oilseed rape. Low yields in Canada, and also in Nordic states within the EU, reflect the use of lower-yielding spring-sown cultivars. In most of Europe, higher yielding winter-sown varieties are grown, taking advantage of the long, cool growing season.

Yields of oilseed rape vary due to many factors such as climate (susceptibility to winter frost and summer droughts), disease, insect pests and soil type. Drought stress conditions typically restrict the areas of oilseed rape production to more northerly latitudes in the EU.

The yield potential of oilseed rape can be improved by both selective breeding and genetic manipulation but also by improving crop management practices to enable realisation of the full genetic potential. Without further increase in yield, the ability to develop very large-scale industrial uses for oilseeds will be limited by the amount of land area available for growing industrial

Farming and Harvesting 51

crops, as intensifying use of existing land is limited by environmental, rotational, agronomic and economic constraints.

2.2.1.2.1 Residual Straw Yields and Fertilise Value. High yielding oilseed rape crops have a harvest index (total seed yield (100% DM)/total above ground dry matter yield) of around 0.3. In lower yielding situations this can fall significantly. At current typical average yields of 3–3.5 t/ha (9% moisture) this gives rise to above-ground straw and pod dry matter yields of around 9–11 tonnes/ha. However, with conventional harvest (leaving an above-ground stubble) and current mechanical bailing equipment, typically only around 60% of this (mostly straw) can effectively be recovered from the field, giving field yields of 5–7 dry tonnes of straw/hectare.

Oilseed rape straw has a nominal fertiliser value. The typical nutrient content of oilseed rape straw is given in Table 2.1, though this will vary depending on fertiliser application and soil nutrient content.

2.2.1.3 Agronomic Practices to Improve the Yield of Oilseed Rape

2.2.1.3.1 Crop Nutrition. In response to lower returns on oilseed crops in Europe in the late 1980s and early 1990s there was a general decline in nitrogen applications to oilseed rape crops, but this has since reversed as economic prospects have improved. Application of around 200 kg N/ha is typical in European situations, but may have been as high as 280 kg/ha N in the early 1980s when some nitrogen was also typically applied in the autumn.

Sulfur availability has been identified as a key factor affecting the yield of oilseed rape. With declining atmospheric deposition, due to cleaning up of sulfur dioxide emissions from fossil fuel burning and other emission sources, and changing practice in moving away from fertilisers containing sulfur (*e.g.* ammonium sulfate) sulfur levels in soils generally have been declining. Oilseed rape is sensitive to sulfur deficiency as it has a high demand (16 kg sulfur per tonne of seed).[1] Many crops now receive a sulfur dressing, though it is commented that many farmers do not apply enough to prevent deficiency from limiting yield potential.[2]

2.2.1.3.2 Disease Control. Booth *et al.* undertook the analysis of UK crops to determine why farm yields in the UK were failing to keep up with

Table 2.1 Typical nitrogen, potash and phosphate content of oilseed rape straw.

Nutrient	Dose addition (kg ton^{-1} straw)
Nitrogen (N)	7.0
Phosphate (P_2O_5)	2.2
Potash (K_2O)	11.5

the potentials identified in variety evaluation trials.[3] The main factor affecting yield potential was found to be presence of disease; in this case light leaf spot (*Pyrenopeziza brassicae*), phoma canker (*Leptosphaeria maculans*) and alternaria (*Alternaria brassicae*).

Mixtures of phoma and light leaf spot infections commonly found on oilseed rape in high disease years can lead to field yield reductions of around 0.86–1.2 t/ha. The highest correlations between disease incidence and yield reflect disease severity on pods, with light leaf spot infection of pods showing the highest correlations. Stem infections accounted for 53% of the variation in final yield. Levels of phoma canker and alternaria infection of pods also show high levels of correlation with final yield. Although there has been an increase in fungicide use in response to better information and advice to growers on the impacts of light leaf spot and other disease, fungicide applications still do not show strong correlations with disease risk. Though there have been good initiatives in this area that have resulted in better targeting of inputs, further advice and support to farmers would help improve performance.

2.2.1.3.3 Consequences of Shortening Rotations between Oilseed Rape Crops. The typical recommended rotational break for oilseed rape production is 3–4 years. However, the frequency of shorter rotations is increasing. In the UK, the frequency with which oilseed rape is grown one year in three has increased from 8% in 1990, to 15% in 1994 and to 23% in 2003. There is increasing evidence of rotations closing to one in two years as growers opt for wheat/oilseed rape rotations in an attempt to optimise economic returns. Surveys of 100 oilseed rape crops in the UK (Judith Turner, Food and Environment Research Agency, York, UK, personal communication) indicate that around 40% of oilseed rape crops are currently grown within three years of the previous rape crop, and that 14% are grown within less than two years of the previous oilseed rape crop, the highest recorded figure for the past two decades.

Reducing the gap to one year between rape crops has been shown to reduce yield by around 0.5 t/ha.[4,5] Shorter rotations increase soil and trash-borne diseases affecting oilseed rape such as club root (*Plasmodiophora brassicae*), *Sclerotinia* infections and Phoma canker. In the case of club root, an effective rotation is the only means of control.[6]

Tightening rotations and increasing the incidence of oilseed rape in the landscape is also likely to increase the inoculum pool for both diseases and pests of oilseed rape such as pollen/blossom beetle (*Meligethes aeneus*), seed weevil (*Ceutorhynchus assimilis*) and pod midge (*Dasineura brassicae*). Increasing incidence of pests and disease is likely to result in erosion of the efficacy of chemical control treatments where the armoury is limited and resistance may develop. Given the known impacts of disease on yield potential this is a real area of concern in trying to expand production.

It is possible to grow rape continuously (as shown by ADAS at High Mowthorpe in North Yorkshire in the UK for 8 years and by The Arable Group (TAG) in the UK for 4 years). In the ADAS trials, although club root

Farming and Harvesting 53

was not a problem, yields declined in later years and relied on high inputs of herbicide to control weeds and pesticides to maintain yield levels. In ongoing trials by TAG (on behalf of the UK Home-Grown Cereals Authority), yields after 4 years of continuous rape were 0.9 t/ha lower than those where rape was grown for the first time in a rotation (3.55 t/ha), and 0.5 t/ha lower than where oilseed rape was grown in a 1 in 2 rotation with wheat. Stem canker and levels of volunteers were higher in the continuous oilseed rape crop. It is questionable whether continuous rape production would be environmentally desirable or favoured economically, due to the high level of pesticide input required to maintain yield, which inevitably would have environmental consequences if undertaken on a large scale.

2.2.1.4 Breeding New Oilseed Rape Cultivars to Improve the Yield

The above highlights the problems posed by intensifying the number of oilseed rape crops on the same land. The alternatives to increase production are to find more land or to increase output from any land in production.

The key objectives for plant breeders to deliver a significant yield increase in oilseed rape have been identified as:[7]

- Optimising the number of seeds set per unit area (optimum population is $130\,000/m^2$).
- Optimising radiation interception during the critical 'pod filling' stage.

Seed set is determined during a critical phase for pod and seed abortion lasting around 19–25 days after mid-flowering. Pod and seed survival are influenced by the amount of radiation intercepted by photosynthetically active tissues per flower and per pod, respectively.[8,9] Radiation interception at mid-flower is also severely reduced by the bright yellow flower canopy (reducing useful photosynthetically active radiation absorption by up to 60%).

Improvements in the existing genetic material could be delivered by bringing flowering forward into cooler conditions (to extend the pod fill period) and increase radiation use efficiency (due to lower temperatures) and by reducing interference from the flowering canopy.

The opportunities for bringing forward flowering are limited by the date of any last frost which would damage embryonic flowers. Non-petal-producing (apetalous) oilseed rape cultivars have been produced and demonstrated to increase yields in Australia[10] but this was not observed with similar approaches in the UK.[11] Agronomically, flower cover can be reduced by lowering seed rates and avoiding very early sowing.

Crops with an optimum number of pods have about 2.5 units of leaf area per 1.5 units of stem;[12] if this could be increased to 3 to 1 by increasing leaf size and duration this would increase radiation use efficiency by around 12%.

Combining the above and assuming water does not constrain growth, Berry and Spink conclude that a conservative yield potential of 6.5 t/ha should be achievable with existing germplasm. The challenge lies in breeders combining the desirable traits into a single cultivar and utilising the best agronomic management to protect the photosynthetic resource. Farm yields of up to 5 t/ha have been recorded in Europe, so the potential of 6.5 t/ha is not unrealistic. Past evidence is that increases in yield have not been at the expense of seed oil content that typically remains around 43–44%.

Agronomic and trait-led breeding could help deliver such developments relatively quickly, but maximum yield potential will still depend on available water resources during the early growth period and pre- and post-flowering.

2.2.1.4.1 Current Developments in Oilseed Rape Breeding. Although capable of high yields, recent hybrid associations of oilseed rape have proven to be variable in their field performance. More recently, restored hybrids have demonstrated improved yield reliability, but as the cultivars are taller than other conventional varieties, and more susceptible to stem canker disease, they have had limited appeal with growers to date. A range of new short-strawed restored hybrids are now appearing on the market. While under ideal fungicide-treated conditions, these cultivars do not appear to offer a significant step in yield improvement over conventional cultivars, they do appear to perform better under dry and difficult autumn conditions, and are able to produce good yields from very low plant populations (down to as low as 9 plants/m^2), as well as offering early maturity. This may help future crops cope with unpredictable weather conditions during establishment and help reduce some of the climate-induced variability seen in oilseed rape yields.

2.2.1.4.2 GM Developments in Oilseed Rape. The regulatory position restricting the ability of European growers to access genetically modified (GM) crops is unlikely to change quickly. However, the technology offers the potential to significantly influence the yield potential of oilseed rape. This can be achieved through ensuring that yield is not lost to pest, disease or weed pressure, or through development of higher yielding cultivars by incorporating new traits into oilseed rape crops.

On a global scale 17% of oilseed rape crops are GM. The vast majority of the commercialised traits relate to incorporation of herbicide tolerance (to specific broad-spectrum herbicides) and to a lesser extent modified oil profiles. In terms of traits under experimental field investigation, the majority are for pest and disease control and further herbicide tolerance traits. Canada leads on genetic modification of rapeseed, and 23% of experimental releases in Canada in recent years have related to investigation of male sterility and restoration traits (used in development of hybrid cultivars). In Europe, most of the experimental traits being investigated are increased oil production, altered oil profiles or herbicide tolerance.

In the near to mid term, even if the technology were made available, current GM technologies in development do not appear to offer significant yield advantages over current conventional and hybrid cultivars. However, the

Farming and Harvesting

development of pest and disease resistance traits could help crops achieve their full potential, by reducing losses.

2.2.1.5 Estimation of Future Yields and Future Production Potential—Conclusions

The yields of most arable crops in developed counties have been increasing steadily throughout the last 40 years due to improved agronomy and management practices. However, yields of oilseed rape have not been seen to increase at the same rates as cereals in recent years (Figure 2.1). In a number of key producing countries, France, Australia, UK and Poland, oilseed rape yields have not shown significant increases in recent years. Up until 1985 the average yield increase in oilseed rape stood between 28 and 65 kg/ha per annum in the UK, France, Poland and Australia, but since then has come to a halt despite continued development of new higher yielding varieties.

In trial studies under 'ideal' conditions in the UK there has been an average annual yield increase of 62 kg/ha/annum between 1978 and 2005 though this reduced to 33 kg/ha from 1987. However, such increases have not been widely reflected in field yields over the same period.

The previous sections identified that the potential realistic yield that could be attained is in the region of 6.5 tonnes/ha. For areas with water retentive soils, the figure for the ultimate yield potential could be as high as 9.2 tonnes/ha. A more realistic goal would be to double the average yield of OSR yield within Europe without increasing the area planted with the crop.

Raising the yield in all EU countries that are capable of producing at least average EU oilseed rape yields currently, to the physiological potential of 6.5 t/ha would produce an additional 12.4 million tonnes of oilseed rape,

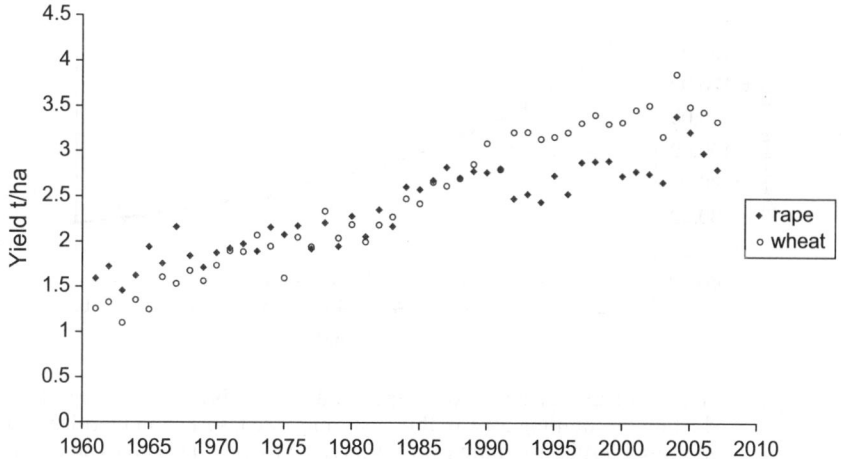

Figure 2.1 Trends in average EU oilseed rape (closed points) and wheat yields (open points) (derived from data obtained from the FAOSTAT database).

increasing current EU production levels by 76%. Even a fraction of this improvement would provide a significant boost to EU productivity.

In addition, if yields in countries currently below the EU average could be brought up to the EU average then this could increase the EU oilseed rape supply by around 1.3 million tonnes. However, in some cases these areas currently present the most difficult growing conditions for oilseed rape *i.e.* drought-prone Mediterranean areas. Taking account of those countries producing below average yields, but with similar climatic conditions to those producing above average yields, then increasing EU-27 production by 0.8 million tonnes without increasing the oilseed area should be a realistic and achievable target. The key countries concerned, Romania, Lithuania, Slovakia, Hungary and Poland, represent newly accessing states to the EU, where agricultural support has been lacking in the past and where access to EU support and investment finance is improving production rapidly. In these cases it is investment in crop establishment and management of inputs that has been lacking, rather than climatic constraints on production.

Increasing production/unit area also helps to offset some of the costs associated with production (see Figure 2.2), a benefit that if passed on helps to reduce costs of raw material supply to industrial users.

2.2.2 Sunflower

2.2.2.1 Introduction

Cultivated sunflower (*Helianthus annuus* L.) has a single stem, a strong and deep central taproot, which can penetrate the soil beyond 2 m, and one large inflorescence with yellow ray flowers. This inflorescence is a capitulum, which is

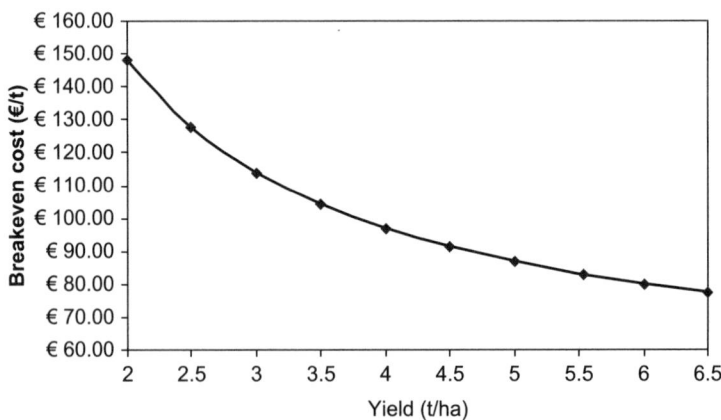

Figure 2.2 Impact of increasing oilseed rape yield on the breakeven costs of production (accounting for variable costs only (seed, sprays and fertiliser) and increased fertiliser requirement for increased yield). This takes no account of additional fixed costs—machinery and depreciation costs, rents and interest payments *etc.*, that are very farm and enterprise specific. Data presented assume no increase in seed or pesticide use with increasing yield.

characteristic of the Asteraceae family.[13] The arrangement of the flowers in the capitulum exhibits two sets of spirals. The fruit is an achene composed of a seed and adhering pericarp. The oil is the main constituent of the seed (about 50% of dry weight on average in France), and is mostly composed of oleic and linoleic fatty acids. The ratio between these two fatty acids is under environmental and genetic control.[14]

Archaeological evidence reveals that sunflower was cultivated in Arizona and New Mexico 3000 years BC.[15] This crop was introduced in Europe in the sixteenth century by the Spanish. Then, sunflower spread through Europe and other continents to become one of the major annual crops in the world grown for edible oil. In Europe, the seed yield of sunflower is poor compared with winter oilseed rape, but the need for fertilisers and crop protection is lower. Sunflower seed yield and oil content depend on their potential values, and the availability of water and nutrients. They are also affected by diseases, pests and weeds. Improvement in agronomic practices and breeding of new cultivars would reduce the effect of these limiting factors.

2.2.2.2 Current Data on Sunflower Cultivated in Europe

Sunflower is the second most important oilseed crop in the European Union (EU),[16] with a total cultivated area of 3 399 000 ha in 2007 and a production of 5 660 000 t. This area cultivated in sunflower in 2007 had been the lowest since 1990, and was 29% less than the maximum area of 1990–2007, which was observed in 1993.[17] Sunflower is mainly cultivated in Southern Europe (Table 2.2). The mean annual seed yield in EU ranged from 1.16 to 1.84 t/ha[17] during 1990–2007. These observed seed yields were much lower than the potential yield of sunflower crop with no tendency to increase. This lack of improvement may partly explain the decrease in the cultivated area of sunflower. However, there are possibilities to increase seed yield through a better control of the limiting factors. There is no statistic data about the amount of crop residues (above ground vegetative biomass) produced, because they are rarely harvested. An estimated value could be calculated using a mean harvest index (ratio between seed yield and total above-ground biomass), which is relatively high in sunflower. Values of harvest index ranging from 0.26 to 0.39 were reported,[18] depending on the water regime, with a mean value of 0.33 indicating that the biomass of crop residues was twice as great as that

Table 2.2 Sunflower production and cultivated area in 2007 in European Union.[16]

Country	Production (tons × 1000)	Area (ha × 1000)
Bulgaria	756	540
France	1300	537
Hungary	1058	470
Italy	300	130
Romania	1125	900
Spain	674	613
Other countries	447	209

of seed yield. However, greater harvest indices were observed in recent cultivars,[19] suggesting that nowadays the relative production of crop residues is lower.

2.2.2.3 The Main Limiting Factors of Sunflower Seed Yield in Europe

Sunflower is drought tolerant through deep rooting.[20] This crop is able to produce satisfactory seed yield under conditions of water stress when other crops grow poorly. However, plant available water is the most limiting factor to sunflower seed yield of dry land agriculture in semi-arid regions.[21]

Diseases are also major limiting factors of sunflower production in most regions. This is attributable to a narrow genetic base and a lack of resistance genes in cultivated sunflower. Diseases caused by fungi are the most widespread and economically important.[22] Sunflower is a known host for over three dozen pathogenic organisms, but fewer than a dozen frequently cause serious economic losses,[23] and just a few are of concern in a particular country or region. In a literature survey of the past four years,[24] a total of 13 pathogens were the subject of a paper, indicating that they have some economical importance. Seven of these pathogens result in diseases of economical importance in the European Union (Table 2.3).

Alternaria spp. cause leaf and stem spots, seedling blight and head rot on sunflower all over the world. Severe *Alternaria* infestations can cause defoliation and yield losses as high as 60 to 80%, especially under warm and humid conditions.[23] Broomrape is an obligate holoparasitic angiosperm that attaches itself to the roots of infected plants, depleting them of nutrients and water, and hence causing important yield losses. It can cause considerable damage: in cases of severe infections, these losses can reach up to 50% and near 100%.[24] Similarly, yield losses ranging from 20 to 90% were reported,[20] depending on the degree of infection. In Bulgaria, Romania and Spain, the parasitic plant *Orobanche cumana* has become a major limiting factor for sunflower crops. Over the past few years, the progression of this parasitic plant, its introduction into new countries and the development of new and more virulent races have all been observed.[25] Charcoal rot may cause premature death of sunflowers grown on light, sandy soil under hot and dry climate.[24] This disease is well known in the southern part of Europe. For instance, in 1983 charcoal rot was reported in

Table 2.3 Sunflower main diseases in European Union.

Disease	Pathogen
Alternaria blight	Alternaria spp.
Broomrape	Orobanche cumana
Charcoal rot	Macrophomina phaseolina
Downy mildew	Plasmopara halstedii
Phoma black stem	Phoma macdonaldii
Stem canker	Phomopsis helianthi
White rot	Sclerotinia sclerotiorum

nearly half of Spanish sunflower fields.[23] During 2004–2007, downy mildew occurred in almost all parts of the world where sunflower was grown, except Australia.[24] This fungus is more prevalent and yield losses more severe in temperate regions compared to sub-tropical regions.[23] Yield losses due to downy mildew vary with the percentage of infected plants in a field, their distribution and with the precocity of the disease.[23] *Phoma* black stem has been recorded on sunflower from many countries in Europe. Most of the time, it is either scarce or of minor importance, as it frequently causes only a superficial stem lesion. Yield losses due solely to *Phoma* stem lesions are generally slight, but when *Phoma* infections are combined with stem-tunnelling insects, affected plants may senesce earlier than normal, a syndrome referred to as 'early death' or 'premature ripening'.[23] *Phoma* black stem is extremely severe only in France where basal stem lesions often result in lodging,[24] and in premature ripening.[26] In this country, the estimated loss of yield ranges from 0.2 to 0.5 t/ha, depending on the region. Phomopsis stem canker was reported for the first time on sunflower in Yugoslavia in 1980. In many fields, 50 to 80% of the plants were infected, causing drastic reductions in yield and quality.[23] This disease can cause significant losses in yield (10 ± 50%) and in oil content (10 ± 15%) when environmental conditions are favourable.[27] Hence, soon after its appearance, stem canker became one of the most limiting factors of sunflower production in many parts of Europe, including Romania, Hungary and France.[24] However, the disease occurrence lessened partly due to unfavourable weather conditions (dry and hot). Sunflower white rot on the stalks and on the heads is considered the most important disease of sunflower in many parts of the world.[24] The disease has been reported from Austria, Bulgaria, Denmark, France, Germany, Hungary and Romania. The severity of yield losses depends on the age of the plant at the onset of disease.[22] Plants infected one week after anthesis may lose up to 98% of their potential yield compared to 12% for plants infected 8 weeks after flowering. Oil concentration is also affected by the disease. Instances of 100% of the plants within a single field having head rot have been recorded, but on a regional basis yield losses are generally from 1 to 5%.

Insects damage plants by feeding on plant tissues.[28] They also cause significant indirect injury to sunflower by transmitting pathogens to plants, or by predisposing injured plants to secondary infection by several pathogens. In Western Europe, insect problems are of less concern compared to eastern Europe.[28,29] Approximately 240 species of insects attack sunflower in the three major sunflower producing countries in Eastern Europe (Bulgaria, Hungary and Romania). Plant bugs, *Adelphocoris* sp., *Lygus* sp. and *Polymerus* sp. infest up to 10% of the sunflower in Hungary, in which seed depredation decreases oil content by 8% and germination by as much as 72%. In Bulgaria, adults of the beetle *Omophlus lepturoides* feed on florets, while their larvae feed on roots, causing wilting damage of plants. Larvae of the dipteran *Napomyza lateralis* burrow in the stalks of sunflower in Bulgaria. Also in Bulgaria, 15 species of insects commonly attack stored sunflower seeds. Insect pests of sunflower in Romania are predominantly species of Coleoptera and Lepidoptera.

In Germany, the jassid, *Aphrodes bicinctus*, feeds in plant terminals. In France, seven species of insects are recorded as pests, including lepidoptera, aphids and leafhoppers. However, insects are not a major problem in this country. Hence, only 40% of the sunflower area received one insecticide in 2006.[30]

Weeds cause substantial losses in sunflower.[20] Some weed species are not controlled by pre-plant or by pre-emergence herbicides. One of the most frequent broadleaf weed species in Spain, the umbelliferous *Ridolfia segetum*, illustrates this problem.[31] In the south-west part of France, sunflower seed yield is affected by weeds in about 5% of the area, inducing a loss of about 0.5 t/ha.

Members of the sparrow family are also important pests of sunflower in Europe.[32] House sparrows and European tree sparrows are the primary species responsible for losses in France, Poland, Hungary and Romania. In southern Hungary, doves are important pests of ripening sunflower. Plant population can also be reduced due to damage by slugs or game animals. In France, slugs, birds or game animals are of more concern than insects. However, only 54% of the sunflower area was sprayed against slugs in 2006, and no treatments are allowed against birds or game animals.[30] Hence, great losses of yields may be observed. The estimated loss of yield due to birds and game animals is 0.3 t/ha in central France and 0.4 t/ha in western France.

2.2.2.4 *Agronomic Practices to Improve the Seed Yield*

2.2.2.4.1 The Main Characteristics of Agronomic Practices in Sunflower. In Europe, water deficit is one of the main limiting factors of sunflower seed yield, along with diseases, because this crop is mainly cultivated without irrigation in semi-arid regions. Irrigation water is often used for the production of crops with greater economic return.[20] For instance, in France only 4% of the area cultivated in sunflower in 2006 was irrigated.[30] Moreover, irrigated sunflowers receive little water and are frequently under a clear water stress.[33] Therefore, the amount of expected available water is often the primary factor influencing management decisions.[21] In the south-west of France, early sowings are dominant because severe deficits in summer are expected.[30] In the Aude region, however, where mid-summer storms are predictable, the date of sowing is delayed. The objective is to postpone the seed filling period, so that it occurs during mid-summer. However, the optimal sowing date does not only depend on the necessity to improve the efficiency of the expected water. The expected soil temperature after sowing must also be taken into account, because warm temperatures are needed for a good stand establishment.[34] Similarly, the plant population density depends on the expected available water, with higher levels recommended for more favourable environments.[20] The amount of N fertiliser also depends on water availability, among other factors, because it is adapted to the expected seed yield. The natural N level in the soil also affects the response of sunflower to N fertilisation.[35] Hence, recommended N fertiliser rates differ with both the plant-available nutrients and the target yield. On average, the amount of N fertiliser applied on sunflower is low compared with other crops (for instance, 38 kg of mineral N ha^{-1} in France in 2006).

The control of diseases is achieved by cultural practices, by genetic resistance and by the use of agricultural chemicals. For instance, the development of resistant hybrids to Phomopsis stem canker was a major factor in the control of the disease.[22] Resistance to Phomopsis is polygenic. Fungicides are also effective to control the disease. To achieve the best disease control, fungicide application should be started before symptom appearance. A two-application schedule, with the first application at the bud stage and the second application at flowering, is suggested. However, foliar fungicide applications do not offer the near total control that can be achieved by genetic resistance. Plant population and nitrogen fertilisation also have an effect: population densities <50000 plants ha^{-1} and N levels <60 kg ha^{-1} are recommended. A rotation of at least 1 year without sunflower will decrease the chance of inoculum from the previously infected crop, because the pathogen is specific to sunflower and does not persist for very long periods in the soil, although ascospores may be blown in from infected debris in nearby fields. Incorporating infected stalks as deeply as possible, or removing the infected stems from the soil surface, should reduce the inoculum the following year.

The control of other diseases can also be achieved by cultural practices, by genetic resistance and by chemicals. Dense canopies, high levels of nitrogen nutrition and a short time between two sunflower crops must be avoided.[22,26] Compared with Phomopsis, genetic control is unfortunately less effective for other diseases. The capacity of some pathogens to evolve is a major reason explaining this difficulty. For instance, resistances to broomrape or to downy mildew have been rapidly overcome.[24,26] There is also a lack of knowledge, and limited efforts have been made to develop genetic resistance. Similarly, the possibility to control diseases other than Phomopsis by chemicals is limited. However, one effective fungicide against Phoma black stem exists, but it is not available for farmers.[26]

The use of chemical insecticides is a primary tool for reduction of pest populations.[28] However, it may seriously reduce the population of pollinators, as well as beneficial parasites and predators of the pest species.

In sunflower crops, weeds are traditionally controlled by pre-plant or by pre-emergence herbicides. Between-row cultivation is used as a complementary tool (for instance, 41% of the area in 2006 in France).

2.2.2.4.2 The Possibilities of Improvement in Agronomic Practices in Sunflower. In order to increase sunflower seed yield, some strategies (for example, those aiming at reducing the impact of slugs, birds and game animals, the development of post-emergence weed control and the development of irrigation) could be readily applied. Other strategies need more time before being applicable. A more accurate adaptation of the choice of cultivars, sowing date, plant density and amount of N fertiliser to the expected water availability would increase the seed yield, but further studies are needed for such improvements. Similarly, alternative strategies to control diseases and insects cannot be readily applied on a large scale. The development of new chemicals could also be an effective way to better control diseases and insects, but they may cause environmental problems, and may be overcome by pathogen changes.[37] Moreover, chemical control also depends on the possible market for a compound, and sunflower diseases often do not provide the acreage necessary to recover the costs of development of new products.[37] Above all, a significant improvement in sunflower yield would occur if the farmers were convinced to follow the recommendations more effectively.

2.2.2.4.2.1 Convincing Farmers to Follow Recommendations. Too often, there is a great difference between recommendations and actual practices. In France, plant population density is often too low and/or heterogeneous within a given field. In the south-west of France, the plant population density is below the recommendations in half of the area. This results from both low seed density in order to reduce the cost of inputs, and poor seedling emergence and survival due to slugs, game damages, climatic conditions and inappropriate seeding depth. The tendency to decrease the input in order to reduce the cost concerns seed density, and also boron. For instance, in the south-west of France there has been a constant decrease in boron applications since 2002: 51% of the area received boron in 2002, 43% in 2004 and only 37% in 2006. Sunflower is also considered by farmers as a crop that is easy to grow. Hence, tools to accurately calculate the amount of N fertiliser, depending on the target seed yield and on the natural availability of this nutrient, are not used frequently. A wider use of these tools would result in higher yields or lower amounts of N fertiliser. This would increase the profitability of the crop and reduce environmental risks.[38,39] The date of sowing by farmers may also be different from the recommendations. For instance, in a survey conducted from 1996 to 2009 in the south-west of France, more than 75% of sunflower crops were sown too late, after 10 April, resulting in a $0.1\,t\,ha^{-1}$ (sowings between 11 and 20 April) to $0.3\,t\,ha^{-1}$ (sowings after 10 May) decrease in yield. The differences between recommendations and actual crop managements by farmers also concern the cultivation of late maturity type hybrids resulting in harvest problems.

2.2.2.4.2.2 Reducing the Impact of Slugs, Birds and Game Animals. A better seedling emergence and growth would limit the effects of slugs, birds and game animals on plant population, and thus on yield. Hence, the improvement of crop management must focus on seed treatments and on sowing practices. Tools to frighten away birds and game animals (sound or visual) and the homologation of treatments that prevent the feeding on seeds by birds would also decrease the damages, while predators and hunters could limit their population more efficiently than they do now. And, last but not least, there is a need for a better understanding of the biology and of the nutrition of birds and game animals, in order to figure out more efficient ways to reduce losses.

2.2.2.4.2.3 Developing Post-emergent Weed Control.

The recent introduction of two herbicide-tolerant sunflower types makes possible a post-emergent weed control option.[40] The CLEARFIELD Production System is based on a non-GMO herbicide-tolerant sunflower and on an herbicide of the imidazolinone chemical family. At present, the herbicide used is IMAZAMOX. It provides contact and residual activity on a number of grasses and broadleaf weeds as well, including nightshade, pigweed, foxtail species, wild oats, volunteer cereals, puncturevine, non-Clearfield wild or volunteer sunflower and broomrape. EXPRESS, the second sunflower production system, facilitates a broad spectrum weed control option for sunflower growers. It provides resistance to EXPRESS herbicide, which is a sulfonylurea. It only offers post-emergence control of broadleaf species. Herbicide-tolerant sunflowers gain a market share very quickly once this trait is incorporated into performing hybrids, because of its high efficiency in weed control. For instance in Romania, sunflower weed control was only partial because most dominant species (Xanthium, Cirsium, Abutilon and Datura) were not controlled by any herbicide before the introduction of the first 'genetically unmodified' sunflower hybrids resistant to imidazolinone or to tribenuron.[41]

Moreover, the post-emergence control strategy with herbicide-tolerant sunflower hybrids makes possible new cultural options. For instance in Andalusia, in several planting date studies it was shown that sunflower planted during the winter yielded about 30% more than when planted in March. This potential increase of yield seemed to be related to the flowering escape from the extremely hot days that often occur in early June. This potential yield increase was threatened by bird damage, herbicide damage from hormonal treatments in neighbouring cereal fields and by weeds due to the slow growth of sunflower during January and February. The introduction of CLEARFIELD sunflowers allows early planting in Andalusia with an effective control of Orobanche cumana and weeds. Hence, herbicide-tolerant sunflowers offer an excellent opportunity to increase yield in different countries due to weed and Orobanche control. This is particularly important for early planting dates in hot Mediterranean countries.

2.2.2.4.2.4 Increasing the Amount of Water Available for Sunflower and Its Efficiency.

Irrigation is usually an effective way to increase seed yield. For instance in France, a limited amount of water (1 or 2 applications) results in a 0.4 to 0.5 t/ha increase in yield on average. In the future, less water will be available for irrigation, because of the constant increase in water demand from other sectors of society and the expected drier climate in southern Europe due to global warming. Hence, crops with little water requirement like sunflower may replace current irrigated crops. This is an opportunity to increase the area of irrigated sunflower, and thus to increase the seed yield of this crop. The efficiency of applied water to sunflower could be improved by optimising the timing of irrigation, resulting in greater yield increase, but further studies are needed to achieve this goal. On the other hand, tillage and management of crop residues could help to save water.[21] These authors also suggested that soil compaction due to an intensive traffic with tractors and other farm vehicles over arable land should be avoided, because of the effects on soil properties regarding water, air movement, nutrient regime, root growth and on yield losses.

2.2.2.4.2.5 Adapting More Accurately the Choice of Cultivars, Sowing Date, Plant Density and Amount of N Fertiliser to the Expected Water Availability. Another means of improvement would be a more accurate adaptation of cultural practices to water availability. In east-central Italy, sunflower yield was strongly dependent on the cultivar cropped,[42] especially under drought conditions. However, the drought-resistance of commercial cultivars, if it exists, is unknown and thus not available for farmers' decisions. The choice of the variety should account for the maturity type but also for differences in leaf area and in stomatal closure, which play an important role in drought resistance.[43] The estimation of a more accurate date of sowing, planting density and amount of N fertiliser (more accurate target yield) is necessary. For instance, in the south-west of France the recommendations of sowing time are based on farm surveys from 1996 to 2009,[44] showing a decrease in yield when the date is delayed after 10 April (Table 2.4). These results are rough estimates, because they compare different fields with possible differences in other cultural operations and soils. In order to take into account possible interactions with other factors (variety, soil depth, *etc.*), and/or to give more site-specific recommendations, many more data would be necessary. The use of crop models to acquire this information was suggested.[30] The SUNFLO model was developed for this purpose.[43] Generally, in sunflower models, varieties only differ in yield components and maturity types. In this new model, varietal parameters are required for crop development, leaf area and its ability to intercept light, response of leaf expansion and stomatal closure to soil water deficit, harvest index and the maximum percentage of kernel in achenes. These parameters are easily measurable, in order to be able to account for the dozens of new varieties appearing each year on the market.[45]

Table 2.4 Sunflower seed yield depending on the date of sowing in the south-west of France.[44]

Date of sowing	Seed yield $(t\,ha^{-1})$
Late March	2.57
1–10 April	2.59
11–20 April	2.50
21–30 April	2.45
1–10 May	2.47
After 10 May	2.32

Results from a survey conducted from 1996 to 2009 on soils with a medium field capacity (300 fields per year).

Sowing date, plant density, irrigation and N fertiliser are also considered. Hence, the SUNFLO model could be used to simulate the effect of all combinations of sowing time, population density and cultivars on seed yield in each soil type and climate, in order to select the best combination of practices. The predicted seed yield would also be helpful to calculate the amount of N fertiliser. The SUNFLO model has already been validated with promising results, but further validations and improvements are still needed before using this tool.

The possibility to recommend site-specific variety, date of sowing, plant density and N fertiliser rate, accounting for water availability, will be a major improvement. These cultural practices also have an effect on the risk of diseases, insect damages and weed problems. Hence, further works are needed to take into account both these risks and the expected availability of water.

2.2.2.4.2.6 Developing Alternative Strategies to Control Diseases and Insects. Biological antagonists could be used to control broomrape,[24] or sclerotinia[22,37] but so far none of them has been applied on a large scale.

Prophylactic methods may also be effective. For instance, one of the most important vectors in broomrape dissemination was the machinery movement along the different growing areas, which could explain the differences in the rates of growth of the disease between areas. This knowledge may help to prevent broomrape dissemination, especially in Andalusia where the new race G which attacks race F resistant hybrids has already been reported.[24]

Alternative insect management strategies include rotating crops, altering planting dates, using sex pheromones to monitor pest populations and conserving/augmenting natural enemies.[28]

2.2.2.5 Breeding of New Cultivars to Improve the Seed Yield

Breeding of new cultivars will help to increase the seed yield by increasing either the potential value or the resistance to water deficit, diseases or insects.

Great increases in seed yield potential have been observed in the last decades. For example in France, the seed yield potential increased by 40% from 1970 to 2000.[46] Meanwhile, no obvious increase in seed oil content has been obtained. Potential seed yields of the most cultivated sunflower cultivars were studied from 1960 to 2000 in France.[19] A part of the genetic improvement was due to an increase in harvest index, resulting from a delay of leaf senescence, of the decrease in leaf N content and of the drop in leaf photosynthesis. Modern cultivars were also more efficient in intercepting solar radiation per unit of leaf area. Considerable efforts have also been devoted to develop drought-tolerant cultivars in some countries, like Romania.[47] However, more appropriate test systems must be developed for breeders to select for drought resistance while saving time, space and cost. A new approach using the modelling of physiological processes was developed to improve the efficiency of conventional plant breeding.[48] This should be helpful for increasing both seed yield potential and drought resistance. In this approach, the parameters of the equations used to describe the phenotypic built up in response to environment is related to genotypic characteristics. Quantitative genetic methods such as quantitative trait locus (QTL) can be used to evaluate the genetic determinism and the variability of the parameters. Using this approach, the impact of the physiological processes on yield variability was studied. The ranking was first biomass allocation and light interception through the plant architecture, and third plant phenology, while the impact of photosynthesis was much less. Interestingly, in this analysis no cultivars were optimised for all the processes described in the model. This indicates that some improvements are still possible. This approach using a model could be used to identify virtual efficient cultivars. This should be useful for increasing the efficiency of breeding programs.

The resistance of sunflower varieties to most of the pathogens must be improved. Recent efforts made by breeders to develop resistant cultivars to sclerotinia or to Alternaria blight were recorded.[24] Resistance breeding has so far been concerned with either major gene resistance (downy mildew, broomrape) or quantitative resistance (Sclerotinia, Phomopsis). With major genes, resistance has been complete but not durable. Although new genes are frequently found, they are rapidly overcome. For quantitative resistance, progress has been rapid in the case of Phomopsis, but it has been slow for Sclerotinia. When resistance is incomplete, breeding programmes search to improve levels of resistance, whereas when resistance is complete, durability is the main objective. A combination of these two objectives, to get a high level of durable resistance, should be the goal in the future.[37] For instance, although new races of broomrape evolve rapidly to overcome the resistance of newly introduced sunflowers, it has been shown that polygenic resistance could occur in some lines. In fact, resistance to broomrape in sunflower is controlled by a combination of qualitative race-specific resistance affecting the presence or absence of the pathogen and a quantitative non-race-specific resistance affecting their number.[24] For resistance that is non-race-specific, the first important step would be to combine the different QTLs for resistance. Horizontal resistance could include some race-specific resistance genes with small effects such that interactions between host and pathogen do not have significant effects on the overall resistance level. Knowledge of the resistance mechanisms underlying the different QTLs would help to decide those that are most likely to have additive effects. This would be long and complex.[37] For resistance that is race-specific, the first tendency is to pyramid genes with the idea of reducing the probability of the mutation of the pathogen. However, the frequency of appearance of multiple virulences could be greater than expected. To get the best use from major genes, they need to be backed up by quantitative resistance.

The breeding of resistant cultivars can also be focused on insects. For instance, the phytomelanin characteristic in the pericarp of achenes has provided a high level of resistance to *H. Nebullela*.[28] In the future, it may be helpful in areas where insects are major limiting factors.

2.2.2.6 Genetic Modification

In sunflower, no significant advances due to GM technology have been reported so far. The main reasons are the difficulty to modify this species, and the high cost to develop GM sunflower compared with the expected economic return. The possible benefit of GM is low, because herbicide-tolerant sunflower types already exist. Moreover, nowadays more complex improvements like drought resistance are out of reach, because they are still a challenge for species that have been genetically modified for decades. However, genetic modification is worth a try to control diseases, because resistance breeding will be long and complex in sunflower.[37]

2.2.2.7 Estimation of Future Yields and Future Production Potential—Conclusions

The average seed yield of sunflower in the European Union is less than $2\,t\,ha^{-1}$. This value is low compared to the potential seed yield as a result of limiting factors, especially drought and diseases. Insects, weeds, slugs, birds and game animals also have significant impact, at least in some regions of Europe. Breeding of new cultivars to increase seed yield potential would be helpful, but there is also a need to increase resistance to drought or diseases and improve cultural practices. Limited amounts of irrigation water would result in large increases in yield in almost all regions where sunflower is cultivated. Attention should be paid to sowing practices in regions where slugs, birds and game animals cause significant reductions of plant population density. The development of post-emergence weed control should also result in increased seed yields, in regions where some weeds are not controlled by pre-plant or pre-emergence herbicides, where broomrape is a major limiting factor or where it makes earlier sowings possible. Other strategies are also possible, such as more accurate adaptation of the choice of cultivars, sowing date, plant density and amount of N fertiliser to the expected water availability, but cannot be readily applied because further studies are needed. In every case, farmers' decisions are the key issue. As long as sunflower is cultivated with low inputs, there will be little improvement in seed yield. There is a need to convince farmers that limited amount of irrigation can be profitable. They should also follow recommendations as accurately as possible. The economics of sunflower cultivation is also crucial. High prices of the seeds increase the economic interest of high levels of inputs. Finally, another important factor is that, since the decline in available water for irrigation is possible in future, sunflower cultivation will be favoured due to the low water requirement of this crop.

2.3 Valorisation of Straw and Leaves through Green Technologies

Rapeseed and sunflower are mainly harvested for their oil while the remainder of the plant such as the straw remains unutilised to a large extent. The residues comprise a substantial proportion of the whole plant and may represent a considerable amount of commercially exploitable biomass. In order to maintain a competitive advantage in the world market and ensure a sustained economic return in the production of oil crops, the potential of these residues in value-added applications should be explored and exploited.

High-value chemicals such as waxes, sterols and polycosanols of substantial commercial value can be extracted through an innovative and clean supercritical carbon dioxide extraction process. This process has already been successfully demonstrated on a pilot scale using wheat straw as a feedstock. This technology could potentially be applied as a first step in a biorefinery to convert by-products of harvesting of oil-rich crops into feedstocks for energy, commodity bioproducts and speciality chemicals. Oilseed straws also constitute an interesting natural composite for agro-materials manufacture. Finally, the potential of producing methane *via* anaerobic digestion is assessed. The anaerobic digestion process is already being applied to a wide range of feedstocks such as energy crops and organic wastes in small- and large-scale units.

2.3.1 Chemicals from Supercritical CO_2 Extraction

The biorefinery concept should exploit the availability of all potential raw materials from the chosen feedstock and the opportunity to capture valuable by-products from straw prior to their conversion to biofuels cannot be ignored. The application of a green technology such as supercritical CO_2 offers substantial benefits within the biorefinery concept as 100% of the required energy and extraction solvent can be obtained from other unit operations. It is envisaged that within this concept the raw material (straws) can be pre-treated by extraction with supercritical CO_2 to recover a range of valuable molecules before being used as a biofuel feedstock. It has been demonstrated that a wide range of molecules, including alkanes, wax esters, sterols and polycosanols, are present and can be selectively extracted using supercritical CO_2 without reducing the calorific value of the straw, which is particularly important if it is to be used for combustion. The removal of these hydrophobic molecules also renders the straw more wettable, which should assist in any subsequent lingo-cellulosic digestion, if that is the chosen route to use this material.

Extraction of a wide range of molecules using supercritical CO_2 is already a commercial reality and more recent applications have demonstrated that large-scale uses of supercritical CO_2 to carry out extraction, cleaning or impregnation can be commercially viable even when the bulk material value is low. Good examples of this are the extraction of seed oils such as sesame seed in Korea, decontamination of cork in Spain and the bulk impregnation of timber in Scandinavia. In this chapter the potential to extract molecules of industrial value from sunflower and oilseed rape straw is examined and the approximate capital and unit operation costs quantified.

2.3.1.1 CO_2 as an Extraction Solvent

Extraction can be carried out using CO_2 in a liquid or supercritical state and the choice largely depends on the solubility of the molecules to be extracted. Liquid CO_2 is significantly less polar than supercritical CO_2 and is applicable only to small non-polar molecules such as mono- and sesquiterpenes. Liquid CO_2 finds commercial use in the extraction of essential oils and oleoresins. Above the critical point ($P_c = 7.38$ MPa, $T_c = 31.1\,^\circ\text{C}$) CO_2 exists as a supercritical fluid

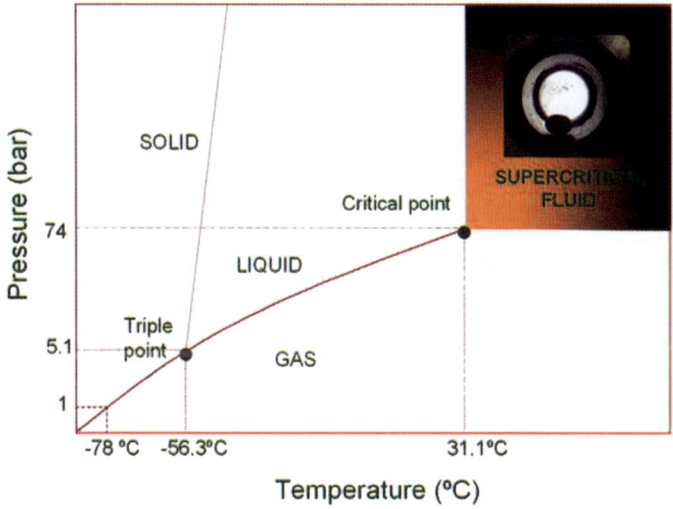

Figure 2.3 Phase diagram of CO_2.

(Figure 2.3) and the manipulation of temperature and pressure produces a highly tuneable solvent that can be used to selectively extract a wide range of molecules and because of this most commercial extraction plants are designed to operate using supercritical CO_2.

Although greater flexibility is achieved by using supercritical CO_2 operating at higher pressures and temperatures, typically 10–40 MPa and 40–60 °C, has a higher energy requirement and capital cost.

For the extraction of cereal, sunflower or oilseed rape straw only the use of supercritical CO_2 can been considered since the molecules that are likely to be obtained are too large and polar to be readily soluble in liquid CO_2. Solubility in supercritical CO_2 is essentially proportional to density[49,50] and this in turn is proportional to pressure and inversely proportional to temperature. Figure 2.4 shows the relationship between density and pressure at selected temperatures and it can be seen that the density rises rapidly once past the critical point.

In addition to the temperature and pressure of supercritical CO_2 extraction efficiency is also directly influenced by other process parameters.

1. Extraction time—equates to total kg CO_2/kg raw material.
2. Extractor design.
3. Raw material pre-treatment.
   a. Particle size.
   b. Particle porosity.
   c. Bulk density.

Extraction time is largely determined at the design stage of any equipment and the circulation pumps sized to meet a pre-determined batch extraction time. Extractor design is also optimised at the same time.

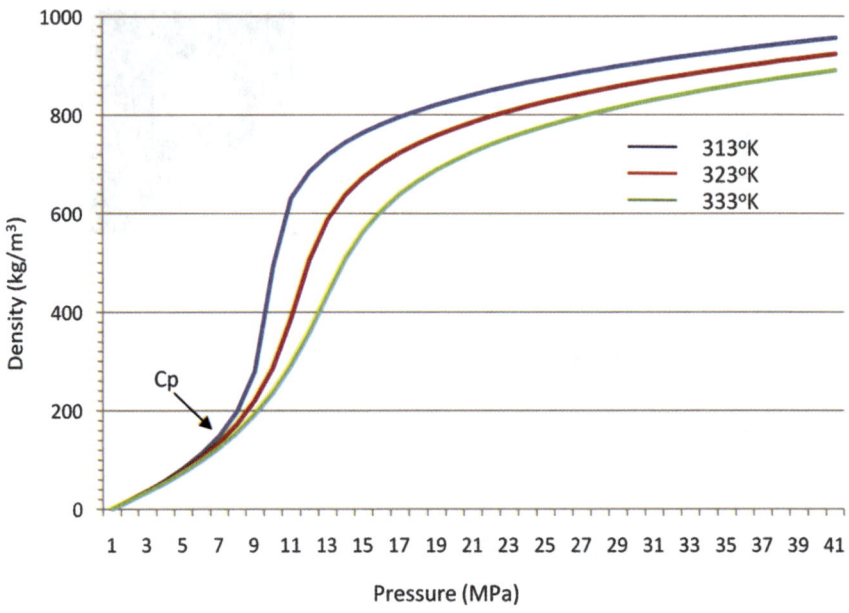

Figure 2.4 Supercritical CO_2 density as a function of pressure and temperature.

A typical extractor arrangement is shown in Figure 2.5. Commercial extraction using supercritical CO_2 is carried out as a batch process. Extraction vessels are loaded with raw material, sealed and after extraction is complete are emptied as quickly as possible before being recharged. The extractors may have an internal basket to assist rapid loading and unloading or the extractor may have hydraulically operated top and bottom lids, which gravity discharge into a hopper and conveyor. The extractor operates within a pumped closed loop. Liquid CO_2 is taken from a storage vessel normally held at 5 MPa and then cooled before entering the primary pump. The compressed CO_2 is then heated to the desired extraction temperature and then passed through the extractors, which may be arranged either in parallel or in series.

The extract is recovered in a series of separators which have sequentially lower pressures resulting in precipitation of the increasingly insoluble extracted components and the CO_2 is finally condensed and returned to the storage tank. The only loss of CO_2 within the extraction circuit occurs when the extraction vessels are emptied; however, this can be minimised by the addition of a recompression stage.

Extraction of molecules that have very limited solubility in supercritical CO_2 can be enhanced by the addition of entrainers. These are co-solvents added as a percentage of the supercritical CO_2 flow that increase the solvating power of the extraction solvent. Ethanol is commonly used as a co-solvent and this can of course be obtained within the biorefinery complex. Previous work on wheat straw[51] has indicated that the use of co-solvents leads to a loss of selectivity.

Farming and Harvesting

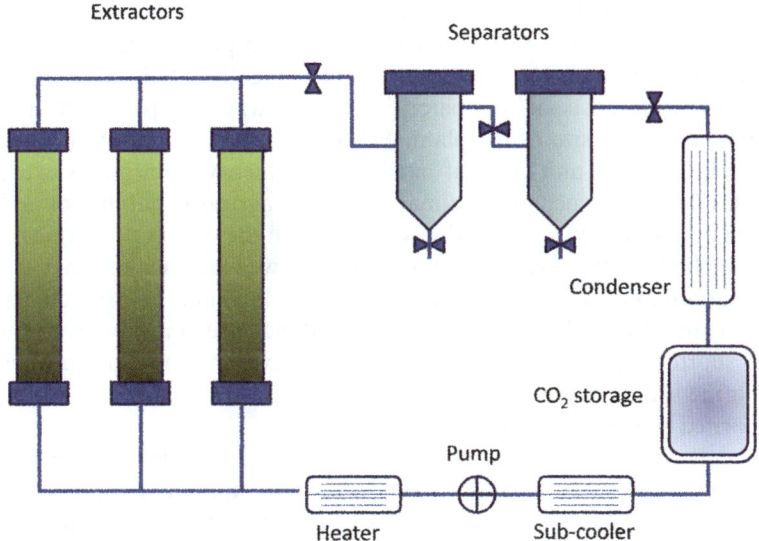

Figure 2.5 CO_2 extraction equipment components.

2.3.1.2 Economic Considerations when Extracting with Supercritical CO_2

Almost all CO_2 extraction plants are now individually designed to deliver optimum performance for their intended application with the lowest possible energy consumption. In designing an extraction plant that would be suitable for a biorefinery application a number of important considerations need to be taken into account.

1. Although continuous extraction would offer significant cost savings no commercial plant exists for this process therefore a high-throughput batch process will need to be considered.
2. Volumes of feedstocks will be high and the loading and unloading of the extraction vessels will be rate limiting.
3. Extract yields will be relatively low (<3% of feed volume) so separator sizes can be scaled down.
4. Maximum load for each extraction needs to be achieved so bulk density of feedstock needs to be maximised. However this should not be so excessive that diffusion of the CO_2 is compromised.
5. Post extraction processing may need to be considered if the residue needs to be delivered in a pelleted form. This is often desirable anyway to reduce dust.

The extraction of biomass prior to its use as a fuel could potentially require extraction plants of a much larger scale than has been built so far. One of the largest commercial CO_2 extraction plants for oil extraction has an extractor volume of $2 \times 3.8\,m^3$; operating at 550 bar and the price for such a plant is currently approximately €5 million without utilities or building. This would have a processing capacity of about 15 tons of seed per day. Larger plants for specific applications such as the decaffeination of coffee are in widespread use.

The operating costs of a supercritical CO_2 extraction plant are inversely proportional to the size of the plant. Figure 2.6 is a compilation of published data[52] reflecting the unit costs of batch and, in theory, continuous extraction processes; however, such cost estimates have an agreed variability of ±30%.

Based on an extraction plant that would match the output of a typical straw pelleting plant (30,000 tons/year or 80 tons/day) we would expect an extraction cost of approximately €350/ton. The lower line in Figure 2.6 represents continuous extraction and an anticipated five-fold increase in productivity (mostly due to shorter use of high pressure volume for extraction) reducing operating costs in the same proportion. However, continuous extraction has so far only been carried out on laboratory scale.

The unit cost of extraction includes all direct and indirect costs and the breakdown of these will again depend on the size of extraction plant. Typical figures are provided by a number of extraction plant manufacturers[53] and Figure 2.7 shows that the depreciation and finance costs of the extraction plant are the largest contributors.

If this proportional breakdown is applied to the overall cost of €350/ton the individual unit costs are as follows:

Plant depreciation	€94.15/ton
Loan interest	€45.85/ton
Labour	€102.90/ton
Electrical energy	€51.45/ton
CO_2 costs	€25.55/ton
Water and steam	€17.85/ton
Engineering spares	€12.60/ton

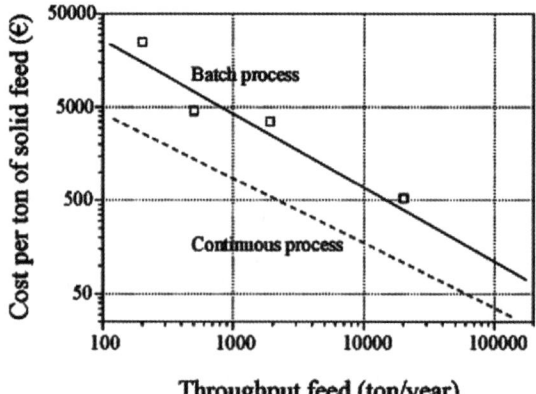

Figure 2.6 Extraction costs (€/ton) relative to throughput.[52]

Farming and Harvesting

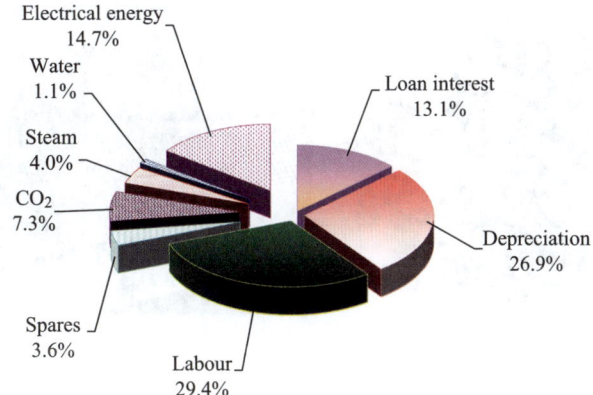

Figure 2.7 Typical breakdown of operating costs for CO_2 extraction.[53]

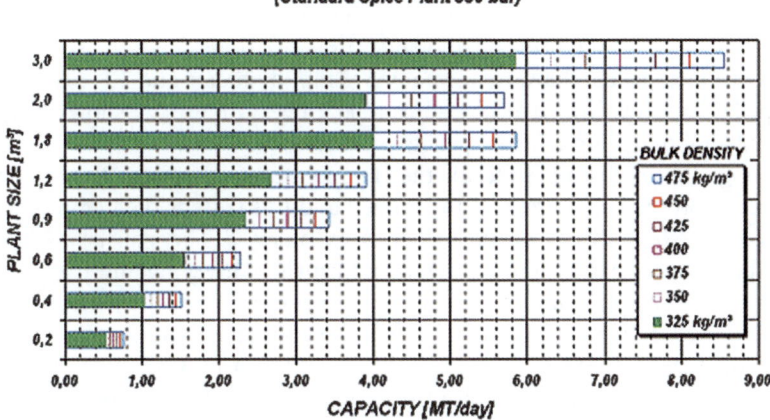

Figure 2.8 Impact of bulk density on plant capacity.[53]

The direct operating costs (depreciation and loan interest removed) are therefore €210/ton raw material input.

2.3.1.3 Straw Densification

The operating costs of extraction are influenced predominantly by plant size/capacity (see Figure 2.8) and by the bulk density of the material to be extracted. Figure 2.8 shows the impact of bulk density on daily capacity of medium-sized plants and this demonstrates the necessity for straw densification prior to extraction. It is anticipated that straw pellet density will be in excess of 500 kg m^{-3} and at this density the pellets will still retain sufficient porosity for efficient extraction.

Figure 2.9 Typical ring die and pellets.[54,55]

Biomass pelletising is a commercially established technology with many operational sites across Europe producing pellets for domestic and industrial boilers or for co-firing. There are two main types of pellet press, flat-die pelletisers and ring-die pelletisers, the latter being the most common. Straw is normally passed through a hammer mill to achieve the desired particle size before pelleting. The milled biomass is then fed to a pellet press where the combined action of die rotation and roller pressure forces material through the die to produce pellets that are normally 6–8 mm diameter and 15–20 mm long. The pellet length is controlled by adjustable knives. Figure 2.9 shows a typical ring die and an example of the pellets produced.

During the pelleting process the temperature is raised to 60–120 °C depending on the pelleting pressure applied and this softens the lignin that acts to bind the straw fibres as the pellet is extruded.[56] The pellets are cooled immediately after pressing to harden them prior to conveying and packaging if appropriate. There is almost no mass loss during milling and pelleting; however, it would be prudent to apply a 98% yield to this process.

Pelleting can be used to increase bulk density to a maximum of 700 kg m^{-3} but although this would maximise the extractor capacity it would be undesirable as the pellet porosity would be too low. A density of 500–550 kg m^{-3} would enable the pellets to be charged directly to the extractor without further milling and at this density the low viscosity of the scCO$_2$ would be able to penetrate the pellet. The pellets would remain intact during and after the extraction process, thereby minimising dust production during subsequent handling.

The capital cost of pelleting equipment will depend very much on the overall throughput and ancillary equipment selected to load and unload the raw materials and pellets. However, it is widely accepted that capital costs will add €18.00/ton for capital equipment based on 100 kt throughput per annum but this does not include the wagon unloading system. A straw pelleting plant will need to acquire straw from the surrounding area and costs therefore need to include straw baling (€12.00/mt) and haulage to the plant (€8.00/mt <50 km). The milling and pelleting of the straw will cost approximately €26.00/mt giving a total cost of €64.00/mt ex plant. If the pellet processing plant is not located in the site of the extraction facility then these will need to be delivered at an additional cost of €8.00/mt based on a distance of 100 km and assuming that 24 mt pellets can be delivered on an average 44 m³ bulk trailer.

2.3.1.4 Raw Material Costs

There is currently within the UK annual availability of about 14 mt of unutilised and harvestable straw of which 2.33 mt is rape straw.[57] This could make a significant contribution to the UK biomass resource as well as reducing GHG emissions. Straw incorporation is seen as a useful means of sequestering carbon and thus mitigating Greenhouse Gas (GHG) emissions. However, straw incorporation is inefficient at sequestering carbon and is estimated to fix only 733 kg CO_2/ha/yr into soil organic matter. In comparison baled wheat straw from 1 ha would replace 2.4 mt of coal, saving over 5000 kg of CO_2. It should not, however, be considered as a waste or free resource. Fertiliser prices have increased dramatically in recent years: and the fertiliser replacement value of straw is now quite significant at £23.39/mt for rape straw (see Table 2.5). In order to encourage any of this potentially available straw biomass into the biomass supply chain the price for baled straw would at least need to cover the fertiliser replacement value plus the baling and handling costs.

Rape straw costs about the same as wheat straw, currently £45.00/mt ex farm, and usually goes for feed or bedding. However, the raw material costs can be treated in a variety of ways depending on the end use of the straw biomass. Because the extraction of straw will not leave any solvent residues and will not reduce either the calorific value or level of mineral nutrients it can be considered as a 'pass through' cost and only the post-extraction treatments and transport included in the raw material costs of the extracts.

2.3.1.5 Potential Straw Extractives

Plant leaf and some stem surfaces are coated with a thin layer of waxy material that has a myriad of functions. This layer is microcrystalline in structure and forms the outer boundary of the cuticular membrane; it is the interface between the plant and the atmosphere. It serves many purposes, for example to limit the diffusion of water and solutes, while permitting a controlled release of volatiles that may deter pests or attract pollinating insects. The wax provides protection from disease and insects, and helps the plants resist drought. As plants cover

Table 2.5 Inherent value of straw related to the nutrient content.

Nutrient value/t	£/kg*	Value £/t
N – 7.7 kg	1.13	8.70
P_2O_5 – 2.2 kg	1.45	3.19
K_2O – 11.5 kg	1.00	11.50
Total		23.39

much of the Earth's surface, it seems likely that plant waxes are the most abundant of all natural lipids. The range of lipid types in plant waxes is highly variable, both in nature and in composition. Table 2.6 illustrates some of this diversity in some of the main components.

In addition, there may be hydroxy-β-diketones, oxo-β-diketones, alkenes, branched alkanes, acids, esters, acetates and benzoates of aliphatic alcohols, methyl, phenylethyl and triterpenoid esters, and many more.

The amount of each lipid class and the nature and proportions of the various molecular species within each class vary greatly according to the plant species and the site of wax deposition (leaf, flower, fruit, *etc.*) and some data for some well-studied species are listed in Table 2.7.

The composition of straw waxes has been extensively studied; however, most work has been carried on the *Triticeae* and in particular wheat (*Triticum aestivum*), barley (*Hordeum vulgare*) and rye (*Secale cereale*). In most studies the

Table 2.6 The major constituents of plant leaf waxes.

n-Alkanes	$CH_3(CH_2)_xCH_3$	21 to 35C – odd numbered
Alkyl esters	$CH_3(CH_2)_xCOO(CH_2)_yCH_3$	34 to 62C – even numbered
Fatty acids	$CH_3(CH_2)_xCOOH$	16 to 32C – even numbered
Fatty alcohols (primary)	$CH_3(CH_2)_yCH_2OH$	22 to 32C – even numbered
Fatty aldehydes	$CH_3(CH_2)_yCHO$	22 to 32C – even numbered
Ketones	$CH_3(CH_2)_xCO(CH_2)_yCH_3$	23 to 33C – odd numbered
Fatty alcohols (secondary)	$CH_3(CH_2)_xCHOH\,(CH_2)_yCH_3$	23 to 33C – odd numbered
β-Diketones	$CH_3(CH_2)_xCOCH_2CO(CH_2)_yCH_3$	27 to 33C – odd numbered
Triterpenols	Sterols, α-amyrin, β-amyrin, uvaol, lupeol, erythrodiol	
Triterpenoid acids	Ursolic acid, oleanolic acid, *etc.*	

Table 2.7 Relative proportions (wt%) of the common wax constituents in some plant species.

	Grape leaf	Rape leaf	Apple fruit	Rose flower	Pea leaf	Sugar cane stem
Hydrocarbons	2	33	20	58	40–50	2–8
Wax esters	6	16	18	11	5–10	6
Aldehydes	6	3	2	—	5	50
Ketones	—	20	3	—	—	—
Secondary alcohols	—	8	20	9	7	—
Primary alcohols	60	12	6	4	20	5–25
Acids	8	8	20	5	6	3–8

Other components present include various diol types and triterpenoid acids.

Farming and Harvesting

level of total 'waxes' varied from 1–3% depending on the variety, straw parts used (leaves, node or internode), crop year and method of extraction. The composition of 'waxes' from wheat straw was first described in 1969[58] and fractionation of the wax carried out using classical separation methods. More recent studies[59,60] have identified a wider range of compounds but extraction was carried out with organic solvent mixtures. The 'waxes' contain a wide range of compounds including alkanes, alkanols, fatty acids, sterols, triglycerides and waxes and these have all been characterised.

The wax fraction from wheat straw has also been extracted using super-critical CO_2[51] and this has demonstrated that the use of this solvent leads to a highly selective extraction of the lipid and wax fraction with minimal content of other compounds such as pigments. The composition of the supercritical CO_2 extract is shown in Figure 2.10.

The composition of barley straw wax has also been studied[61] and found to be very similar to that of wheat straw wax. These cuticular 'waxes' protect the plant against microbial attack and in barley straw have been found to exist in three distinct layers,[62] each having different degrees of order and composition. The composition of 'waxes' from the cuticular wax layers of rye leaves has also been reported;[63] the total wax mixture from both sides of the leaves was found to contain primary alcohols (71%), alkyl esters (11%), aldehydes (5%) and small amounts (<3%) of alkanes, steroids, secondary alcohols, fatty acids and unknowns. A homologous series of alkyl resorcinols was also identified.

Straw 'waxes' from other plant families have also been examined. Flax (*Linum usitatissimum*) straw was first examined in 1931[64] and a more detailed study has recently been reported.[65] However there has been very little work carried out on oilseed rape (*Brassica napus*), or sunflower straw (*Helianthus annuus*) other than examination of the structural components. The total

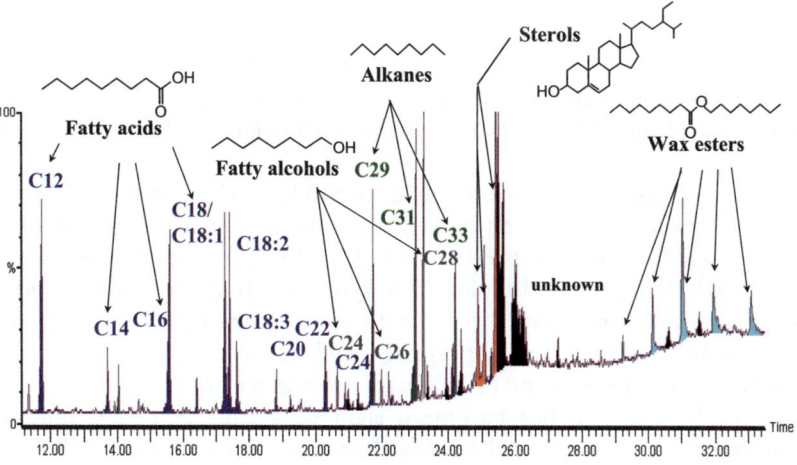

Figure 2.10 Composition of wheat straw wax extracted with supercritical CO_2.

Table 2.8 Alkane yields from sunflower and rape straw.

Raw material/solvent	Hexane	Ethanol	SCO_2
Sunflower straw	0.33%	1.99%	0.33%
Rape straw	0.71%	1.83%	—

benzene/ethanol extractable components from rape straw have been reported[66] and this indicates a yield of 1–2%. Analysis of surface leaf waxes of oilseed rape[67] has indicated that they may be particularly rich in alkanes. Recent extraction trials carried out at the University of York have indicated that yields from sunflower and rape straw are in the range 0.3–2% of the biomass depending on the extraction solvent used (Table 2.8).

Analysis of these extracts indicates that both contain complex mixtures of fatty acids, alkanes and sterols with other minor components. The rape straw extract contained higher levels of odd-numbered alkanes than the sunflower extracts. The extraction yield and composition of the hexane extract is almost identical to that obtained when using supercritical CO_2 (30 MPa and 50 °C) and this is not surprising given that the polarity of both solvents is almost identical.[68] These yields provide only a 'snapshot' of the potential yields as significant variation in both yield and composition can be expected between varieties, crop year and harvest conditions.

It should be recognised that the extraction of these 'waxes' is largely from the surface of the straw and there may be plant protection residues present on the straw. Non-polar residues would be soluble in supercritical CO_2 and given that the extract will represent a 30–50-fold concentration the levels might be significant. A survey of all possible residues is outside the scope of this study but should be considered in future work.

2.3.1.6 *Potential Applications of Straw 'Waxes'*

Waxes are produced commercially in large amounts for use in cosmetics, lubricants, polishes, surface coatings, inks and many other applications. Many of these are of mineral origin but three in particular are derived from plant sources.

Beeswax—Beeswax is recovered as a by-product when honey is harvested from the honeycomb and then refined. It contains a high proportion of wax esters (35 to 80%). The wax esters consist of C_{40} to C_{46} molecular species, based on 16:0 and 18:0 fatty acids some with hydroxyl groups in the ω-2 and ω-3 positions. In addition, some diesters with up to 64 carbons may be present, together with triesters, hydroxy-polyesters and free acids.

Jojoba—The jojoba plant (*Simmondsia chinensis*) grows in the semi-arid regions of Mexico and has become a significant crop. Jojoba is unique in producing wax esters rather than triacylglycerols in its seeds and the waxes consist mainly of 18:1 (6%), 20:1 (35%) and 22:1 (7%) fatty acids linked to 20:1 (22%), 22:1 (21%) and 24:1 (4%) fatty alcohols. Therefore, it contains C_{38} to

C_{44} esters with one double bond in both the acyl and alkyl moiety. As methylene-interrupted double bonds are absent, the wax is relatively resistant to oxidation.

Carnauba—The carnauba palm (*Copernicia cerifera*) has a thick coating of wax on the leaves, which can be extracted when dried. It contains mainly wax esters (85%), accompanied by small amounts of free acids and alcohols, hydrocarbons and resins. The wax esters constitute C_{16} to C_{20} fatty acids linked to C_{30} to C_{34} alcohols, giving C_{46} to C_{54} molecular species.

In addition there are also other plant waxes that are commercially produced in smaller quantities such as citrus wax and apple wax. These are mostly used in cosmetic or personal care applications.[69]

There is so far no commercial production of straw 'waxes'; however, within the groups of molecules so far reported in wheat and barley straw, potential applications have been identified[70] for the alkanes, alkanols and sterol wax fraction.

Waxes are currently extracted with organic solvents and can be derived from plant or animal sources. Using supercritical CO_2 as the extraction solvent extracts can be produced with no organic solvent residues and furthermore can be certified as organic.[71] These waxes can be used as replacements for a wide range of existing products, such as cosmetics, as their physical properties can be varied by selecting the temperature and pressure of the supercritical CO_2 used in the extraction.[51] Importantly the waxes from straws have been shown to have a microcrystalline structure and this property is important for many cosmetic uses.

Most straw waxes appear to contain a range of straight-chain and branched alkanes and it has recently been shown[72] that similar odd-numbered alkanes are important semiochemical molecules. Ladybirds secrete these molecules as they inhabit plant surfaces and aphids are able to detect these and avoid these surfaces. Application of wheat wax extracts has been shown[73] to reduce aphid foraging on important food crops. Extraction of these molecules from straw offers a low-cost and sustainable source of these semiochemicals.

Polycosanols are used in the treatment of various chronic diseases such as diabetes and hypercholesterolelia and work by inhibiting the production of cholesterol in the liver. Recent work[74] has shown that the polycosanol content of wheat straw (164 mg/kg) and sugar cane leaves (181 mg/kg) are remarkably similar, reinforcing the potential of wheat straw as an alternative source of this important group of molecules. Sterols are also used to reduce plasma cholesterol and LDL by interfering with the intestinal absorption of cholesterol originating in the diet. Combinations of polycosanols and sterols are claimed to act synergistically[75] and development of the use of these molecules from straw extracts could be delivered as either dietary supplements or incorporated into high-fat foods such oils or emulsified spreads.

Given the lack of data on the yield and composition of lipophilic extracts from sunflower or OSR straw it is not possible to determine at this time the relative economic contribution that each group of molecules could make to the overall biorefinery. Because sunflower and OSR belong to other genera it is highly likely that other groups of molecules may also be identified that could have commercial value.

2.3.1.7 Existing Biorefinery Concepts Incorporating SFE

There are currently no industrial biorefineries that are using an integrated SFE step as part of their processes. However, there are recent patents that demonstrate that such integration is being considered as an addition to existing processes. The most advanced proposed process[76] envisages capturing the CO_2 generated from the production of ethanol and using this to extract the lipids from the corn germ as a raw material for biodiesel.

It is possible that such an integrated extraction process could be used within the SUSTOIL concept both to extract valuable molecules from alternative feedstocks such as straw and husk and to recover residual oil from the seed press cake.

2.3.1.8 Conclusions

A range of valuable products can be extracted from straw and although the absolute value of by-products that may arise from the extraction of straw cannot be quantified at this time it is possible to arrive at a base cost for such extracts that will enable market consideration.

The following assumptions are used:

1. There is no raw material cost as this will be recovered through end use as a biofuel or by reincorporation back into the land.
2. All transport, baling and pelleting costs will be considered part of the extraction cost – these will in total be £69.00/mt assuming that the residue is re-pelleted.
3. Extraction costs based on 80 mt/day will be £200/mt.
4. Yields will be in the range of 1–3% m/m.

Using these assumptions the crude extracts will be £9.00–£27.00/kg, which is higher than bulk lipids and waxes but well within the acceptable range of costs for the specialised uses that have been identified from the work with wheat straw wax.

2.3.2 Biomethane

Energy production from renewable resources has become a very attractive alternative to the conventional fossil-based fuels. The major biofuels dominating the booming bioenergy market are biodiesel and bioethanol. Biogas is another important biofuel gaining much attention recently. Biogas has been generated primarily in small farms, wastewater treatment plants, plants treating the organic fraction of the municipal solid wastes (OFMSW) and landfills. Other types of biomass such as energy crops and agricultural wastes are

Farming and Harvesting

emerging as potential feedstocks to give a boost to biogas production on a large scale and make it a competitive alternative biofuel. In what follows, the main aspects of anaerobic digestion are presented, as an alternative process to exploit the residues of oil-rich crops.

2.3.2.1 Application of Anaerobic Digestion for Bioenergy Production

Biogas is a mixture of methane and carbon dioxide produced during anaerobic digestion of organic matter. Anaerobic digestion is a multi-step microbial process of degradation of the organic matter, but it is usually described as a four-step process. In the first step, hydrolysis, the complex organic matter (carbohydrates, proteins *etc*.) is transformed into soluble compounds (sugars, amino acids *etc*.) by extracellular enzymes. During the second step, acidogenesis, these products are broken down to simple organic acids and alcohols. The latter are further converted to acetate, hydrogen and carbon dioxide during the acetogenesis step. Finally, during methanogenesis, biogas is produced from acetate or hydrogen and carbon dioxide.

Anaerobic digestion is a well-established technology applied to convert several types of feedstocks to biogas on a large or small scale. Feedstocks with a high biomethane potential include agricultural (livestock manure, agricultural residues, animal mortalities, energy crops), industrial (wastewater, sludges, by-products, slaughterhouse waste, spent beverages, biosolids) and municipal (sewage sludge, organic fraction of municipal solid waste) residues. Table 2.9 lists typical values of the biomethane potential of various feedstocks.

The biogas plants in Europe are very diverse in terms of feedstock, size and technology depending on the regional framework conditions. In some regions, the agricultural biogas plants use liquid manure and energy crops from their own production. Other regions have concentrated on the use of agricultural residues and organic wastes. The size of the biogas plants in Europe ranges from very small plants (15 kWel) up to large plants (several MW).[77]

Large-scale biogas plants, commonly called centralised anaerobic digestion (CAD) plants, have been developed to use feedstock from a variety of sources. The anaerobic digesters may operate at mesophilic or thermophilic conditions and the typical retention times vary from 12 to 20 d. The location of CAD

Table 2.9 Anaerobic potential of different substrates.

Substrate	BMP ($L\ CH_4/g\ VS$)	Methane yield ($m^3\ CH_4/ton_{ww}$)
Slaughterhouse waste	0.57	150
OFMSW	0.5–0.6	100–150
Energy crops	0.3–0.5	30–100
Straws, sugar beet tops	0.2–0.4	36–145
Pig manure	0.29–0.37	17–22
Cow manure	0.11–0.24	7–14

plants is very crucial and they usually serve either a single large farm or several farms within a radius of about 10 km.

Small-scale biogas plants have been established in many farms. In this case, the digesters are constructed as simply as possible in order to be economic. They are heated containers, shaped like silos, troughs, basins or ponds, and may be placed underground or on the surface. They may be batch type (much simpler to construct and maintain) or continuous type. On-farm digesters usually operate at a mesophilic range of temperatures at a typical retention time of 10–30 d.

Another criterion to classify the anaerobic digesters is the solid content of the feeding mixture; if it is up to 12–15% or more, the anaerobic system is characterised as wet or dry respectively. Wet anaerobic systems involve bioreactors of the continuous stirred tank reactor (CSTR) type and are considered suitable when the agricultural wastes are mixed with liquid manure. For the dry anaerobic systems, the plug-flow type digesters are preferred and are operated at high temperature (50–55 °C). Since the volume of the mixture to be digested is low due to the high solid content, the digester's volume can be reduced and the thermal energy consumption needed to maintain the thermophilic conditions is also reduced, making the operation at high temperatures more economic.

In the case of agricultural feedstocks, the anaerobic digestion is limited by the slow hydrolysis of the lignocellulosic material. Lignocellulose is a complex where cellulose is embedded in an amorphous matrix of hemicellulose and lignin, which is rather difficult for common enzymes produced anaerobically to access.

Various pre-treatment methods have been proposed to improve the anaerobic digestibility of the agricultural wastes by disintegrating the lignocellulose structure. These methods are mainly physico-chemical, based on the use of a chemical agent, application of high pressure and temperature, or biological, which is less efficient but more amenable to the environment. Biological methods can be applied in combination with physicochemical pre-treatment or alone. Addition of commercially available enzymes hydrolysing cellulose, hemicellulose and lignin into readily fermentable compounds[78] is an efficient method. Wen *et al.*[79] used a fungi culture in order to produce the enzymes from animal manure, reducing the cost of lignocellulosic hydrolysis. Research is also focused on developing appropriate methods for improving the hydrolysing enzyme activity levels of bacteria under anaerobic conditions.[80] Well-known bacteria used to produce enzymes are *Clostridium thermocellum*, *Bacteroides cellulosolvens* and *Ruminococus albus*.[81]

Another interesting approach is the two-stage anaerobic digestion process that has already been applied to the anaerobic treatment of several agro-industrial residues.[82–86] Typically, the first stage involves solid hydrolysis and acidogenesis, while in the second stage the dissolved organic matter is converted to biogas by methanogens. In a two-stage process hydrolysis and methanogenesis can be optimised separately to suit the needs of each kind of bacterial community.[87] It was reported that the separation of these processes can provide high stabilisation of organics[88,89] and great pathogen reduction.[90] Moreover, the separation of hydraulic and solids retention time through proper handling of the operational mode of the bioreactors[86] can enhance hydrolysis and improve the overall process efficiency.

Biogas needs purification to improve its quality since it contains hydrogen sulfide, ammonia in small quantities and traces of hydrogen, nitrogen, carbon monoxide, saturated or halogenated carbohydrates and oxygen. Its energy content is determined by the methane content (1 kWh per m^3 of biogas with 10% of methane). Biogas can be converted to electricity and heat in the combined heat and power (CHP) engines on site or, preferably, injected to the natural gas grid and dispersed for the production of electricity and heat where necessary. In this way, the CHP application is improved, otherwise heat would be lost or used inefficiently in the biogas plant. Another option is to use it as a vehicle fuel after upgrading. In countries like Italy and Germany, where the network of natural gas for vehicles stations is extensive, this option of biogas utilisation can have an immediate positive impact.

Several demonstration projects have been realised to show the sustainability of the biogas plants based on agricultural feedstocks. A biogas plant in a pig breeding farm near a small village in Austria, Reidling, was set up with a two-stage continuous stirred tank reactor system in 2003.[91] The digesters had a volume of 2000 and 1850 m^3 and fed on a mixture consisting of pig manure (30%), energy crops like maize and residues from vegetable processing (70%). The agricultural feedstocks were provided by local farmers who utilised the digestate as a fertiliser in their farms. Rates of 14.5 eurocents per kWh electrical energy and 2 eurocents per kWh of thermal energy were paid to the farm with the biogas plant. The plant was operated successfully and achieved more than 98% of the theoretical capacity. The sale of the thermal energy has been increasing since 2003 because more and more people joined the local heat grid. The net profit from the energy sale (both electrical and thermal) in the first years amounted to 6–7 eurocent per kWh produced (IEA).

Another example is in a village in Germany, Juhnde.[92] The village's energy demand was covered by exploiting the biomass coming from the local agriculture and forestry. For this purpose, the project involved an anaerobic system co-digesting liquid manure and energy crops, a boiler fed on wood chips and a heating network for 145 houses. The biogas plant produced 5000 MWh of electricity per year, which supersedes the electricity demand of the village. The heat sold covered close to 99% of the heat demand. Heat was produced both by the CHP (85%) and the boiler (15%). The heat losses in the grid were approximately 22% while the demand of the plant for heat was lower than 10% of the total heat produced. Various aspects, such as infrastructure, biomass production, social awareness and willingness to participate, are important to the sustainability of such bioenergy schemes. Therefore, detailed feasibility studies and economic plans are necessary before a demonstration project is realised.

The technology of dry thermophilic anaerobic digestion adapted from digestion of OFMSW to ferment agricultural crops has been applied in Nustedt, Germany.[93] A vertical plug-flow digester of 1200 m^3 was operated continuously and fed on 12 500 tons/year of agricultural crops (maize, sunflower, rye and grass) and 1200 tons/year of solid manure. The biogas yield amounted to 145 m³/ton on average and the potential of electrical output was 750 kW. The cost of the total investment was approximately €3 M.

Many more examples of case studies can be found showing the sustainability of the anaerobic digestion process.

2.3.2.2 Economic Considerations

The cost of the production of biogas from agricultural feedstocks involves the crop production, transportation of the crops or the residues to the biogas plant, the investment and the operation of the biogas plant. The costs of the crop cultivation can be categorised into land rental, equipment, labour, fuel, seeds and fertilisers. The investment of the biogas plant ranges from 3000 to 5000 euro per KWh$_{el}$[94] depending on the complexity of the technology applied to design the system (usually a two-stage digester system and a gas storage tank), the size of the system and the price of the auxiliary equipment such as pumps, stirring mechanism, CHP *etc.* as well as the materials of construction (steel or concrete). The operational costs range between 2 and 4.5 eurocent per KWh$_{el}$ and mainly include the maintenance of the CHP and the biogas plant and labour costs.[94]

On the other hand, the benefits from biogas come mainly from the sale price of the energy. The sustainability of green energy production from biogas is determined to a large extent on the remuneration policy of the local government. In European countries (Austria, Belgium, England and Wales, France, Germany, Italy, Poland, Slovenia and Spain) there are two main systems: feed-in-tariffs and certificates. The feed-in-tariffs system is based on the electricity (produced from biogas) fed into the grid. Although there is a great differentiation of the regulations for the tariff calculation from country to country, what usually influences the tariff rate involves the capacity (kW$_{el}$) of the biogas plant, the capacity of the electrical grid, the feedstock (agricultural crops, residues, organic wastes *etc.*), the innovation of the technology applied, the efficient utilisation of the produced redundant heat, the period of the electrical feeding *etc*. The green certificate system is based on the green MWh$_{el}$ generated divided by the carbon dioxide saving rate. This rate is defined as the ratio of the carbon dioxide produced during the bioelectricity generation (emissions from fuel consumption for the energy crops, transportation of the feedstocks, but also from the processing of the feedstocks, the combustion of the biogas *etc.*) to the carbon dioxide produced in the conventional electrical system. Other sources of benefits include the treatment of organic wastes such as manure together with the agricultural feedstocks and the use of the digestate as a fertiliser replacing the mineral fertilisers and reducing the cost of the crop cultivation.

2.3.2.3 Conclusions

Anaerobic digestion, as a mature technology already applied successfully to convert agricultural wastes and crops to biogas, can easily become an

Farming and Harvesting

important segment of the biorefinery scheme. Since the sustainability of the process is very reliant on the income coming from the sale of the bioenergy, the CHP unit should operate during all the year, depending on the availability of the feedstocks. However, the rapeseed and sunflower crops, taken as case studies in this chapter, are seasonally cultivated. As a result, the waste stream consisting of the residues after the oilseed harvesting should be mixed with other feedstocks (for example other agricultural wastes, manure *etc.*) to secure the constant feeding of the biogas plant and the continuous bioenergy generation.

2.3.3 Biomaterials from Thermocoupling

In most oilseed crops production, stalks are a by-product for which no added value has really been considered. The main use of this lignocellulosic part is soil enrichment; the stalks are not harvested. For soy and rapeseed, this is really understandable as these stalks are not rigid enough to be sorted during harvesting but in the case of sunflower, as it is for linseed for example, a real industrial use of stalks could be investigated as their fibre quality is quite good and they contain a high amount of pectin in the pith.

2.3.3.1 Structure and Composition of Sunflower Stalk

When sunflower reaches maturity, stalks account for around 25% of the total dry matter of the plant.[95] When compared to seeds production,[96] if they were harvested, sunflower stalks would represent a potential of 4.7 Mt/year in Europe (Table 2.10).

2.3.3.1.1 Structure. Sunflower stalk is made of two very distinct parts: the pith and the bark or husk (Figure 2.11).

The outer part is constituted of a wood-like layer on which long fibres are encrusted. The whole bark is covered by a thin, brown skin. Dry and easily breakable, the bark is easy to separate from the pith and accounts for around 90% of stalk mass.

The pith is white, more or less elastic and not dense (around 0.035 while bark density is around 0.43). Very porous, its inner structure is well organised in cells (Figure 2.11).

Table 2.10 Average composition of sunflower mature plant.[95]

Part	Composition (% DM)
Root	8
Stalk	25
Leaves	18
Capitulum	19
Seeds	30

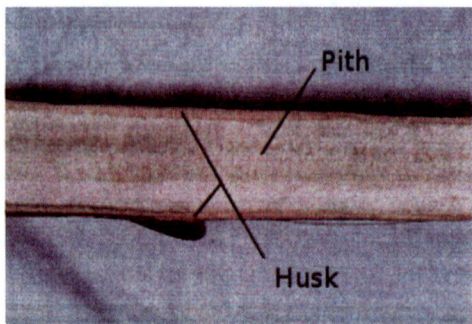
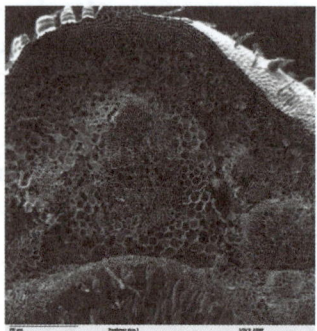

Figure 2.11 Longitudinal cut of a stalk (left), SEM micrograph of stalk section (right).

Table 2.11 Composition in hemicellulose/cellulose/lignin of various plant fibres.[97]

	Hemicelluloses	Cellulose	Lignin
Hard wood	25	48	24
Soft wood	25	43	29
Wheat straw	34	38	14
Kenaf			
- whole stalk	23 (pentosans)	39	10
- depithed stalk	16	45	8
- pith	22	37	14
Sorghum			
- canes	24 (pentosans)	27	9
- depithed canes	22–34	31–44	7–14
- pith	32	37	9
Corn			
- whole stalk	27 (pentosans)	46	17
- depithed stalk	26 (pentosans)	48	16
- pith	26 (pentosans)	48	13
Sunflower			
- depithed stalk (D.W.: 88%, ash: 3.8%, proteins: 1.4%)	32	41	17

2.3.3.1.2 Composition. The composition of fibres of sunflower stalk is very like that of other annual plants such as corn and paper sorghum (Table 2.11). They are characterised by high cellulose and hemicellulose contents. Acid hydrolysis indicated that they were of galacto-arabino-xylan type.[97]

Sunflower pith has a high pectin content and low levels of hemicelluloses and lignin (Table 2.12), which distinguishes it from the stem piths of corn, sorghum and kenaf (Table 2.11). Almost all the pectins from the pith are extracted in ammonium oxalate. Their main characteristics are a high anhydrogalacturonic

Farming and Harvesting

Table 2.12 Chemical composition of sunflower pith (%w/w on a dry basis).

	Content (%)	Characteristics
Dry weight (D.W.)	88	
Ash	16.6	
Proteins	0.9	
Lipid fraction	4	Waxes: 0.3%
Total pectins	17.6	
• extracted in water at 20 °C	4.7	• D.W.: 86%; ash: 19.9% D.M.: 4% GA.A.: 46% Ca: 9.3%
• extracted in ammonium oxalate	11.7	• D.W.: 85%; ash: 3.7% D.M.: 9% GA.A.: 82% Ca: 2.3%
• extracted in HCl	1.2	• D.W.: 90.1%; ash: 14.1% D.M.: 8% GA.A.: 43% Ca: 12.7%
Polysaccharides extracted in NaOH	4.4	• D.W.: 89%; ash: 49.4% • Uronic acid: 17% • Total sugars: 24.5% Sugars: xyl. (25%), glu. (24%), gal. (20%), rham. (16%), arab. (10%), man. (6%)
Lignin	3.2	
Cellulose	45.4	

D.M.: Degree of methylation, GA.A.: gallic acid content, Ca: Calcium content, arab.: L(+)-arabinose, gal.: D(+)-galactose, glu.: D(+)-glucose, man.: D(+)-mannose, rham.: L(+)-rhamnose, xyl.: D(+)-xylose

acid content and a very low acetyl content.[98–100] They have an average molecular weight ranging from 39 500 to 52 000.[95,99]

2.3.3.2 Review of Possible Industrial Uses of Sunflower Straw and Pith

Apart from the bioenergy production potential, other applications have also been considered for the two main components of sunflower stalks.

2.3.3.2.1 Separation of Pith and Straw. As pith and straw densities are significantly different, 35 kg m^{-3} and 430 kg m^{-3} respectively, sunflower stem can be easily split into two parts through crushing and separation in a cyclone.

For example, the comparison of the composition of the different fractions resulting from the particle size analysis after crushing through a 15 mm grid is presented in Figure 2.12.

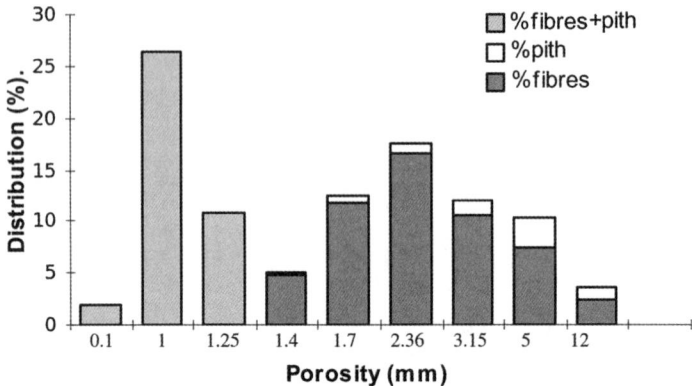

Figure 2.12 Distribution of pith and fiber fractions from crushed sunflower stalks.[3]

With a simple sieving device, 70% of the pith is easily separated and can be completely sorted, for example through fluidisation.

2.3.3.2.2 Potential Application of Sunflower Pith

2.3.3.2.2.1 Low-density Materials. The low density of the pith can be directly exploited in the manufacture of light materials without any requirement for additives or mould-drying process. The mechanical characteristics of the materials obtained seem to depend on the amount of water used in the forming process (Table 2.13). The water appears to solubilise a fraction of the free carbohydrates and pectins in the pith. This may account for the superior mechanical properties of the materials based on sunflower pith compared to those manufactured from piths of other plants (Table 2.14).

2.3.3.2.2.2 Pectin Extraction. Pectins find many applications in the food and cosmetic industries due to their texturing properties. The high pectin content of the sunflower pith and head makes it an interesting potential source for pectin production.

Pectins from pith account for 45% of the total pectin content of the sunflower.[100] Their content varies from 5 to 20% with the plant growth.[98] Their main characteristics are a high anhydrogalacturonic acid content (77–85%) and a low acetyl content (2.3–2.6%).[99] They form firm gels in the presence of calcium but these are highly sensitive to pH.[101]

2.3.3.2.3 Potential Use of Whole Stalk

2.3.3.2.3.1 Adsorption Properties. Biosorption is becoming a potential alternative to the existing technologies for the removal and/or recovery of toxic metals from wastewater. The major advantages of biosorption technology are its effectiveness in reducing the concentration of heavy metal ions to very low

Table 2.13 Mechanical properties of pith-based light materials according to water content.

Materials shaped under zero pressure	Density	Flexion to breakage			Compression by 1 mm		Cohesion	
		Breaking strength/thickness ($kg\,mm^{-1}$)	Slope ($kg\,mm^{-2}$)	Strength (kg)	Residual sag (mm)	Thickness (mm)	Strength/thickness ($kg\,mm^{-1}$)
8 g pith + 0 g H_2O	0.11	0.1	0.1	1.0	0.17	8.9	0.2
8 g pith + 8 g H_2O	0.33	1.2	0.4	5.2	0.31	3.0	3.4
8 g pith + 16 g H_2O	0.32	1.7	0.5	4.8	0.26	3.0	5.4
8 g pith + 24 g H_2O	0.30	1.5	0.4	4.6	0.30	3.2	5.3
8 g pith + 32 g H_2O	0.22	1.5	0.8	3.4	0.30	4.2	1.8
Pith rinsed in water	0.26	0.5	0.2	2.7	0.31	3.3	0.6
Raw stalk	0.27	0.2	0.1	2.1	0.43	—	—
Packing polystyrene	0.02	0.5	1.2	0.7	0.26	—	—

Table 2.14 Comparison of pith-based materials from various plant sources.

Materials shaped under zero pressure (8 g pith + 8 g H_2O)	Density	Flexion to breakage		Compression by 1 mm		Cohesion
		Breaking strength/ thickness (kg mm^{-1})	Slope (kg mm^{-2})	Strength (kg)	Residual sag (mm)	Strength/ thickness (kg mm^{-1})
Sunflower	0.33	1.2	0.4	5.2	0.31	3.4
Kenaf	0.11	0.3	0.2	1.6	0.39	1.5
Sorghum	0.16	0.5	0.2	2.1	0.26	4.0
Corn	0.18	0.6	0.4	2.4	0.25	2.5

levels and the use of inexpensive biosorbent materials. Metal adsorption and biosorption onto agricultural wastes is a rather complex process affected by several factors. Mechanisms involved in the biosorption process include chemisorption, complexation, adsorption-complexation on surface and pores, ion exchange, microprecipitation, heavy metal hydroxide condensation onto the biosurface and surface adsorption.[102]

The high content of minerals in sunflower stalk (Table 2.10), associated with the cellular structure of the pith, makes stalks a good candidate biosorbent. It has been tested with additional chemical modification on Hg and Cu solutions[103,104] or for colour removal from textile wastewater.[105]

2.3.3.2.3.2 Fibrous Materials. The whole sunflower stalks have been considered to make paper or particle boards.

For paper applications, whatever the pulping conditions were (twin-screw extrusion[97] or soda-antraquinone[106–108]), pith appeared to decrease the paper quality. However, considering the economical interest due to the cheap raw material, pulping of sunflower stalks is promising for low-quality pulp, for example cardboard production.

The same conclusion about depithing was stated for particle board applications using phenol-formaldehyde resins[109] or tannin-modified UF resin.[110] In this case, the mechanical properties of the particle boards are sufficient for many applications, confirming the potential value of the sunflower stalks fibres. An alternative to depithing is to mix the sunflower stalks with poplar wood particles in three-layer boards.[111]

2.3.3.3 Composites Made from Residues of Whole Plant Oil Extraction

A new integrated approach of the sunflower biorefinery, based on the aqueous extraction of oil from the whole plant, provides the conditions for complete industrial utilisation of the crops.

Aqueous extraction could be an interesting alternative to classical oil extraction because:

- it is an energy-efficient one-stage process using twin-screw extrusion;
- the liquid phase is an oil-in-water emulsion stabilised by sunflower proteins, which can be directly used in industrial applications or become the medium for the transformation of oil into fatty acids and fatty esters;
- it is a 'soft' process for the oil cake that allows further possible use of the resultant fibrous materials to make biodegradable materials.

2.3.3.3.1 Aqueous Extraction of Sunflower Oil. Oil extraction from seeds in aqueous conditions has been evaluated.[112] In the twin-screw extrusion extraction process, presented in Figure 2.13, the use of fibres is necessary to create a dynamical plug (in the counter screw as seen in the seventh module of the screw profile of Figure 2.13) rigid enough to ensure the liquid/solid separation. Two types of fibres have been tested, wheat straw and depithed sunflower stalks, for this purpose. It has been made obvious then that the extraction from the whole plant could be really promising. In these conditions, with seeds and fibres added separately, the lipid extraction yield in optimal conditions is more than 70%[112] (Figure 2.14).

A complete matter assessment of the fractionation process is presented in Figure 2.15 when the extraction is performed directly on whole sunflower plant.

This novel process involves the treatment of the obtained liquid phase to recover the oil. After treatment, three phases are obtained (Figure 2.16).

Higher hydrophobic phase and lower hydrophobic phase are both oil-in-water emulsions accounting for the 16% and 8% approximately of the total mass of the filtrate respectively. Their water content is about 80%. Their dry matter consists mostly of lipids, corresponding to 36% and 8% of the oil in whole plant respectively. For both hydrophobic phases, the remainder contains proteins, phospholipids and minerals. The presence of proteins is indicative of their role as natural surface-active agents in the emulsions. They represent 20%

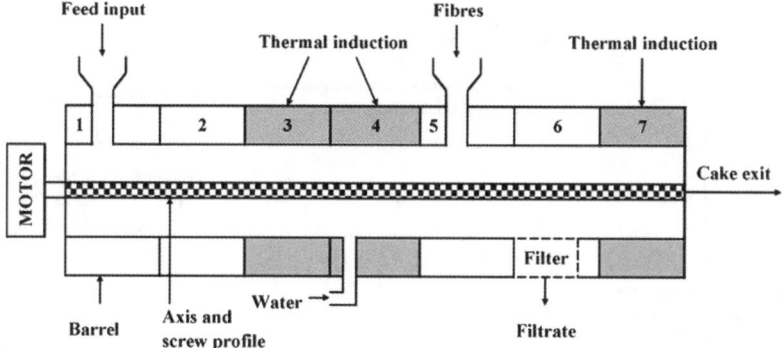

Figure 2.13 Schematic modular barrel of the Clextral BC 45 twin-screw extruder used for aqueous extraction of oil from whole sunflower seeds ($\theta c = 80\ °C$).

1	2			3			4			5			6	7	
T2F 66	C2F 50	C2F 33	DM 10×10 (45°)	C2F 25	BB 5×5 (90°)	C2F 33	C1F 33	C1F 25	BB 5×5 (90°)	C1F 33	C1F 33	C1F 33	C1F 25	C1F 15	CF1C -15	C1F 25

T2F, trapezoidal double-thread screw; C2F, conveying double-thread screw; C1F, conveying simple screw; DM, monolobe paddle-screw; BB, bilobe paddle-screw; CF1C, reversed screw. The numbers following the type of the screw indicate the pitch of T2F, C2F, C1F and CF1C screws and the length of the DM and BB screws.

Figure 2.14 Example of screw configurations for aqueous extraction of oil from whole sunflower seeds.

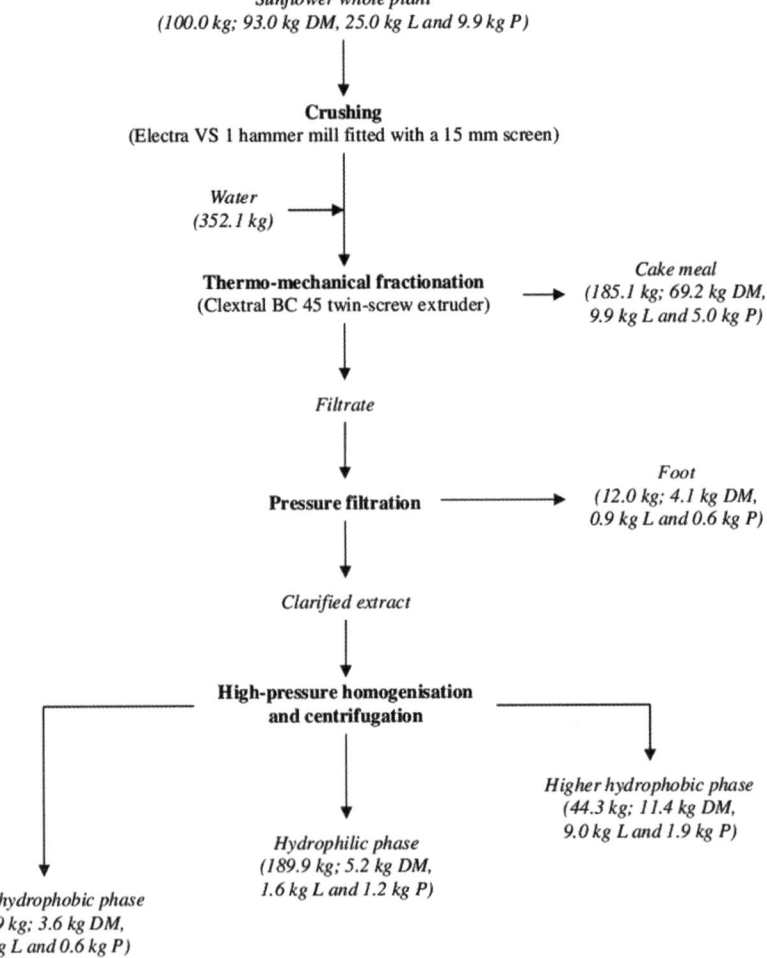

Figure 2.15 Matter assessment for thermo-mechanical fractionation of sunflower whole plant conducted with the Clextral BC 45 twin-screw extruder. DM: dry matter, L: lipid, P: protein.

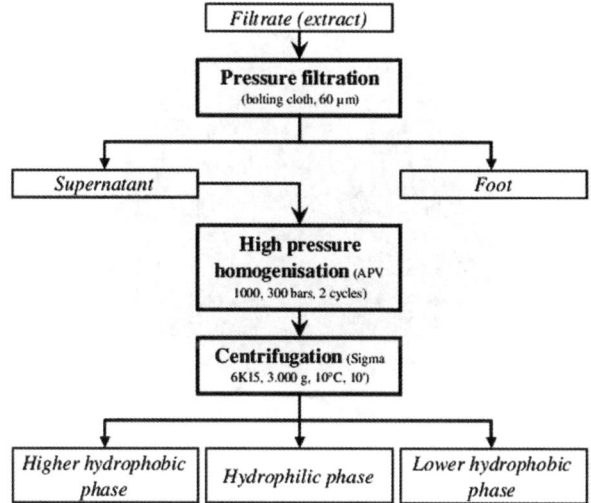

Figure 2.16 Flow-sheet of the filtrate treatment after the aqueous extraction process.

of the proteins in the whole plant for the higher hydrophobic phase, and 6% of the proteins in the whole plant for the lower hydrophobic phase. Pectins and non-pectin sugars are constituents of the lower hydrophobic phase, indicating that part of the hemicelluloses is also extracted during the aqueous process. These characteristics in the chemical composition of the lower hydrophobic phase also explain why it is denser than the other two liquid phases.

The hydrophilic phase (aqueous phase) is the largest (about 70% of the filtrate) since the liquid/solid ratio for the extraction is close to 4. It represents almost half of the injected water. It is an aqueous extract of the water-soluble constituents from the whole plant: (i) proteins from kernel, (ii) pectins from pith and head and (iii) hemicelluloses from stalk. The dry matter content is 3%, corresponding to 6% of dry matter, 12% of proteins and 9% of pectins in the whole plant. Lipids account for 31% of the dry matter. The presence of a high quantity of lipids indicates that the separation between the three liquid phases is not optimal. The hydrophilic phase appears to be incompletely separated from the higher hydrophobic phase. However, taking into account the low percentage of the dry matter, these losses are minimal; they account for only 6% of the oil in the whole plant (Figure 2.16).

2.3.3.3.2 Biomaterials Production from Cake Meal.
This fractionation process results in a fibrous cake meal suitable to make biodegradable biomaterials through direct hot-pressing (Figure 2.17).

Depending on the extrusion operating conditions (*i.e.* screw profile, temperature, solid and liquid input rate), the composition of the oil cake can change. In regular conditions, the cake contains most of the lignocellulosic parts of the stalk and the seed husk and some non-negligible amounts of protein and oil from seeds (Table 2.15). The protein content was 7% of the dry matter, corresponding

Figure 2.17 Fibrous cake meal from aqueous extraction of sunflower whole plant.

Table 2.15 Chemical composition of the sunflower whole plant and cake meal.

	Sunflower whole plant	Cake meal from trial 4
Minerals	6.46 ± 0.19	5.90 ± 0.09
Lipids	26.83 ± 0.43	14.33 ± 0.06
Proteins	10.65 ± 0.17	7.25 ± 0.02
Cellulose	23.93 ± 0.55	32.00 ± 0.37
Hemicelluloses	7.83 ± 0.09	14.83 ± 0.03
Lignin	9.13 ± 0.03	10.71 ± 0.14
Pectins	6.97 ± 0.24	–

to 51% of the proteins in whole plant. Besides, DSC measurements indicated that the denaturation of proteins was almost complete in the cake meal after thermo-mechanical fractionation in the twin-screw extruder.[113]

The cake meal would be suitable for animal feeds or for pellets production for energy. New utilisation of the cake meal as a mixture of lignocellulosic fibres and proteins could also be considered. Thus, this natural composite can be processed into biodegradable and value-added agromaterials by hot-pressing. During moulding operation, macromolecular structure of the proteins is completely transformed due to their thermal properties. The simultaneous effects of pressure and temperature resulted in their complete denaturation. The reorganisation of their structure gives the panels their mechanical resistance while fibre entanglement ensures their structure.

The mean apparent density of the material depends on the pressing conditions (Table 2.16). It varies between 0.74 for panel 1 and 1.13 for panel 3. The mechanical properties are evaluated by the flexural strength at break (σ_f) and the elastic modulus (E_f). These values increase simultaneously with

Table 2.16 Hot-pressing conditions and mechanical properties in bending (σ_f, stress at break; E_f, elastic modulus) of the moulded test specimens.

	Panel				
	1	2	3	4	5
Thermo-pressing conditions					
Temperature (°C)	160	160	160	180	200
Pressure (kg cm^{-2})	107	320	320	320	320
Time (s)	30	30	60	60	60
Flexural properties					
t (mm)	6.01 ± 0.39	3.97 ± 0.19	3.82 ± 0.12	3.88 ± 0.18	3.89 ± 0.15
D	0.74 ± 0.05	1.09 ± 0.05	1.13 ± 0.04	1.09 ± 0.05	1.04 ± 0.04
σ_f (MPa)	0.4 ± 0.1	5.0 ± 0.2	6.0 ± 1.0	11.3 ± 0.9	11.5 ± 0.5
E_f (GPa)	0.10 ± 0.03	1.07 ± 0.15	1.28 ± 0.19	2.11 ± 0.16	2.22 ± 0.17

Table 2.17 Comparison of the mechanical properties of various experimental and industrial materials obtained by hot pressing.

	Isorel®	Laminated board	Maize cob	Cake meal[1]
t (mm)	12.0	0.9	3.2 ± 0.1	3.9 ± 0.2[96] 3.9 ± 0.1[97]
d	0.30	0.67	1.20 ± 0.04	1.09 ± 0.05[96] 1.04 ± 0.04[97]
σ_{max} (MPa)	1.1	14.7	11.7 ± 1.0	11.3 ± 0.9[96] 11.5 ± 0.5[97]
E_f (GPa)	0.02	1.45	1.78 ± 0.11	2.11 ± 0.16[96] 2.22 ± 0.17[97]

temperature, pressure and pressing time. The highest stress at break (11.5 MPa) and highest elastic modulus (2.22 MPa) are therefore obtained for panel 5 at a temperature of 200 °C, pressure of 320 kg cm^{-2} and pressing time of 60 s. The corresponding mean apparent density is 1.04.

Due to their promising mechanical properties in terms of bending compared with those of other experimental materials (Table 2.17), panels (Figure 2.18) could be used as inter-layer sheets for pallets in the handling and storage industry or for the manufacturing of biodegradable containers (composters, crates for vegetable gardening) by assembly of panels.

Additional results, as yet unpublished,[114] show that:

- the amount of protein is the key factor determining the mechanical properties of the panel, as seen in DMA analyses;
- the panel water contact angle is relatively high (between 80 and 90°) but decreases with time: residual fatty compounds create a lipophilic surface;
- panels absorb around 7% of water when stored at 85% RH and their thickness increases by 10% in the same conditions.

Figure 2.18 Photograph of the cake meal based materials.

Hot-pressed panel from the cake after the whole sunflower oil extraction has some interesting properties due to the lignocellulosic fibres of the straw and the adhesive properties of the sunflower protein. In addition, the remaining oil content of the cake gives the panel some relative resistance to moisture when compared to other 100% biobased materials.[115]

2.3.3.4 Conclusions

Among all major oilseed by-products, sunflower stalks seem to be the most promising to make fibrous materials.

These stalks are made of a light pectic-based core called pith and a lignocellulosic husk. For direct industrial use, separation of both parts seems to be necessary. The pith can be used to produce light materials or as a source of pectin while the outer part can be used for fibrous materials, cardboard or fibre board. However, when compared with other alternative sources of fibres, like wheat straw for example, the fibre quality of sunflower stalks does not seem to be especially high.

The aqueous extraction of sunflower oil from the plant can be an interesting alternative for a complete valorisation of sunflower. In a one-stage twin-screw extrusion process, oil- and water-soluble compounds can be extracted.

The raffinate contains mainly the lignocellulosic fibres of the stalk and a large part of sunflower protein (around 50%). These proteins have some interesting adhesive and plastic properties found in panels made directly from this extraction cake without addition of any kind of resin. They are denser than usual panels (around 1), and are processed at high pressure (320 kg cm^{-2}) and high temperature (200 °C) but have high mechanical resistance (about 2 GPa of bending modulus) and show good behaviour towards moisture.

References

1. S. P. McGrath, F. J. Zhao and P. J. A. Withers, *Proceedings – The Fertiliser Society*, 1996, **379**.
2. S. P. McGrath and F. J. Zhao, *J. Agri. Sci.*, 1996, **126**, 53.
3. E. J. Booth, I. Bingham, K. G. Sutherland, D. Allcroft, A. Roberts, S. Elcock and J. Turner, *HGCA Research Review*, 2005, **53**, 58.
4. K. Sieling and O. Christian, *Eur. J. Agron.*, 1997, **7**, 301.
5. TAG 2007, Impact of previous cropping on winter oilseed rape. Interim annual project report, 2007 results. Available to view on Home-Grown Cereals Authority website crop research section (project number RD-2003-2922), www.hgca.com/publications/documents/cropresearch
6. D. B. Turley, in *Proceedings 16th BCPC International Plant Protection Congress 2007, Glasgow*, pub: BCPC, 2007, pp. 170–171.
7. P. M. Berry and J. H. Spink, *J. Agri. Sci.*, 2006, **144**, 381.
8. N. J. Mendham, P. A. Shipway and R. K. Scott, *J. Agri. Sci.*, 1981, **96**, 389.
9. P. Leterme, in *Colza: Physyologie et Elaboration du Rendement*, CETIOM, Paris, INRA/CETIOM, 1988, pp. 124–129.
10. M. S. S. Rao, N. J. Mendham and G. C. Buzza, *J. Agri. Sci.*, 1991, **117**, 189.
11. M. J. Fray, E. J. Evans, D. J. Lydiate and A. E. Arthur, *J. Agri. Sci.*, 1996, **127**, 193.
12. G. D. Lunn, J. H. Spink, D. T. Stokes, A. Wade, R. W. Clare and R. K. Scott, Canopy management in winter oilseed rape. *HGCA Project Report No OS49*. Pub: Home-Grown Cereals Authority, London, 2001.
13. G. J. Seiler, in *Sunflower Technology and Production*, ed. A. A. Schneiter, Agronomy Monograph 35, ASA, CSSA and SSSA, Madison, WI, USA, 1997, p. 67.
14. D. J. Connor and A. J. Hall, in *Sunflower Technology and Production*, ed. A. A. Schneiter, Agronomy Monograph 35, ASA, CSSA and SSSA, Madison, WI, USA, 1997, p. 113.
15. E. D. Putt, in *Sunflower Technology and Production*, ed. A. A. Schneiter, Agronomy Monograph 35, ASA, CSSA and SSSA, Madison, WI, USA, 1997, p. 1.
16. Anonymous, *Statistiques des oléagineux & protéagineux, huiles & protéines végétales. De la production à la consommation. France – Europe – Monde*, PROLEA, Paris, France, 2007.
17. http://faostat.fao.org/site/567/default.aspx#ancor
18. H. Quinones, V. Texier, M. Cabelguenne and R. Blanchet, in *Le tournesol et l'eau, adaptation à la sécheresse, réponse à l'irrigation*, ed. R. Blanchet and A. Merrien, Les points science du CETIOM, CETIOM, Paris, France, 1990, p. 56.
19. P. Debaeke, A. M. Triboi, F. Vear and J. Lecoeur, in *Proc. 16th Int. Sunfl. Conf.*, Fargo, ND, USA. Int. Sunfl. Assoc., Paris, France, 2004, p. 267.

20. F. P. C. Blamey, R. K. Zollinger and A. A. Schneiter, in *Sunflower Technology and Production*, ed. A. A. Schneiter, Agronomy Monograph 35, ASA, CSSA and SSSA, Madison, WI, USA, 1997, p. 595.
21. G. Petcu and E. Petcu, *Helia*, 2006, **29**, 135.
22. S. Masirevic and T. J. Gulya, *Field Crop. Res.*, 1992, **30**, 271.
23. T. Gulya, K. Y. Rashid and S. M. Masirevic, in *Sunflower Technology and Production*, ed. A. A. Schneiter, Agronomy Monograph 35, ASA, CSSA and SSSA, Madison, WI, USA, 1997, p. 263.
24. F. Virányi, in *Proc. 17th Int. Sunfl. Conf.*, Cordoba, Spain, Int. Sunfl. Assoc., Paris, France, 2008, p. 1.
25. G. Gagne, P. Roeckel-Drevet, B. Grezes-Besset, P. Shindrova, P. Ivanov, C. Grand-Ravel, F. Vear, D. T. De Labrouhe, G. Charmet and P. Nicolas, *Theor. Appl. Genet.*, 1998, **96**, 1216.
26. C. Seassau, E. Mestries, P. Debaeke and G. Dechamp-Guillaume, in *Proc. 17th Int. Sunfl. Conf.*, Cordoba, Spain, Int. Sunfl. Assoc., Paris, France, 2008, p. 199.
27. F. Raducanu, E. Petcu, C. Raducanu, M. Stanciu and D. Stanciu, in *Proc. 17th Int. Sunfl. Conf.*, Cordoba, Spain, Int. Sunfl. Assoc., Paris, France, 2008, p. 219.
28. C. E. Rogers, *Field Crop. Res.*, 1992, **30**, 301.
29. L. D. Charlet, G. J. Brewer and B. Franzmann, in *Sunflower Technology and Production*, ed. A. A. Schneiter, Agronomy Monograph 35, ASA, CSSA and SSSA, Madison, WI, USA, 1997, p. 183.
30. F. Flénet, P. Debaeke and P. Casadebaig, in *Proc. 17th Int. Sunfl. Conf.*, Cordoba, Spain, Int. Sunfl. Assoc., Paris, France, 2008, p. 13.
31. F. López-Granados, J. M. Peña-Barragán, M. Jurado-Expósito and L. García-Torres, in *Proc. 17th Int. Sunfl. Conf.*, Cordoba, Spain, Int. Sunfl. Assoc., Paris, France, 2008, p. 477.
32. G. M. Linz and J. J. Hanzel, in *Sunflower Technology and Production*, ed. A. A. Schneiter, Agronomy Monograph 35, ASA, CSSA and SSSA, Madison, WI, USA, 1997, p. 381.
33. F. Castillo-Llanque, C. Santos, I. J. Lorite, M. García-Vila, M. Tasumi, J. Estévez and P. Gavilán, in *Proc. 17th Int. Sunfl. Conf.*, Cordoba, Spain, Int. Sunfl. Assoc., Paris, France, 2008, p. 299.
34. G. Sin, M. Botea and L. Drăgan, in *Proc. 17th Int. Sunfl. Conf.*, Cordoba, Spain, Int. Sunfl. Assoc., Paris, France, 2008, p. 329.
35. D. Laureti, S. Pieri, G. P. Vannozzi, M. Turi and R. Giovanardi, *Helia*, 2007, **30**, 135.
36. J. Fernández-Escobar, M. I. Rodríguez-Ojeda and L. C. Alonso, in *Proc. 17th Int. Sunfl. Conf.*, Cordoba, Spain, Int. Sunfl. Assoc., Paris, France, 2008, p. 231.
37. F. Vear, in *Proc. 16th Int. Sunfl. Conf.*, Fargo, ND, USA, Int. Sunfl. Assoc., Paris, France, 2004, p. 15.
38. F. Montemurro and D. De Giorgio, *J. Plant Nutr.*, 2005, **28**, 335.
39. D. De Giorgio, V. Montemurro and F. Fornaro, *Helia*, 2007, **30**, 15.

40. L. C. Alonso, in *Proc. 17th Int. Sunfl. Conf.*, Cordoba, Spain, Int. Sunfl. Assoc., Paris, France, 2008, p. 53.
41. A. Popescu, in *Proc. 17th Int. Sunfl. Conf.*, Cordoba, Spain, Int. Sunfl. Assoc., Paris, France, 2008, p. 483.
42. D. Laureti, A. Del Gatto and S. Pieri, *Helia*, 2007, **30**, 141.
43. P. Casadebaig, PhD thesis, INP Toulouse, 2008.
44. Anonymous, *Tournesol Régions sud 2010*, CETIOM, Paris, France, 2010.
45. P. Casadebaig, P. Debaeke and J. Lecoeur, *Eur. J. Agron.*, 2008, **28**, 646.
46. F. Vear, H. Bony, G. Joubert, D. Tourvielle de Labrouhe, I. Pauchet and X. Pinochet, *Oléagineux, Corps Gras, Lipides*, 2003, **10**, 66.
47. E. Petcu, M. Stanciu, D. Stanciu and F. Raducanu, in *Proc. 17th Int. Sunfl. Conf.*, Cordoba, Spain, Int. Sunfl. Assoc., Paris, France, 2008, p. 345.
48. J. Lecoeur, R. Poiré-Lassus, A. Christophe and L. Guilioni, in *Proc. 17th Int. Sunfl. Conf.*, Cordoba, Spain, Int. Sunfl. Assoc., Paris, France, 2008, p. 429.
49. J. Chrastil, *J. Phys. Chem.*, 1982, **86**(15), 3016.
50. R. Hartono, G. A. Mansoori and A. Suwono, *Chem. Eng. Sci.*, 2001, **56**, 6949.
51. F. E. I. Deswarte, J. H. Clark, J. J. E. Hardy and P. M. Rose, *Green Chem.*, 2006, **8**, 39.
52. G. Brunner, *J. Food Eng.*, 2005, **67**, 21.
53. Natex Prozesstechnologie, http://www.natex.at (cited 6 November 2008).
54. http://www.stovesonline.co.uk/what-are-wood-pellets.html (cited 12 December 2008).
55. http://www.uk-energy-saving.com/wood_pellets.html (cited 12 December 2008).
56. G. Evans, Techno-Economic Assessment of Biomass "Densification" Technologies, NNFCC Project 08-015, 2008, p. 38.
57. J. Kilpatrick, Addressing the land use issues for non-food crops, in response to increasing fuel & energy generation opportunities, NNFCC project 08-004, 2008, p. 27.
58. A. P. Tulloch and R. O. Weenink, *Can. J. Chem.*, 1969, **47**, 3119.
59. R. C. Sun and J. Tompkinson, *J. Wood. Sci.*, 2003, **49**(1), 1611.
60. R. C. Sun and J. Tompkinson, *Cellulose Chemistry & Technology*, 2001, **35**(5), 471.
61. R. Sun and X.-F. Sun, *Separ. Sci. Tech.*, 2001, **36**(13), 3027.
62. S. K. Wiśniewska, J. Nalaskowski, E. Witka-Jeżewska, J. Hupka and J. D. Miller, *Colloids and Surfaces B: Biointerfaces*, 2003, **29**(2), 131.
63. J. Xiufeng and J. Reinhard, *Phytochemistry*, 2008, **69**, 1197.
64. W. H. Gibson, *Chem. Eng. Res. Des.*, 1931, **9a**, 30.
65. A. P. Tulloch and L. L. Hoffman, *J. Am. Oil Chem. Soc.*, 1977, **54**(12), 587.
66. F. Karaoosmanogalu, E. Tetick, B. Guerboy and A. Seanli, *Energy Sources, Part A: Recovery, Utilization, and Environmental Effects*, 1999, **2**, 801.
67. http://www.lipidlibrary.co.uk/Lipids/waxes (cited 18 December 2008).

68. Y. Ikushima, N. Saito, M. Arai and K. Arai, *Bull. Chem. Soc. Jpn*, 1991, **64**(7), 2224.
69. E. Flemming and U. Hehner, *US Pat.*, 5 885 561, 1999.
70. F. E. I. Deswarte, J. H. Clark, A. J. Wilson, J. J. E. Hardy, R. Marriott, S. P. Chahal, C. Jackson, G. Heslop, M. Birkett, T. J. Bruce and G. Whiteley, *Biofuels, Bioprod. Bioref.*, 2007, **1**, 245.
71. Soil Association Certification – clause 8.03.07a.
72. Y. Nakashima, M. A. Birkett, B. J. Pye, J. A. Pickett and W. Powell, *J. Chem. Ecol.*, 2004, **30**, 1103.
73. G. Powell, J. Hardie and J. A. Pickett, *Entomologica Experimentalis et Applicata*, 1997, **84**, 189.
74. S. Irmak and N. T. Dunford, *J. Agr. Food Chem.*, 2005, **53**, 5583.
75. C. K. Dartey, *Eur. Pat.* 1108364, 2001.
76. Semo Milling LLC, patent WO2008/020865, Power production using grain fractionation products.
77. Biogas regions – Biogas development in your region, Catalogue with overview of 40 shining examples of biogas plants in eight European countries, http://www.biogasregions.org/shining_examples.php
78. M. von Sivers and G. Zacchi, *Biorec. Technol.*, 1995, **51**, 43.
79. Z. Wen, W. Liao and S. Chen, *Biores. Technol.*, 2004, **96**, 491.
80. W. S. Adney, C. J. Rivard, S. A. Ming and M. A. Himmel, *Appl. Biochem. Biotechnol.*, 1991, **30**(2), 165.
81. V. S. Bisaria, in *Bioconversion of Waste Materials to Industrial Products*, ed. A. M. Martin, Elsevier, London, 1991, p. 210.
82. D. Verrier, F. Roy and G. Albagnac, *Biol. Wastes*, 1987, **22**, 163.
83. V. C. Kalia, A. Kumar, S. R. Jain and A. P. Joshi, *Biores. Technol.*, 1992, **41**, 209.
84. P. Weiland, *Wat. Sci. Tech.*, 1993, **27**(2), 145–151.
85. A. Mtz-Viturtia, J. Mata-Alvarez and F. Cecchi, *Resour. Conservat. Recycl.*, 1995, **13**, 257.
86. K. Stamatelatou, K. Dravillas and G. Lyberatos, *Wat. Sci. Tech.*, 2003, **48**(4), 235.
87. S. Ghosh, in *Biogas, Technology, Transfer and Diffusion*, ed. M. M. El-Halwagi, Elsevier Appl. Sci. Publ., London, 1984, p. 400.
88. S. Ghosh, J. P. Ombregt and P. Pipyn, *Water Res.*, 1985, **29**, 1083.
89. B. G. Yeoh, *Wat. Sci. Tech.*, 1997, **36**(6–7), 441.
90. K. M. Lee. C. A. Bruunner, J. B. Farrel and A. E. Eralp, *J. Water Pollut. Control Fed.*, 1989, **61**, 1421.
91. http://www.iea-biogas.net/Dokumente/casestudies/reidling_final.pdf
92. http://www.iea-biogas.net/Dokumente/casestudies/biogas_village.pdf
93. L. De Baere, Dry continuous anaerobic digestion of energy crops, http://www.ows.be/pub/Dry AD of energy crops.Papenburg.March07.pdf
94. M. Laaber, R. Madlehner, E. Brachtl, R. Kirchmayr and R. Braun, Aufbau eines. Bewertungssystems für Biogasanlagen – "Gütesiegel Biogas", Final report, 2007.
95. V. Vandenbossche Marechal, PhD thesis, INPT, France, 1998.

96. Prolea Statistiques des oléagineux et protéagineux. Huiles et protéines végétales, 2007.
97. V. Marechal and L. Rigal, *Ind. Crops Prod.*, 1999, **10**, 185.
98. S. Campbell, F. Sosulski and M. Sabir, *Can. J. Plant. Sci.*, 1968, **58**, 863.
99. M. Iglesias and J. Lozano, *J. Food Eng.*, 2004, **62**, 215.
100. M. Lin, F. Sosulski, E. Humbert and R. Downey, *Can. J. Plant. Sci.*, 1975, **55**, 507.
101. W. Kim, F. Sosulski and S. Campbell, *J. Food Sci.*, 1978, **43**, 746.
102. A. Demirbas, *J. Hazard. Mater.*, 2008, **157**, 220.
103. A. Hashem, A. Abou-Okeil, A. El-Shafie and M. El-Sakhawy, *Polym-Plast. Technol.*, 2006, **45**, 135.
104. A. Hashem, *Polym-Plast. Technol.*, 2006, **45**, 35.
105. W. Shi, X. Xu and G. Sun, *J. Appl. Polym. Sci.*, 1999, **71**, 1841.
106. P. Khristova, S. Bentcheva and I. Karar, *Bioresour. Technol.*, 1998, **66**, 99.
107. P. Khristova, S. Gabir, S. Bentcheva and S. Dafalla, *Ind. Crops Prod.*, 1998, **9**, 9.
108. F. Lopez, M. Eugenio, M. Diaz, J. Nacimiento, M. Garcia and L. Jimenez, *J. Ind. Eng. Chem.*, 2005, **11**, 387.
109. P. Khristova, N. Yossifov and S. Gabir, *Bioresour. Technol.*, 1996, **58**, 319.
110. P. Khristova, N. Yossifov, S. Gabir, I. Glavchev and Z. Osman, *Cellul. Chem. Technol.*, 1998, **32**, 327.
111. I. Bektas, C. Guler, H. Kalaycioglu, F. Mengeloglu and M. Nacar, *J. Compos. Mater.*, 2005, **39**, 467.
112. P. Evon, V. Vandenbossche, P. Pontalier and L. Rigal, *Ind. Crops Prod.*, 2007, **26**, 351.
113. P. Evon, V. Vandenbossche, P. Pontalier and L. Rigal, *European Biomass Conference Berlin*, Germany, 2007, p. 2094.
114. P. Evon, PhD thesis, INPT, France, 2008.
115. A. Rouilly and L. Rigal, *J. Macromol. Sci., Polym. Rev.*, 2002, **42**, 441.

Website Sources

Food and Agriculture Organization of the United Nations:
  http://faostat.fao.org/
Eurostat:
  http://epp.eurostat.ec.europa.eu/portal/page/portal/eurostat/home/
UK National Statistics:
  http://www.statistics.gov.uk/hub/index.html
Vermont Agency of Agriculture:
  http://www.vermontagriculture.com/energy/anaerobic/forage.html
IEA Bioenergy:
  http://www.iea-biogas.net/
Anaerobic Digestion Systems website:
  http://www.anaerobic-digestion.com/index.php

CHAPTER 3
Primary Processing

WIM MULDER,[a] PAULIEN HARMSEN,[a] JOHAN SANDERS,[a] PATRICK CARRE,[b] BIRGIT KAMM,[c] PETRA SCHÖNICKE[c] AND GEERTJE DAUTZENBERG[d]

[a] Wageningen UR, Bornse Weilanden 9, 6708 WG Wageningen, The Netherlands; [b] CREOL, 11, rue Monge, F-33600 Pessac, France; [c] Research Institute Bioactive Polymer Systems e.V., Kantstrasse 55, D-14513 Teltow, Germany; [d] Biorefinery.de GmbH, Stiftstrasse 2, D-14471 Potsdam, Germany

3.1 Introduction

This chapter describes the state of the art in primary processing of oil-bearing plants such as rapeseed, sunflower and olive. The seeds contain the majority of the oil in the kernels whereas the olive fruit contains the oil in the meat. Based on these differences, oilseeds and olives are processed in different ways (see Section 3.3). Nowadays, vegetable oils and meals are commodity products with a standard quality. Production techniques are mature and since World War II no major changes were carried out in the different processing steps such as pre-treatments, oil expelling and solvent extraction. Different oil-bearing cultivars were bred and processing techniques were optimised in relation to oil production and oil quality. Reduction of crushing costs was obtained by increasing the size of the processing plants.

Currently, a broad interest exists to develop novel biorefinery schemes by which better utilisation of the crop is being realised. The most important component in the meals is protein, which enables it to be used as animal feed. Looking at the oil production process and taking into account both the oil and

Primary Processing

proteins, developing novel biorefinery schemes gives opportunities to increase the value of the proteins and the meals for example, and even make them suitable for human consumption. Also the quality of meals can be improved by simply removing most of the fibres. This could be achieved by dehulling the seeds before oil isolation. As well as improving the meal quality, the hulls themselves could have interesting applications. Until now the focus has been on oil recovery, and almost no information is available concerning the relation between the different processing steps and the possibilities to exploit other seed components, such as levulinic acid and proteins. To investigate the possibilities of novel biorefinery schemes, the different processing techniques will be discussed further.

3.2 Pre-treatment Processes

3.2.1 Dehulling

Valorisation of hulls after separation of the seed coats from the meat could be regarded as a first and indispensable step. In the cases of rapeseed and sunflower, the two main domestic oil crops in Europe, industrial dehulling is still poorly developed. The outlet of the meals is restricted to cattle feed. To explore more valuable utilisation of meals at an industrial scale, such as the exploitation of proteins, which is an important fraction of the meals, as long as water extraction remains in the domain of the research and development, it will be necessary to dehull the seeds before oil extraction in order to extract the proteins, or other valuable components, and improve the valorisation of these crops.

3.2.1.1 Dehulling Equipment

Different dehulling equipment can be used to separate the hulls from the seeds. The choice of the equipment is dependent, amongst other things, on the type of seeds that have to be dehulled.

Rapeseed
The French institute, CETIOM, in the 1970s patented a centrifugal propeller in which rapeseeds are projected against a target with controlled speed in order to separate meats and hulls. Seeds are fed in the centre of a rotating disk bored of radiating channels. The seeds are propelled against the wall of the device due to centrifugal forces and the impact makes the seed shatter. The speed of the disk and flowrates are both adjustable.

Secondly, an abrasive dehuller can be used.[1] Abrasion is the most common way to produce white rice. The seeds are introduced in a rotating device having a rough surface. The coats of the seeds are removed by abrasion. By adjusting the rotating speed and the throughput of seeds, it is possible to adjust the percentage of biomass that is removed.

A third method is using a roll dehuller. The seeds are briefly passed through co-rotating rolls with a carefully adjusted gap. This method has been described by Schneider in 1979.[2] The seeds have to be sorted before treatment in order to remove the small ones and hydrothermal conditioning is carried out to improve the hullability. A 0.6 mm space between rolls would permit a better efficiency to be reached. No data were provided concerning the percentage of fat of the hulls thus obtained.

Hulls can also be separated from meats after oil extraction by air transportation. This is called tail-end separation.[3] The separation is based on supposed differences in behaviour between meats and hulls. The meals must be milled prior to this treatment. According to the available information tail-end separation is not very efficient.

By use of radiation, infrared and/or microwaves it is possible to provoke a quick vaporisation of water at the surface of the seeds and the vaporisation of this water can loosen the hulls/meat linkage enabling a weak constraint such as a compressed air flux to separate the two fractions. This process has been developed for the removal of fruits like walnuts, chestnuts and almonds.

Sunflower Seeds

The most popular decorticator for sunflower is proposed by Bühler Cie. It consists of a rotating blade that propels the seeds by centrifugal forces against a wall. It functions on one impact. It is designed for partial removal of hulls. Further, good dehulling is not possible without increasing the force of the impact. The kernel oil content is so high in sunflower seeds that under the violence of the shock, some oil can be transferred to the hulls and is lost. Moreover, increasing the rotating speed of such a dehuller increases the production of fines that are difficult to separate from the hulls. Another monoshock kind of dehuller is air-jet impact, where the propelling of the seeds is done by a strong current of air. Multi-shock dehullers such as the ripple mill improve the dehulling quality because they carry out several impacts on the achene with milder violence and enable an enhanced separation of hulls without generating an excess of fines.

The Impco Company proposes a cutting apparatus in which the seeds are opened by a scissor effect provided by two co-rotating cylinders fitted with grooves, which have differential rotating speeds. The separation of hulls and meats is pursued in a secondary stage where rotating beaters shake the cut seeds in order to separate them. The sorting of the hulls is done by sieving and air classification. This apparatus has been designed for cotton seeds but, according to the manufacturer, it can also be used for sunflower seeds.

The principal aspect of a compression–decompression apparatus is to place the achenes in a tight chamber into which compressed air is introduced. Once the pressure is stable, decompression is provoked enabling the air in meats and hulls to expand quickly in order to crack the hulls and remove them. This apparatus was set up for the decortication of striated sunflower for human food. Its advantage is that it does not damage the kernel with percussions. With oil-rich seeds, this process is not efficient and is energy consuming.

3.2.1.2 Hull Separation

Hulls have a much lower density than the seeds and the meats. Therefore, most of the separators remove the hulls in a flow of air, with adjustable force, to carry away the hulls without taking the meats. The drawback of this principle of separation is that one needs to remove the fines prior to the aspiration of hulls. Otherwise, the fines, which are mainly meat fragments, will end up in the hull fraction. The eviction of fines can be done by sieving or by air sorting adjusted to aspirate only the fines. The first solution is adopted for sunflower while the second one functions with rapeseed. In the case of sunflower, it is possible to purify the hulls by removing the fines after aspiration in a rotating sieve with the help of brushes. The best way to obtain hulls without an excess of oil is to carry out dehulling without producing a lot of fines. The separation of meats from the non-dehulled seeds is required if a total dehulling is desired. To achieve this goal, it is necessary to eliminate the small seeds in order to process them separately. When all the seeds have a minimum size, it is possible to obtain a good separation of the whole seeds and the meats by size separation on a sieve. In the case of sunflower, a specific gravimetric separator can be used for the same function. Electrostatic separation has been proposed by Sket Cimbria for the separation of hulls. The dehulled particles are charged with electrostatic charge and subsequently they are passed in a chute within an electromagnetic field of the opposite charge attracting the seeds particles. Since hulls have a large surface compared to their mass, they can bear more charge per unit of mass and are able to be moved by the electrical field. This technology requires less energy than the regular one, which involves large ventilators.

3.2.1.3 Rapeseed Dehulling

The composition of rapeseed seeds, hulls, kernels and meals is shown in Table 3.1 (source CETIOM).

The percentage of hulls in rapeseed is 12–13% according to CETIOM. Other sources place the hulls percentage in a wider range (12–20%). Rapeseed hulls contain around 3% of seeds, oil and protein, and more than 70% of fibres, of which 95% contains lignin. The removal of this material could enable the production of an improved meal quality thanks to a dramatic reduction of its fibre content. Provided that it can be desolventised without damaging the protein solubility, this meal could be used after extraction of soluble sugars

Table 3.1 Composition (as percentage of dry matter) of rapeseed and rapeseed-related products.

	Seeds	Kernels	Hulls	Meal (seeds)	Meal (dehulled seeds)
Dry matter [%]	91.9	93.4	14.3	93.3	93.5
Oil [%]	47.7	53.3	12.0	1.0	1.0
Proteins [%]	21.8	24.5	15.2	40.7	44.2
Fibres [%]	8.1	2.7	32.3	15.2	6.4
Ash [%]	4.7	4.7	6.6	8.5	9.7

or protein extraction for human food or non-food applications. CETIOM studied the composition of the oil stemming from the hulls under industrial conditions. It was noticed that the oil extracted from hulls after 11 days of storage had a very high acid value, was dark in colour and was of a very low quality. The oil quality just after dehulling has been investigated. Acidity was 3.3% *vs.* 1.0 and 1.4 for the press and extraction oils, para-anisidine index 4.4 *vs.* 2.1 and 2.8, tocopherol 850 ppm, *vs.* 916 and 1180 ppm. Although low, this quality was not considered too low to exclude valorisation by extraction. Liu et al.[4] have studied the composition of the fats of Canadian and Australian rapeseed hulls. According to their results, rapeseed hulls contain waxes and polar compounds that can precipitate during cold storage. These waxes are not the same as in sunflower. The average chain length is shorter and, as a consequence, the standard winterisation treatment is not convenient. These waxes can cause some trouble in rapeseed oil after refining although dewaxing of rapeseed oil is not a common practice in Europe.

In Europe, dehulling is marginally practised and is not developed in the large plants that function with solvent extraction. The Teutoburger Ölmühle in Germany has developed an original technology to dehull rapeseed and valorise the oil obtained by this process. Within this factory, knowledge was developed enabling the cold pressing of rapeseed meats. The seeds are dried before dehulling, and subsequently the meats are cold pressed, cooked and passed to a second expeller. The hulls are fermented to produce biogas after oil recovery by expelling. The oil from the hulls is used by cogeneration to produce energy for the process (drying of the seeds and cooking of the cake of the first expelling before the second one).

In France, a small-scale oil mill has performed dehulling of rapeseed in Châlon sur Saône. The technology used for dehulling was the one developed by CETIOM. Hulls were not extracted but were sold as feed for rabbits. This mill was closed because of a low profitability due to its small size.

The main reasons why dehulling is not being carried out on large scale are:

- Low profitability caused by the loss of oil in hulls. The oil content in the hulls was rather high due to the presence of meats in the hulls. Around 18% of oil in hulls was considered a reasonable level to expect;
- Low valorisation of the hulls;
- Difficulties encountered in processing the meats.

To improve the economic value of dehulling at an industrial scale, the effect of the percentage of dehulling on the value of the hulls and the meals was investigated (Table 3.2).

It was expected that by decreasing the percentage of removed hulls lower oil losses would be obtained and therefore the economical value of the meals and hulls would improve. The values of dehulled meals are calculated according to their protein content. It is supposed that their value increases when the protein content increases. The meal with 20% dehulling had 41.8% protein, the 10% dehulled meal had 38.8% protein while the standard meal contained 35.1%

Table 3.2 Economical evaluation of dehulling. The price of the oilseeds (2008) was €320/t.

		Non-dehulled seeds	10% dehulled seeds	20% dehulled seeds
Oil	price [€/t]	741	741	741
	content [%]	41	40	38
	value [€/t seeds]	300	294	282
Meal	price [€/ton]	137	155	163
	content [%]	57	48	40
	value [€/t seeds]	78	74	66
Hulls	price [€/ton]	–	105	134
	content [%]	–	11	20
	value [€/t seeds]	–	11	27
Total value		378	380	374

protein. The value of hulls is based on the value of the seeds as a source of oil. The fat content of 20% dehulled seed hulls was 18.4% and the oil content of the 10% dehulled seed hulls was 14.4% while the seeds are supposed to contain 42% of oil. Based on the 1980 market prices, it was concluded that efficient dehulling was not realistic to be implemented on a large scale. The gap between the total value for non-dehulled materials and dehulled materials in today's prices is even broader than in 1980, being unfavourable with respect to dehulling.

Dehulling can become important when a good technical performance is possible. So far, this has not been feasible under industrial conditions. To improve the economic features of dehulling, the following possibilities can be investigated:

- Improve the dehulling in order to get better technical performances in terms of oil losses and hulls purity. If possible, this option has the advantage of not being based on market values;
- Recovery of 60% of the oil left in the hulls. By expelling the hulls, it is possible to decrease their oil content from 14.4 to 6%. The recovery costs are easily compensated by the benefits;
- Improvement of the value of oil. According to the OPTIM'OILS study, it is possible to obtain press oil after dehulling with a better quality regarding micronutrients. Better oil quality is possible provided that mild refining could be carried out. This proposal could take advantage of research in progress and generate more value than all the other ones;
- Improvement of the value of the meal. Low fibre content is supposed to improve the digestibility of the meal by animals. In consequence, the value of the feed should be increased. According to past experience amelioration will be difficult since the dehulled meals must compete with soybean meals that are abundantly and widely used while rapeseed meal must deal with anti-nutritional factors like glucosinolates;
- Define a better valorisation for the hulls. The rabbits market is not very important and as a consequence the development of the technology to a

larger scale was rather uncertain. This seems the most difficult option due to the fibrous constitution of the hulls. They have to compete with numerous sources of ligno-cellulosic material and they need to be pelletised for transportation.

For the ulilisation of rapeseed hulls there are several possibilities:

- As animal feed;
- For the production of energy;
- Use as substrate for growing fungus.

The use of rapeseed hulls in animal feed was preferred in France. L17 dehydrated alfalfa is a low grade of alfalfa that can be used as feed for rabbits and horses. The value of dehydrated alfalfa is rather high compared to rapeseed meals. In December 2008, the prices of these raw materials were €155/t for alfalfa L17 and €143/t for rapeseed meal that has 35% of proteins. That difference can be partly explained by the cost of production and dehydration and partly by the harvest conditions of 2008 with a strong reduction of the production due to bad weather in spring compromising harvests. That price is equally biased because the main part of the marketing of the 2008 harvest was done in the first month of the campaign when the prices of feedstuffs were high. The December prices concern resale of dehydrated alfalfa bought in spring. It is difficult to think that it could be possible to substitute L17 dehydrated alfalfa by rapeseed hulls at this price because the market for L17 dehydrated alfalfa is small and the competition could question the economic viability of the alfalfa crop, a problem that the growers are not likely to observe without reaction. On the other hand, it is possible to envisage that rapeseed hulls could take a share in the dehydrated alfalfa market in a future where agriculture will be committed to limit its impacts on climate change, and therefore the consumption of fossil energy subsidised by the European Community. The €33/t subvention that the EU grants to dehydrators is maintained until 2011 but should be 'decoupled' after 2011. In this new context, 80% of the current dehydration will disappear according to the union of the dehydrators. In Europe, the production of dehydrated alfalfa represents around 4 Mt when the production of rapeseed represents 18 Mt. Even if all rapeseed was dehulled at 10%, it would not be enough to replace the dehydrated alfalfa. The presence of omega-3 fatty acid in the rapeseed hull could bring an advantage to this fodder since this nutrient is taking more importance in cattle feed. To market a valuable dehydrated alfalfa substitute, it would be necessary to recover a part of the oil left in the hulls by expelling or by extraction and to improve the protein content by addition of some meal. Of course, this product needs to be pelletised to be used in the same condition as dehydrated alfalfa.

With 4700 kcal/kg (5.4 kWh/kg) the hulls at 13% of fat could be used as fuel for the energy supply of the oil mill. In the case of rapeseed, the heat needed for the process is 350 kWh/t; theoretically 65 kg of hulls would be enough for this supply. For sunflower it is necessary to burn 8.25 kg of hulls to supply the equivalent of heat provided by 6.25 kg of sunflower hulls. The yield of hull burner is approximately 75%. The caloric value of 1 kg of rapeseed hulls is therefore 4.05 kWh/kg. To cover 350 kWh/t, the oil mill must burn 86.5 kg of rapeseed hulls. By choosing cogeneration, it is possible to use 50% more fuel, so that 130 kg of hulls could be valorised for each ton of seed to crush. In France the electricity produced by cogeneration from combustion of biomass is bought at €0.06/kWh. The crushing of 1 t of seed will generate 0.173 MWh of electricity and 0.3 MWh of heat. The price of the substituted heat can be evaluated at the level of coal because, like coal, it requires special burners to be properly used. The coal kilowatt hour has an approximate value of €0.01. The combustion of the hulls in these conditions gives 1.34 MWh of electricity and 2.71 MWh of heat, therefore €107 of value. This valorisation has the advantage of requiring neither transport nor storage of the hulls. It can make the oil mill more than auto-sufficient in energy and able to deliver better quality meals. However, it requires high capital investment and we are not able to estimate these capital costs. On the other hand, this disadvantage could be compensated by fiscal measures or direct help from the public powers in the context of reduction of the greenhouse gas emission. The limit of that possible valorisation is the nitrogen content of the hulls. Because of its high level, it is impossible to imagine the combustion of that material without a treatment of the smoke. Considerable investment would be required and would decrease the economical feasibility of that valorisation.

As previously mentioned, the hulls can be used to produce biogas by methanogenesis. This biotransformation occurs in the absence of oxygen and is due to the degradation of biomass by microorganisms. Around 90% of the energy of the biomass is converted and 10% is used for the metabolism of the microorganisms. This solution requires a long residence time in digesters and seems difficult to use with large amounts of biomass. The investment for a large oil mill seems to be important and, compared to combustion, this solution appears not to be competitive.

Pyrolysis is the transformation of biomass in liquid or gaseous fuel induced by heating at temperatures between 450 °C and 550 °C. Most of the organic matter is vaporised then condensed to produce a liquid with characteristics of a viscous oil. Some incondensable matter is usable as gas. Depending on the conditions (concentration in O_2, temperature), the proportion of gas/liquid resulting from these transformations is variable. A number of processes have been proposed to use lingo-cellulosic biomass as a commodity fuel but the industrialisation of these processes is still not evident.

In China, during the Wuhan 12th international rapeseed congress,[5] a new process including dehulling was presented by the Wuhan University. They proposed to use the rapeseed hulls as substitute for dehydrated alfalfa and as support for the cultivation of fungus. In France Phelipaea ramosa is an invasive species, mainly located in the Charente and Vendée regions, that penalises the rapeseed cultivation. Around 50 000 ha of arable land are unfit to grow rapeseed. This use has still to be validated and needs to be authorised as a plant protection compound under 91/414/EEC. By spreading rapeseed hulls on the ground, it could be possible to stimulate the germination of broomrape before the real sowing and decrease the parasite pressure on the crop. In the future, if the presence of that parasite increases, a few thousand tons of rapeseed hulls could be required for that use. Currently, there is no product to fight the parasite.

The most promising possibility to valorise rapeseed hulls is as a dehydrated alfalfa substitute. Because of the CAP evolution, the volume of available dehydrated alfalfa is likely to decrease after 2012 generating an opportunity to sell off rapeseed hulls.

3.2.1.4 Sunflower Dehulling

The composition of rapeseed seeds, hulls, kernels and meals is shown in Table 3.3 (source CETIOM).

The percentage of kernels and hulls in sunflower is variable according to the cultivar, the size of the seeds and their oil percentage. Most of the values are between 22 and 28%. Breeders have done considerable work to increase the oil composition of the seed. Part of this change comes from a reduction of the hull percentage in the seeds. In general, the varieties with high oil content have a low hulls percentage. The consequence of this change is a reduction of the ability of the seeds to be dehulled. Hulls in sunflower contain 2–3% of seed oil, and 10% of the seeds' protein and lignin. The architecture of the hulls has been studied by Agnès Beauguillaume (PhD thesis), who explained the hullability of the seeds by the repartition of two kinds of sclerenchymas. Type 1 clumps are wedge shaped and have a fibro-vascular bundle at their base while type 2 clumps are not vascularised and wide. In cultivars with favourable ability for dehulling, type 1 clumps are more frequent. Their structure gives more rigidity to the hulls that are more likely to break under the impact of the dehuller. A greater abundance of type 2 structures gives more elasticity to the hull that will deform under the chock instead of cracking. This character is genetically determined and could be improved by selection. A peculiarity of sunflower is the presence of waxes in the oil. Waxes are concentrated at the surface of the hulls where they play a role in protecting the seeds against water, therefore inhibiting moulds which attack the seeds. Denise[6] indicates that the dehulling enables the production of crude oils with 40 to 300 ppm of waxes whereas non-

Table 3.3 Composition (as percentage of dry matter) of sunflower seed and sunflower-seed-related products.

	Seeds	Kernels	Hulls	Meal (seeds)	Meal (dehulled seeds)
Dry matter [%]	92.8	90.5	–	88.8	90.5
Oil [%]	48.0	61.3	2.5	2.2	1.2
Proteins [%]	16.7	20.6	6.2	31.9	52.6
Fibres [%]	17.3	2.4	57.6	28.1	6.2
Ash [%]	3.5	3.6	3.2	7.1	9.2

Primary Processing 111

dehulled seeds contain 600 to 900 ppm (up to 1200 ppm for oils coming from the USA). Winterisation reduces the wax content of oils down to 50–150 ppm. It is possible to explain the differences between the information from the two authors by the type of dehulling that they are taking into consideration. Another characteristic of the sunflower hulls is the silicon content that makes them a very abrasive product that causes some wear in the presses. The removal of hulls is likely to reduce maintenance costs associated with the press spare parts; however, it will generate a constraint on the dehulling equipments, especially if some pneumatic transportation is used.

Industrial dehulling of sunflower is a common practice but the dehulling is generally limited by two factors:

- The valorisation of the hulls by combustion: the oil mill cannot valorise more energy than it needs to process the sunflower. The order of magnitude of the energy requirement is 350 kWh/t of seeds for the crushing. The part required for heat consumption is 85 (300 kWh). The theoretical calorific power of sunflower hulls is 5.1 kWh/kg. Therefore, around 60 kg of hulls are necessary for the heat supply of the mill. It means that the removal of 6% of the seeds mass is enough. On this basis, if one supposes that the seeds contain 25% of hulls, only 25% of these hulls can be valorised.
- The hulls oil content. As it has been indicated previously, the industrial dehuller and the inability for dehulling of the modern sunflower seeds make it impossible to produce a large dehulling percentage without enrichments of the hulls by oil. Therefore, it is necessary to limit the dehulling percentage in order to avoid oil losses.

The marketed meals have 33% of proteins *vs.* 29% for the non-dehulled. It is necessary to remove about 34% of the hulls to obtain that increase in proteins content.

One can note that more than 25% is necessary for heat production, which can be explained by the energy yield of the burner and the boilers that are not optimal when the combustible is biomass. Improvement of the performance of dehulling by use of a multi-shock dehuller is possible but 100% dehulling is not feasible with the existing cultivar. It would be possible to select some varieties easy to dehull if one wanted such dehulling for the production of pure kernel meal but this would mean having to pay a premium to get the right raw material. At the global scale, a reorientation of breeding is necessary to supply cultivar having the necessary ability to be dehulled. Breeding studies[7] have demonstrated that hullability is a heritable character and has a weak negative correlation with oil content. Inbreds and hybrids producing achenes with quite small hull content, high oil content but a good hullability were found to be the most promising for improving the quality of sunflower seed meal. Such genotypes are rare so a recurrent selection program was recommended to increase the frequency of favourable genes. In the current state of the market an oil mill that would carry out the best possible dehulled sunflower meal would have to use a diagram with a separation of the seeds according to their size prior to the dehulling. The seeds would have to be dried before dehulling and the dehulling would be carried out with a multi-shock dehuller. The dehulled seeds are then sorted by sieving to separate kernels from non-dehulled seeds and both fractions are passed onto an air classifier to remove the hulls. In the case of the small size fraction, the hulls have to be sieved to recover the fines. The small seeds are generally difficult to dehull and the simplest way to treat them is to crush them separately without dehulling. A dehulling is possible but the quality is inferior to that of the bigger seeds. According to the works done by CETIOM on the subject, the protein content in the meal from such a dehulling scheme is 40%. Around 66% of the hulls have to be removed to obtain that result. This means that 100 kg of seeds having 25% of hulls are going to give 16.5 kg of hulls.

Increasing the removal of hulls will decrease the amount of raw material entering the oil mill. In consequence, that could decrease the energy necessary for the crushing but we have seen that some drying is necessary to improve the seeds hullability in consequence; the 50% ratio of valorisation for crushing needs can be accepted as a rough estimation. An important characteristic of the hulls is their low density. According to the cultivar, the range of the density of the hulls is 0.15 to 0.20. This low density generates flow problems because in bulk storage the hulls are bridging. Moreover, low density makes it impossible to ship the hulls in their state to distant customers. If pelletising presses are to be employed, the structure of the fibre will be damaged and could limit its applications. Another problem already mentioned is their abrasive quality that interferes with air transportation. A simple possibility to valorise more hulls in the boiler is to use a high-pressure boiler in order to run turbines activating an electrical generator and use the turbine extraction steam at 10 bars in the processing areas. The excess of energy will become easy to export by the electric grid. The limit is that to produce 1 kWh of electricity, it is necessary to valorise 2 kWh of heat. The steam needs for the plant are not sufficient to valorise all the biomass extracted due to the potential dehulling.

The cost of a hulls boiler is about seven times more expensive than an oil boiler (information from De Smet). According to available information, the peculiarity of the combustion of the hulls is that the gasification of the fuel occurs quickly and the particle size must be small. The firing and air distribution system must be adapted to this particular raw material. The fly ashes born by the flue gas have a very small particle size, are very concentrated and are very hydroscopic with a high tendency to bind and form a concrete-like material. In consequence, the design of the boiler must avoid the presence of features allowing deposits of ashes such as horizontal tubes for heat recovery and manage the self-cleaning of these tubes during operation without damages. Some devices like a soot blower or mechanical rapping and shaking system are necessary to keep the exchange surfaces clean. The exhaust requires filters that are able to remove small particle sizes, which means extra costs for the cleaning of the fumes.

As the calorific value of sunflower is 4.1 kWh/kg and the possible valorisation of heat is limited to 300 kWh/t of crushed seeds, only 11% of the seed mass can be valorised by this process (44% dehulling, 36% of proteins in the meal).

Primary Processing

One ton of hull can supply 1353 kWh of electricity and 2706 kWh of heat. On the same basis as for rapeseed oil, a ton of hull is valued at €108.

In the case of 100% dehulling, it would be interesting to supply steam to another plant in order to valorise all the hulls. In the case of biodiesel production, the crushing plant has a semi-refining plant near to it and, possibly, an inter-esterification plant that could use this available energy. However, it is unlikely that the needs are going to match the supply. *A priori*, even if the seed quality becomes very favourable to dehulling, it seems more realistic to imagine that the plant will process two meal grades, a good-quality one with the seeds having a good hullability that will be used in high-added-value products and a lower grade meal resulting from the crushing of the seeds of poor hullability. The quantity of hulls extracted would be adjusted according to the energy needs and the quality of the meal to be marketed. In consequence, it seems unavoidable to produce a small percentage of hulls that will be not absorbed by the production of energy. A small percentage in a plant that crushes 1000 t/day can represent a few tens of tons, at least 100 cubic metres, a volume that the plant manager cannot neglect and for which a reliable customer is necessary.

The exportation of the sunflower hulls to distant customers is possible because it is a source of combustible biomass available in significant quantities with low water content with a regular composition. Compressing this material would allow efficient storage and transportation. The compression of hulls could be done in a pelletising press after conditioning by steam.

Compared to wood, sunflower hulls contain much more nitrogen and this nitrogen during combustion will produce nitrogen oxides that are likely to be much more important. According to a German study cited in a review of the French Agency for Energy and Environment Preservation, the emission of pollutants resulting from the combustion of biomass depends on their composition and it seems that wheat straw produces 3.3-fold more NO_x than wood. It is not possible to predict the real emission of sunflower hulls in NO_x, but it is certain that they will be higher than straw because of their higher content of nitrogen.

The regulation for industrial boilers (>300 MWh) limits the concentration of NO_x to 200 mg/m^3 (300 mg/m^3 for power comprised between 100 and 300 MWh). The wheat straw already produces 330 mg/m^3, therefore it is a problem for sunflower hulls as they contain almost twice the protein content.

Compared to rapeseed hulls, the sunflower hulls are poor in protein and oil. If sunflower hulls have a little more protein than wheat straws, they have thrice times their content of lignin. This high lignin content is even higher than sawdust. These figures make sunflower hulls a very bad fodder, in fact more a litter material than a fodder. The wheat straw is sold for €50/t provided that it can be compressed to a 0.3–04 density. Sunflower hulls have a density of 0.16. Several kinds of treatments have been studied in order to increase the digestibility of sunflower hulls:

- Mechanical treatments (milling, pelletising);
- Physical treatment (steam, extrusion);

- Treatment by alkalis (NaOH, Ca(OH)$_2$, NH$_4$OH) or oxidants (NaClO, SO$_2$, Cl$_2$);
- Enzymatic treatments or biologic digestion by unicellular organisms able to digest lignin.

Due to the insignificant increase in value realised by these treatments, their industrialisation has not progressed.[8]

Due to their high lignin content and the hydrophobic compounds found at their surface, sunflower hulls could be used as construction materials. The presence of sunflower hulls in particleboard panels decreased their mechanical resistance and dimensional stability, however up to 50% of incorporation was possible. Sunflower hulls have been studied to build bricks or tiles that are very light, solid and insulating. The incorporation of hulls in concretes enables a reduction of the density (15–20% of sunflower hulls decrease the density to 1 *vs.* 2.4) and improves the insulating properties of the material. ($\lambda = 0.1$ W/m °C *vs.* 1.6 to 2.1 W/m °C for ordinary concrete).

A process has been proposed by the Laboratoire de Chimie des Agro-ressources in Toulouse in which diluted soda is used to dissolve the hemicelluloses, and then the pH is decreased to 5.4 in the presence of ethanol to obtain the precipitation of solute than can be filtered and dried. Yields of 17% were obtained in laboratory conditions. The carbohydrate thus obtained is glucuronoxylan (polymer of glucuronic acid and xylose), which has promising uses in numerous applications such as food, cosmetics, paints and glues.

In Argentina large amounts of seeds have been studied.[9] In order to find high-value applications. Due to their high C/N ratio (83), sunflower hulls cannot be composted alone and the addition of nitrogen is necessary to obtain acceptable compost. Inorganic nitrogen addition was not an efficient means to improve the decomposition of sunflower hulls. Co-composting of sunflower hulls with alfalfa or vetch gave better results, even with relatively high C/N ratios (64). The article does not present economical aspects of the treatment and is proposed as an alternative to combustion, which is a pollutant activity in urban areas. In Europe, composting can be considered as a low-added-value valorisation, more a way to eliminate wastes than a way to produce value.

Whatever the pollution problems stemming from the SF combustion, it seems to be the best way of using these materials in large volumes. For large oil mills, the cost of equipment to reduce the emission of pollutants could be affordable in a context of high energy prices and with support of the public bodies to reduce the carbon footprint of the biofuel production.

3.2.1.5 Conclusions – Dehulling

Dehulling is a step that must be addressed if one wants to achieve a better valorisation of the oilseeds. The main reason why dehulling is practically not realised in the case of rapeseed and only to a weak percentage for sunflower is the practical problems related to extracting hulls without generating oil losses.

Primary Processing

Also, the crushing of pure kernels meets technical problems, first for expelling soft materials, second to remove the hexane residues of the meal without decreasing the protein solubility. These three hurdles must be overcome to allow the valorisation of the proteins from oilseeds.

Concerning sunflower, it would be necessary to work upstream in order to select varieties with good hullability. Some existing cultivar could be used in order to produce seeds that are easy to dehull and to get pure kernels and pure hulls without requiring complicated equipment.

In the case of rapeseed, it seems possible to imagine a good valorisation of hulls to replace dehydrated alfalfa, for which production is likely to diminish after 2011. In this case, the presence of oil could be valorised due to its high content in omega-3 fatty acids for cattle feed. The presence of omega-3 improves the quality of animal products and improves the health of consumers.

3.2.2 Thermal Pre-treatment

Before thermal treatment, seeds are usually flaked, which increases the surface area of the material and improves the heat transfer and the water evaporation. The flaking is conducted to improve the oil availability for solvent extraction. Small seeds like rapeseed are difficult to flake properly because some seeds can escape the rolls. Flaking is adjusted by the space between the rolls. When the spacing is too narrow, the flakes become too soft and are difficult to press. When the flakes are too thick, the oil becomes less extractible. To work properly, flakers require the absence of large materials like corn seeds or vegetable residues. These large particles can stay over the passage between the rolls and block the seeds or they can force the rolls to open the space and let unflaked seeds pass. For this reason, the seeds are passed through cleaners and crackers that eliminate and grind these large pieces of material.

A side effect of flaking is that it disrupts some oil bodies and when enzymes are in contact with their substrates for a certain period some hydrolysis occurs. In terms of loss of value for the crusher, the action of D phospholipase is important, as it cuts phospholipids between the phosphatidic acid and the polar group of the molecule. In the case of phosphatidylcholine the transformation results in the loss of the hydratable character of the molecule and makes it mandatory to carry out an acidic pre-treatment before degumming. A treatment of the flakes by superheated steam has been proposed to inactivate the enzymes quickly after flaking but this treatment has not been adopted by the industry. The treatment is carried out in a vertical cooker heated by indirect steam. Heat is transferred to the material by forced convection thanks to a powerful stirrer. The mechanical consumption of energy of the conditioner is far from negligible and is of the same order of magnitude of the expeller. Hydrothermal pre-conditioning improves the oil availability through protein denaturation and reduces oil viscosity due to the temperature increase. Other positive effects include the inactivation of different enzymes (lipase, myrosinase, lipoxygenase), which enhances product quality.[10] Depending on the process chosen, conditioning temperatures range from 75 to 110 °C, moisture ranges from 8 to 10% and conditioning duration reaches up to 90 minutes.[11]

Oil in the seeds is contained in 'oil bodies' that separate the lipids from the rest of the cell content. Oil bodies are coated by specific proteins called oleosins which have an amphiphilic structure that helps to stabilise the oil bodies. The mechanical extraction of oilseeds without pre-treatment is not efficient because it has to break the oil bodies only by mechanical means. Heat by coagulation of oleosins destroys the oil bodies and facilitates the extraction. With rapeseed, the performance of the expelling is strongly correlated to the temperature of cooking (Table 3.4).

The consequence of intense cooking on quality is a decreased protein solubility that makes the feed value of cake less interesting for monogastric livestock (Table 3.5). On oil, the main marker of heat treatment is the para-anisidine index that takes into account the secondary products of oxidisation. Unfortunately, this index is perturbed by the presence of phenolic compounds that are present in crude oils, and rapeseed oil is rich in these products. It is worth noting the effect of the cooking on the presence of phosphorus in oil. Cold pressing drives phosphorus to a very low presence, meaning that membranes are weakly disrupted by the treatment. On tocopherol, one can observe a similar trend. In the case of very intensive extraction, more tocopherol is extracted. Phospholipids and tocopherol have a synergetic effect on oxidative stability. Phospholipids are scavenging oxidising ions like iron and protect

Table 3.4 Effect of cooking temperature on oil recovery of rapeseed (source: Creol).

	Cold pressing	Medium cooking at 105 °C	Intensive cooking at 125 °C
Flowrate [kh/h]	230	525	525
Oil in cake [%]	13.3	12.2	7.7

Table 3.5 Effect of cooking on the extraction of rapeseed compounds (source: CETIOM/Creol).

		Two-step expelling	
	Intensive cooking	Cold pressing	Second expelling
Oleic acid [%]	0.8	0.4	3.9
Peroxide value [meqO$_2$/kg]	0.4	1.1	1.6
Para-anisidine value [IPA]	8.3	0.7	20.3
Index total ox. [IPA + 2 IP]	9.1	2.9	23.5
Tocopherols [mg/kg]	790	740	930
Phosporous before water degumming [mg/kg]	360	7	1620
Phosporous after water degumming [mg/kg]	11	8	200
Iron [mg/kg]	1	0.4	22

Primary Processing

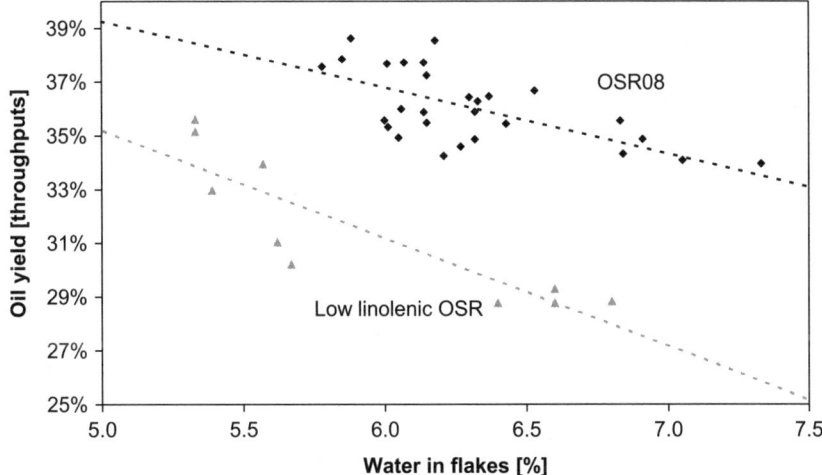

Figure 3.1 Effect of the water content in flakes in the yield of the oil recovery.

tocopherols from oxidisation. In the case presented above, the high iron content is explained by the treatment carried out before the second expelling, which was extrusion carried out with an experimental apparatus.

An important effect of the pre-treatment on the expelling is the removal of some water from the product to be extracted. Water plays the role of plasticiser in the de-oiled matrix. By increasing the water content of the cake, the softness of the cake also increases causing a reduction of the pressure that can be carried out on the material. By drying the flakes, it is possible to adjust the mechanical work applied on the cake and the oil content of the cake.

Oil yield is determined from the ratio between oil output and flakes input. The data presented in Figure 3.1 have been collected during a trial where cultivars were harvested in 2008 and were crushed under similar conditions and compared cultivar with low linolenic acid harvested in 2005. Clearly, it can be seen that with increasing water content, the oil yield decreases.

3.2.3 Microwave and Radio Frequency

The main purpose of oilseed pre-treatment is to damage the oil cells and adjust the oil viscosity and moisture content to facilitate oil extraction. Thermal pre-treatment can also be effected by using dielectric heat generation (*i.e.* microwave and radio frequency) instead of hot steam.[12,13] Microwaves can be considered as electromagnetic radiation with an oscillation frequency in the range of 300 MHz to 300 GHz with radio frequency in the low regions and microwave in the higher regions.

The heating effect is based on an interaction between the electromagnetic field and dielectric effects *e.g.* by polar molecules and ions. The molecules will rotate and oscillate, causing internal friction leading to heat generation. For microwaves this effect is limited to a small area near the surface of the dielectric material due to low penetration depth. Heating in the radio frequency field follows the same principle, but because of the lower frequency and the higher penetration depth heating with radio frequency energy causes a more homogeneous heat distribution compared to microwave application.

3.2.4 Pulsed Electric Field

The application of pulsed electric fields is a non-thermal food processing technology. An external electric field can induce critical electrical potential across the cell membrane leading to breakdown and local structural changes of cell membranes, thereby increasing the permeability.

Many reports have demonstrated the advantages of pulsed electric field application for high pressing efficiency as well as high purity of juices. A few reports describe the impact of pulsed electric field on the recovery from oilseeds and the impact on high-fat plant cells.[14,15] Pulsed electric field had a positive effect on oil yield and content of functional food ingredients.

3.2.5 Enzymatic Pre-treatment

The use of enzymes can enhance oil extraction from fruits and oilseeds. There is a difference in effect of enzymatic treatment when seeds and fruits are treated:

- In seeds, oil is in intracellular vacuoles linked to other macromolecules and its extraction is enhanced by the hydrolytic action of carbohydrases. Exogenous enzymes are employed to increase the amount of oil recovered. So when enzymatically treated oilseeds are processed, the increase in oil is due to cell wall rupture. In practice, enzymatic pre-treatment is not widely employed since hexane extraction allows efficient oil recovery at competitive cost;
- The extraction of oil from fruits is accomplished by the addition of hot water to the ground fruit, mixing of the paste and separation of the three phases (*i.e.* solid, aqueous and oily phases). When enzymatically treated oil fruits are processed, the higher amount of oil recovered is due to cell wall rupture and rupture of the interphase of lipoproteic membranes (dispersing the colloidal system formed during grinding). So enzymatic treatment is particularly useful for pastes which form emulsions that need to be disrupted. In the case of olives, industrial enzymes are pectinases, hemicellulases and cellulases. Their addition enables an 86% oil yield *vs.* 82% without enzymes when oil is extracted by a centrifuge decanter. The throughput of the decanter can be increased up to 40% thanks to enzymes. With a traditional press, the paste extracted with enzyme shows faster release of the oil over time and gives higher total yield. Moreover, enzymatic preparation has a positive effect on the oil quality by decreasing the acidity and improving the extraction of phenolic compounds.[16]

3.3 Novel Oil Recovery Processes and Valorisation of Waste Streams

3.3.1 Introduction

A universal approach to recover oil from oilseeds is shown in Figure 3.2.

The bulk of the oil in oilseeds is recovered by a pressing/expelling step (batchwise or nowadays continuously by screw presses). Often the pressing stage is followed by an extraction step with organic solvents, water or supercritical fluids to recover the residual oil from the press cakes. In order to increase the oil yield, the seeds or fruits can be pre-treated by preconditioning.

Olives are in contrast to oilseeds fruits and the oil is recovered usually in a different way. Some local, small producers of olive oil do this by squeezing and pressing the olives. The biggest part of olive oil production, however, is produced by large companies using large centrifuges to separate the oil from the wastewater. In practice, both a two-phase and a three-phase centrifugation process can be used (Figures 3.3 and 3.4).

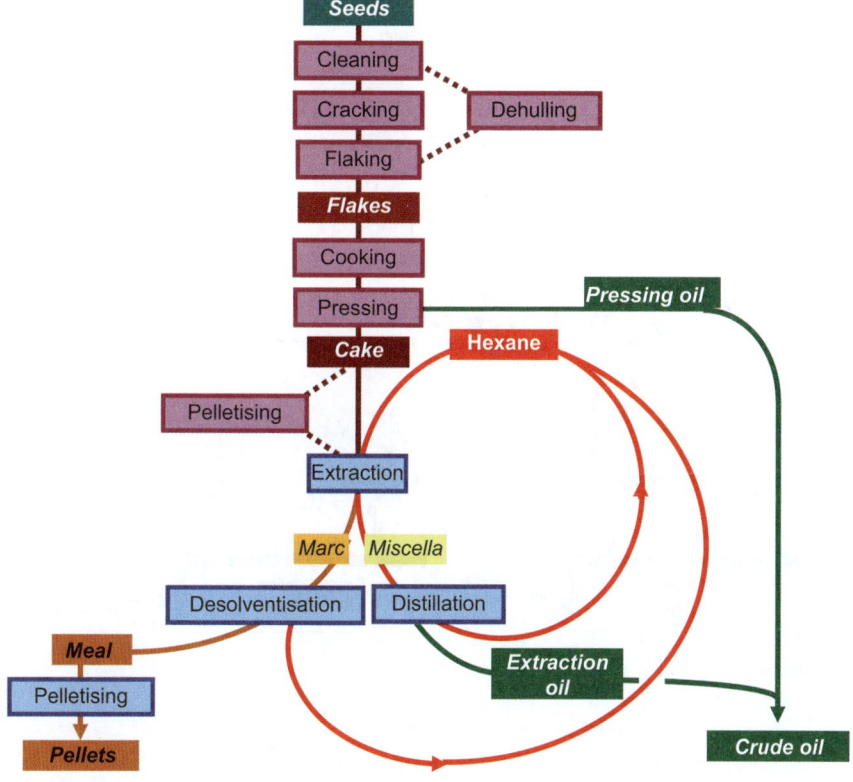

Figure 3.2 Universal process of the oil recovery from oilseeds.

A three-phase centrifugation system was designed at the beginning of the 1970s. By means of this system, the oil, vegetation water and solid phase of the olive can be separated in a continuous process. The main inconvenience of the three-phase system is the generation, during a short period of the year (November–February), of large quantities of olive mill wastewater. The solid waste stream produced by the three-phase system is called 'orujo'.

A new two-phase centrifugation system for oil extraction was developed during the early 1990s. Although this is called the ecological system because it greatly reduces wastewater generation and its contaminant load, it still produces a solid and very humid by-product called 'alperujo' or 'alpeorujo'.

The different possibilities of oil recovery from oily biomass are described in this chapter. The most suitable oil-recovery process depends on the oil content

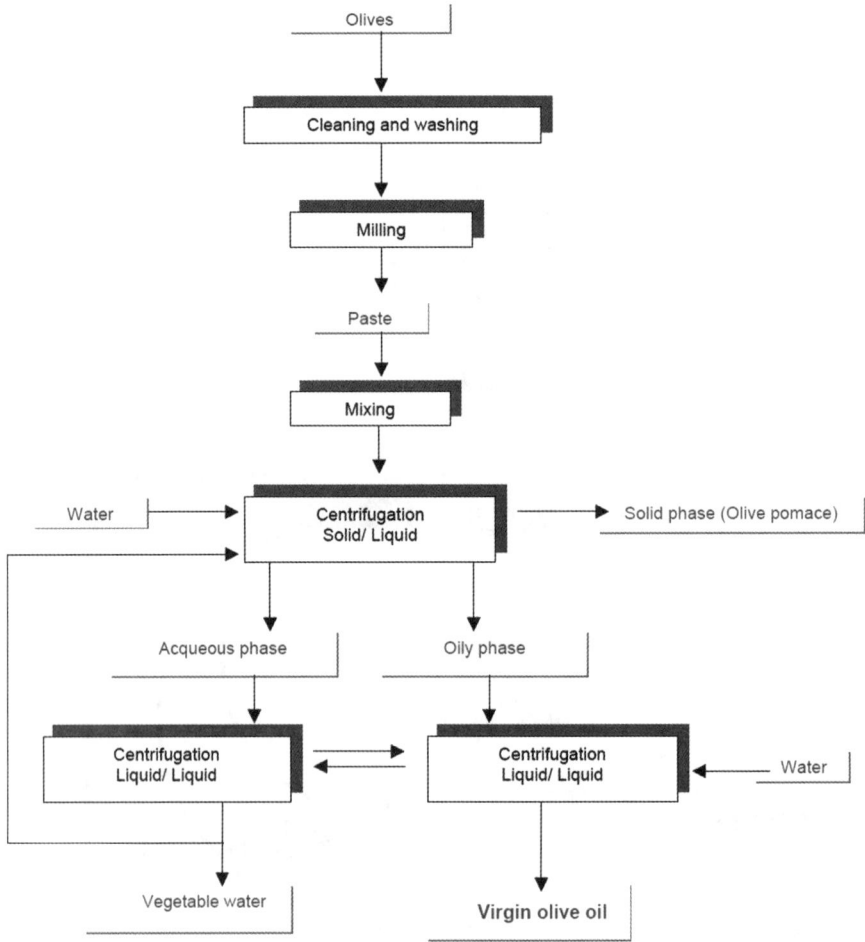

Figure 3.3 Olive oil production using a three phase centrifugation process.

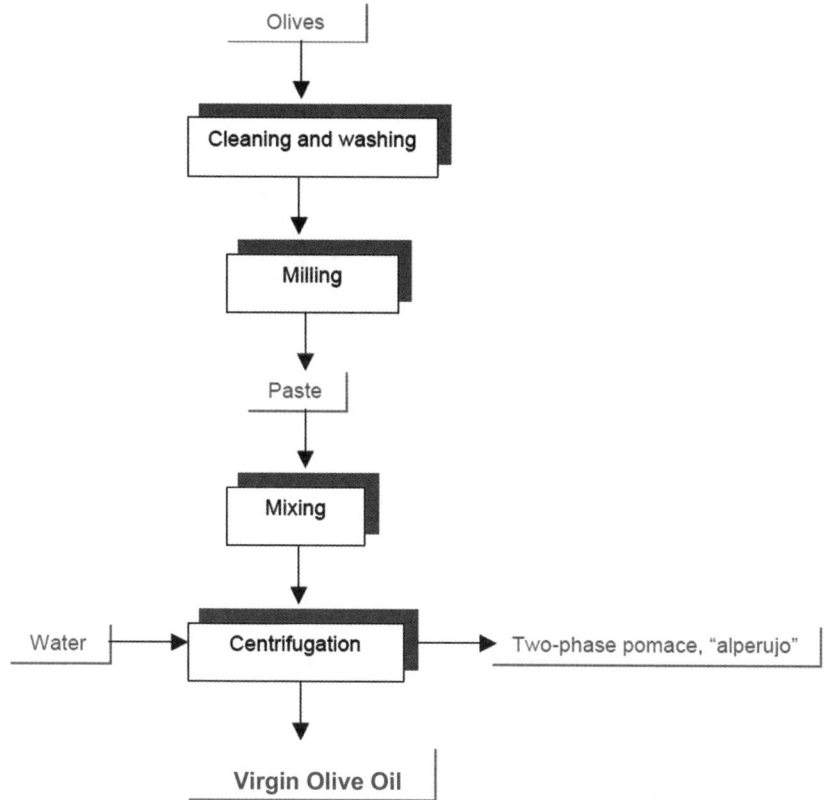

Figure 3.4 Olive oil production using a two phase centrifugation process.

of the seed or fruit. The oil content for sunflower seed, rapeseed and olives shows that seeds contain more oil compared to fruit pulp (Table 3.6).[17] Based on more recent analysis of more than 5000 samples of rapeseed and more than 700 samples of sunflower more detailed information is available. This set of data is slightly different from the values found in the literature.

For materials with a relatively high oil content, a three-step process is carried out, which consists of pre-treatment, pressing and solvent extraction. This combined process is generally used on high oil sources (oil content > 35%). For safety reasons, the solvent extraction step requires high equipment costs. The trend is to build very large plants to cut costs by economies of scale. In France, the size of the most recent plants exceeds 600 000 t/year.

3.3.2 Oil Extraction from Olives

The olive can be structurally separated into three distinct anatomical parts: epicarp (skin), mesocarp (pulp or flesh) and woody endocarp (stone) containing the seed. All three influence the quality of the end product.[18,19]

Table 3.6 Oil content in rapeseed, sunflower and olives.

	Type	Oil content (literature) [%]	Oil content (source: CETIOM) [%]
Rapeseed	seed	42	46
Sunflower seeds	seed	53	48
Olive pulp	fruit	22	–

- The epicarp or skin is the protective tissue of the olive (1–3 wt%). Classes of compounds in the epicarp include alkanes, aliphatic alcohols, aliphatic aldehydes, wax esters and polycyclic triterpenes.
- The mesocarp constitutes the major part of the olive comprising 70–80% of the whole fruit. Around 70–75% of the mesocarp weight is water and 15–30% is oil, dependent on the maturity of the olives. Classes of compounds in the mesocarp include triglycerides, sugars, glucosides and phenols.
- The endocarp or stone represents 18–22% of the olive weight. The seed or kernel contains a relevant amount of oil or triglycerides (22–27%).

The land area occupied by olive orchards has increased in recent years, largely in response to the worldwide rise in olive oil consumption.[20] The world production of olive oil has risen from 2460 thousand tonnes between 1996 and 2000 to 2770 between 2000 and 2004. The estimation for olive oil production for 2005–2006 is 2580 thousand tonnes. Production across the 25 countries of the European Union, where Spain (45%), Italy (31%) and Greece (22%) are the largest producers, is expected to account for almost 75% of the world's olive oil.

The by-products from olive oil extraction depend on the technology applied.[21] A three-phase and a two-phase centrifugation procedure can be distinguished.

3.3.2.1 Olive By-products

Olive Leaves
Olive leaves refers to a mixture of leaves and branches from the pruning of olive trees as well as the harvesting and cleaning of olives prior to oil extraction. The production of leaves from pruning has been estimated to be 25 kg per olive tree.

The leaves can be used as animal feed or as a calorific source. Olive leaves are a well-known source of antioxidant compounds and are marketed as herbal teas.

Olive Cake
Olive cake consists of olive pulp, skin, stone and water. Different terms may be given depending on factors such as composition and oil content (crude or extracted olive cake), stones or moisture (fresh or dry). Olive cake from the two-phase system has a higher moisture and lower oil content as a result of a

Primary Processing

more efficient and environmentally friendly centrifugation process, compared to olive cake from the three-phase system (also called olive husk).

Olive Mill Wastewater

Olive mill wastewater or alpechin is a very polluting liquid made of the olive vegetation water plus the water added in the different steps of oil production. With the three-phase system large quantities of wastewater are produced during a short period of the year (November–February). With the development of the two-phase system the generation of wastewater is greatly reduced.

However, due to the presence of organic acids, lipids, alcohols and polyphenols the wastewater remains an environmental problem. In recent years many management options have been proposed for the treatment and valorisation of olive mill wastewater. Most of these methods aim to reduce the phytotoxicity in order to reuse it for agricultural purposes, but more recently alternative methods have also been proposed.[22]

Other By-products

Olive stones may be a single by-product when they are well separated from pulp either before or after oil extraction.[20] This separation results in another by-product: the olive pulp. Stones are valuable products because of their high calorific power. They are currently used as an energy source, but other uses are also proposed (*e.g.* soilless substrate or activated carbon).[22,23]

Oil extraction also produces watery waste, which is sometimes dried and transformed into a concentrate known as 'olive molasses'.

Finally, olive cake is increasingly used as fuel in olive oil mills and in power generation, which results in a new by-product called olive cake ash. An estimated 140–160 g of ash are produced per kg of olive cake.

Valorisation of Olive Cake

The solid waste stream from the two-phase centrifugation system, alperujo, is characterised by a high moisture content (50–70%), slightly acidic pH, a high concentration of organic matter (mainly fibres) and richness in minerals (especially potassium). The concentration of lipids varies between 3 and 18%, depending on the extraction yield, and it also has a high concentration of polyphenols. It has a strong odour and a doughy texture, making management and transport difficult. Roig and co-workers wrote a review on the valorisation of this waste stream.[22] They described various treatments including physico-chemical treatments, direct soil application and biotechnological transformations. The most interesting ones are listed here:

Drying Followed by Second Oil Extraction

Alperujo is usually treated with a second centrifugation to extract the residual oil. The resulting by-product of this second extraction is dried, and then subjected to chemical extraction with hexane in order to produce an extra yield of oil. The recently discovered problems concerning the detection of PAHs (polycyclic aromatic hydrocarbons) in this oil as a result of drying the alperujo before oil chemical extraction has obliged manufacturers to perform a further purification step, which greatly increases production costs. This new resulting waste of the chemical extraction ('orujillo') can be used for the co-generation of electrical power.[21]

Solid State Fermentation
It is possible to improve the nutritional properties of alperujo through solid fermentation.[24] This process, performed by microorganisms in a solid medium, has been successfully exploited for production of animal feeds, fuel and enzymes. The protein content in olive husk was increased from 5.9% to 40.3% with this method.

Extraction of Valuable Products
Alperujo is a potentially rich source of a great range of phenols with a wide array of biological activities. Olive mill waste is rich in polyphenols and typically contains 98% of the total phenols in the olive fruit. Hydroxytyrosol, tyrosol, oleuropein and caffeic acid are the major phenolic components. A large number of scientific articles demonstrate the antioxidant, cardioprotective, antimicrobial, antihypertensive and anticarcenogenic activities of these compounds, which could be used in the pharmaceutical, cosmetic and food industries. Some authors have developed new technologies for the improvement of their extraction methods.[25,26]

Due to the replacement of the three-phase by the two-phase extraction system a new solid waste stream (olive cake or alperujo) with specific characteristics has been created. It presents more disposal problems compared to the olive cake or husk from the three-phase system. Most of the research is focussed on the valorisation of this waste stream.

Due to the low residual oil content it is probably not so interesting to recover this residual oil in order to increase the oil yield. More focus lies on the recovery of valuable products such as polyphenols, or possibilities of using the by-products as soil conditioners or fertilisers to close the cycle of residues and resources.[22]

3.3.3 Oil Extraction from Rapeseed

Rapeseed oil has the third largest production volume in the world, after soybean and palm oil. In order to obtain oil from rapeseeds, a combined process of thermal and mechanical pre-treatment, mechanical expression and extraction is applied.

3.3.3.1 Thermal Pre-treatment

Thermal pre-treatment or cooking is mainly carried out to inactivate the naturally occurring enzymes present in the seed, which otherwise would negatively affect the quality of the oil. Also, denaturation of proteins within the cells takes place, cell membranes become permeable and, due to the increase in

Primary Processing

temperature, the viscosity of the oil is reduced. These factors facilitate the flow of the liquid out of the solid matrix during pressing, therefore improving the performance of this process. During this thermal pre-treatment, the flakes or seeds are heated as fast as possible up to 80–105 °C, accompanied by a rise in moisture content up to 10–12% due to injecting steam or spraying water. The material is subsequently dried at 105–110 °C down to a moisture content of 3–7% and is immediately pressed.[27]

3.3.3.2 Enzymatic Pre-treatment

Enzymatic hydrolysis has been shown to be a useful option for pre-treatment of oilseeds such as rapeseed.[28] In a study by Srivastava the enzyme-induced microstructural change in rapeseed was studied to further understand the mechanisms to release extra oil.[29] Rapeseed was pre-treated by enzymatic hydrolysis with crude enzyme (mainly cellulase and protease) and dried. Oil content was determined by Soxhlet extraction. They concluded that enzymatic pre-treatment of oilseed enhances oil recovery by two mechanisms:

- Cell walls are biodegraded by the enzyme action facilitating oil extraction;
- Enzymes break up the complex lipoprotein and lipopolysaccharide molecules, releasing extra oil for extraction that is otherwise not extractable.

3.3.3.3 Pre-treatment by Microwave and Radio Frequency

Microwave and radio-frequency energy was applied for the thermal pre-treatment of oilseeds to increase oil yield obtained by pressing and at the same time preserve or improve the oil quality.[12,13] A micro-structural change was detected for the microwave and radio frequency pre-treated material. The oil drops in the rapeseed cells were spread to numerous very small droplets that could be extracted more easily by pressing due to easier flow through the existing capillaries of the seed tissue. Important parameters include the temperature reached during treatment and the velocity of energy input. The duration of the treatment seemed of lower importance. Radio frequency pre-treatment seemed to be more suitable compared to microwave due to a more homogeneous heating and lower loss of moisture. The lipid acid composition was not affected by the pre-treatments.

From a study by Valentova it was concluded that irradiation of rapeseeds by microwaves was more gentle with respect to final oil quality compared to conventional heat treatment before oil processing.[30]

3.3.4 Oil Extraction from Sunflower Seeds

Sunflower seeds represent an important source of food-grade oil. Their protein fraction is characterised by a well-balanced amino acid pattern and they are recognised as a potential source of proteins for human consumption. Furthermore, sunflower oil is richer in vitamin E than many other edible oils, and it is important to preserve this antioxidant fraction.

The utilisation of defatted sunflower meal in the human diet requires dehulling of achenes (*i.e.* dry fruit containing the seed) and the removing of chlorogenic acid, a polyphenolic compound.[31] The oil as well as the defatted protein meal are important products in the human diet and the search for food-grade extractions of oil from sunflower seeds is a topic of many studies.

The industrial process for sunflower oil production consists of four successive stages: particle size reduction, pressing, extraction of residual oil using hexane and refining.[32,33] A flow diagram of sunflower seed processing is shown in Figure 3.2.[34] The extraction yields are close to 100% with very good quality. However, the use of hexane to remove oil from the press cake is an increasingly controversial issue and could be prohibited due to its carcinogenicity.[35] As a consequence various other routes have been considered and a selection of these are described here.

3.3.4.1 Supercritical Extraction

Supercritical extraction of sunflower seeds has the advantage of preserving the antioxidant fraction, in particular with CO_2 because of its relatively low critical temperature (31 °C), non-toxicity, non-flammability, good solvent power and ease of removal from the product.

The extraction of dry, milled and dehulled sunflower seeds by supercritical CO_2 extraction on laboratory scale was described by Salgin and co-workers.[36] The effects of pressure, temperature, supercritical CO_2 flowrate and mean particle size on the extraction yield were investigated. The experimental results showed that extraction yield increased linearly with time in the early stages of extraction, and the extraction process was limited by the solubility of oil in the supercritical CO_2. Solubility is a function of solvent density, which in turn is a function of pressure and temperature.

3.3.4.2 Extrusion

Extraction of sunflower by extrusion was described in numerous articles.[37–40] From the same research group aqueous extraction of residual oil from press cake by extrusion was reported.[41] The separation of liquid and solids was assisted by the introduction of fibres (*i.e.* wheat straw and sunflower depithed stalk).

It appeared that the use of fibres was essential to enable the liquid/solid separation. The process efficiency was limited and the best oil extraction yield was only 55% of the total lipid content of the seed. The residual oil content of the cake meal was approximately 30%, partly due to incomplete cell lysis within the seed. Direct expression in the twin-screw extruder was more efficient (oil extraction yield close to 70%) compared to the extraction using water.

Primary Processing

During the aqueous extraction process, the oil was extracted in the form of oil-in-water emulsions, stabilised by phospholipimaradonads and proteins. Demulsification by alcoholic extraction made it possible to isolate the oil extracted during the process. This process also allowed the production of a protein extract.

In a more recent study, Evon and co-workers suggested that a co-rotating twin-screw extruder could also be used for the aqueous extraction of residual oil from press cakes produced after pressing of the whole seeds to improve the total oil yield.[35] Again, wheat straw was essential to enable liquid/solid separation. During the aqueous extraction process, the oil was extracted in the form of oil-in-water emulsions. However, the contribution of the aqueous extraction stage for the total oil yield was limited (<5%), probably due to insufficient particle size reduction, low ratio of water to the press cake, and thermomechanical denaturation of the proteins during the expression stage. Another factor was the technological limits of the twin-screw extruder that did not enable full separation of liquid and solid phases.

3.3.4.3 Enzymatic Pre-treatment

Pressing and solvent extraction is the most widely used technique for the processing of high oil-bearing seeds in the industry, owing to its high efficiency in oil recovery. An enzymatic pre-treatment step could be incorporated with some modifications to the current system. Its application would require additional steps of incubation at intermediate moisture (15–40%) and further drying before pressing and solvent extraction. Since the costs of enzyme treatment and subsequent drying would constitute major costs of the treatment, both moisture and enzyme parameters should be optimised.

In a study by Dominguez the combined effects of moisture, enzyme/sunflower kernel ratio and treatment time were studied.[42] The use of an enzymatic treatment for high oil content seeds not only enhanced the oil extractability in the pressing stage, but also made the residual oil in the cake more easily extractable by solvents, achieving increases of up to 10%.

3.3.5 Pressing and Pressing-related Processes

In this section the methods applied for pressing or expressing of oil from materials are described. The techniques described include screw pressing, extrusion and gas assisted pressing extrusion.

3.3.5.1 Screw Pressing

Recovery of oil from oil-containing materials is usually done with continuous screw presses such as expellers. An expeller press is a screw-type machine, which presses oilseeds through a caged barrel-like cavity. Raw materials (usually pre-conditioned) enter one side of the press and cake products exit the other side.

Cake, fed in at one end of the barrel, moves down the screw because friction between the cake and barrel or baffles prevents the cake from turning in the barrel.

The cake compacts because:

- The screw's root diameter progressively increases;
- Its pitch progressively decreases;
- The diameter of the barrel decreases;
- Cake outflow is throttled;
- Combinations of these actions occur.

Expressed fluid flows out through openings in the barrel. Barrels made of perforated sheet metal are used when pressing relatively soft fruits. In the case of seeds, the cage is made of bars of special profile that allow liquid to escape with relatively low solids.

In general, the worm assembly is made of screw elements separated by rings that can be smooth or conical. The flight is discontinuous, so that the main pressure is carried out on the cake between two screw units and without excessive pressure on the barrel. According to the hardness of the cake it is possible to adjust the screw profile to increase or reduce the retention of the press. Material with high fibre content must be processed without conical rings, while soft material requires the presence of cones. Dehulled seeds are difficult to expel because of the lack of rigidity of the incoming material. Both rapeseed kernels and sunflower have low fibre content and high oil content. Cold pressing of these materials is not possible with a regular screw press.

Screw presses are used to recover oil from plant matter with high oil content (>25%). Usual rest oil values in a single (hard) pressing are 8–10% for rapeseed and sunflower. Fines expelled with the oil are separated from it and recycled to the press. The energy consumption of expelling is 10 to 20 kWh/t of seeds for pre-pressing and 30–40 kWh/t in the case of hard pressing. In order to minimise wear and problems caused by recycling of fines, high-oil-content feeds are usually pre-pressed rather than fully pressed. Pre-pressing removes two-thirds of the oil and generates cake containing 15% to 18% oil. Then, 95% to 98% of the residual oil is recovered by subsequent solvent extraction, discussed in the next section, resulting in a final residual oil content of 2% in the case of sunflower and 3.5% in the case of rapeseed (source: CETIOM). The throughput can be doubled when full-press expellers are used as pre-presses, but it is preferable to use units specifically designed for pre-pressing.[43] Full presses have the disadvantage of high energy consumption for attaining low residual oil content and losing valuable product.

3.3.5.2 Extrusion

In an extruder, oil can be extracted from material by mechanical lysis of the oil-containing cells and thereby separation of solid and liquid phases. Co-penetrating and co-rotating twin-screw extruders are most common. A very wide choice of screw elements is available. The screw elements affect different functions such as conveying, heating, cooling, shearing, crushing, mixing, chemical reaction, liquid/solid extraction, liquid/solid separation and drying.

Twin-screw extrusion was proposed as a good alternative to traditional oil extraction processes by Dufaure. Lipids are industrially extracted by a two-step process of mechanical pressing (70–80% yield), followed by extraction with hexane (98% yield). The raw material needs to be pre-treated before pressing (*e.g.* drying, dehulling and cooking). In addition, the oil collected during pressing contains solid particles in suspension that have passed through the filter and need to be removed by decanting, filtration or centrifugation. The new process proposed by Dufaure gave an 80% oil yield of good quality. High pressing temperature and low moisture content improved the oil extraction. The oil yield was further increased to 90% by adding acidic alcohol during extrusion.[37–40]

3.3.5.3 Gas-assisted Pressing

With gas-assisted pressing the gas, in general CO_2, is passed through the oilseed before or during pressing in order to achieve lower residual oil contents. It is assumed that the effect of gas-assisted pressing is mostly based upon dissolution of CO_2 into the oil, thereby changing the physical properties of the oil in such a way that the pressing efficiency increases. Dependent on the type of gas and conditions applied, the gas becomes a supercritical fluid. CO_2 becomes supercritical under relatively 'mild' conditions: above 31 °C and 73 bar. Supercritical fluids can diffuse through solids like a gas, and can dissolve materials like a fluid.

The aim of gas-assisted oilseed pressing is to improve the efficiency of mechanical expression in presses in order to reach sufficiently low residual oil contents at justifiable energy costs without the need of a subsequent solvent extraction step.

The effectiveness and feasibility of gas-assisted pressing in hydraulic batch presses is proven by several experimental investigations, especially when CO_2 is used. However, for industrial development the pressing process must be continuous. Compared to conventional hydraulic pressing the pressing times and CO_2-saturation times in continuous pressing processes are relatively short. The main goal at present is to develop a continuous gas-assisted pressing process.

Around 1960 organic solvents (*e.g.* hexane, ether) or gases were used in continuous screw presses in order to enhance the oil yield. The solvents were added either before or during the pressing stage. Besides the addition of liquids, displacement of oil by steam injection was also possible, but this had the disadvantage of increasing the moisture content in the press. These developments of using solvents or gases for oil-yield improvement led to the idea of using compressed gases like CO_2 instead.[44–48]

A group from the University of Twente and the Eindhoven University of Technology introduced the term GAME: Gas Assisted Mechanical Expression. In the GAME process oil-containing material is saturated with CO_2 at around 10 MPa, after which the resulting CO_2–oil mixture is expressed from the seeds by mechanical expression. The CO_2 displaces part of the oil from the cake, to an extent that is related to the solubility in the oil. To accomplish the full increase in yield, the oil in the seeds has to be completely saturated with CO_2, which requires 30–60 min. either in a hydraulic press or in a pressure vessel with sufficient residence time.[49–56]

Voges and co-workers studied theoretically and experimentally the predominant effects of gas-assisted pressing, which can be divided into dissolution-related and gas-flow-related effects. It became clear that the dominance of one mechanism over another depends on processing conditions, particularly residence time and dimension of the press cake.[57]

3.3.6 Solvent Extraction

3.3.6.1 Organic Solvents

Continuous solvent extraction systems are very suitable for processing oil-containing materials to a meal with a very low residual oil content. Modern solvent-based processes usually consist of extraction by successive counter-current washes with an organic solvent (*e.g.* hexane, petroleum ether) of the previously cracked, flaked, ground or pressed oil-containing material. The extracted meal is then carried by a sealed conveyor for solvent recovery in enclosed vessels. Hexane is removed from the oil in rising film evaporators and vacuum distillation.[33]

The major drawbacks of organic solvent extraction are the hazardous nature of common organic solvents and the energetic effort required for complete removal from the product and the organic residue.

3.3.6.2 Supercritical Fluids

A gas becomes a supercritical fluid above its critical point. Supercritical fluids can diffuse through solids like a gas, and dissolve materials like a liquid. In most cases carbon dioxide is applied because of some advantages. It becomes a supercritical fluid under relatively 'mild' conditions: above 7.38 MPa and 31 °C. It is non-toxic, non-flammable, has good solvent power and it is easy to remove from the product just by releasing the pressure.

According to the literature the use of supercritical fluids is often limited to the recovery of high-value natural products such as high-quality vegetable oils without traces of organic solvents. According to Del Valle there is still no commercial application for commodity oils such as those of canola seed, corn germ, rapeseed, soybean or sunflower seed; however, there is renewed scientific and commercial interest in speciality oils to be used in cosmetics and other high-value applications.[58] However, there are examples of industrial use of supercritical fluid extraction (*e.g.* www.natex.at), where 15 ton/day of sesame seed is extracted. More information on supercritical fluid extraction of rapeseed and sunflower is given in the next sections.

Wageningen UR has developed a new clean process to isolate valuable components from solids such as oil-containing materials. In contrast to other conventional processes, the new method concerns a continuous process that can be controlled easily and leads to higher extraction yields.

Within the process supercritical CO_2 is used as solvent in combination with extrusion. An extruder is adapted in such a way that it can be employed as a high-pressure vessel in which the continuous extraction can take place. The high pressure is maintained by creating two material plugs in the beginning and end of the extruder. In between the material plugs, the supercritical carbon dioxide extracts the desired components in a counter-current manner. The equipment is able to handle pressure up to 250 bar. However, the maximum process pressure depends on the material to be processed. Next, the dissolved product is separated from the carbon dioxide by decreasing the pressure after which the carbon dioxide is reused and the pure product remains as solid or liquid. The scale of the trials was 6.5 kg/h.

Advantages of this process include high extraction yields, low energy costs and flexible processing. In addition, no organic solvent is used.

3.3.6.3 Aqueous Extraction

Aqueous extraction processing, which uses water as an extraction and separation medium, has regained considerable interest during the past decade as an environmentally friendly approach for extracting oil from oilseeds. In general aqueous extraction processing is less efficient in achieving high oil yield. It is based on the insolubility of oil in the extraction medium rather than its dissolution as applied in hexane extraction. Studies have reported *e.g.* soybean oil extraction yields from 65–75% *via* aqueous extraction processing compared to >95% *via* hexane extraction.[33,59] Much of the oil extracted with aqueous extraction processing is present in the form of a stable cream (*i.e.* oil-in-water emulsion).

However, aqueous extraction processes can be improved by treatments that enhance the dissolution of proteins or carbohydrates such as increasing the temperature or by using specific enzymes. Enzymes can be applied to break down the proteins in the cell wall and pseudo-membranes surrounding the oil bodies to eliminate/reduce barriers to oil extraction. This process is called enzyme-assisted aqueous extraction processing or aqueous enzyme extraction. Several cell-wall degrading enzymes can be used during aqueous extractions such as cellulases, hemicellulases, pectinases and proteases. For each system the optimal temperature and pH value has to be determined.

Enzymes can degrade walls of oil-bearing cells but also break the colloidal system in olive paste, thereby releasing maximum oil and also improving oil quality. During the extraction process, enzymes present in the olive fruit are in general deactivated during the extraction process or crushing step, and additional enzymes need to be added to the olive paste during the mixing step to replace the deactivated enzymes. A number of studies have focused on maximising oil yield by optimisation of enzyme use or enzyme combinations.[60]

Thermoplastic extrusion facilitates the accessibility of enzymes to proteins surrounding oil bodies. Oil extraction yields of 88% have been reported for extrusion-aided, enzyme-assisted aqueous extraction;[61] different proteases showed different oil extraction efficiencies. A comparison between aqueous and solvent extraction was made by Rosenthal.[33]

3.3.7 Residual Oil Recovery

Recovery of residual oil from press cakes or other oil-containing by-products is strongly dependent on the properties of the material and the efficiency of the oil-extraction process.

The ideal situation is an oil recovery method or process that is so efficient that the oil content of the resulting by-products is low and that a subsequent solvent extraction is not necessary. An example of such an approach is the gas-assisted mechanical expression process or GAME. The research focus at the moment is to make the GAME process continuous.

Another option is to improve or optimise the pre-treatment of the oil-containing material. When the oil availability is improved by increasing the particle size or disruption of the oil bodies, the subsequent pressing and extraction will become less energy-consuming. However, it is essential that the quality of the oil is not deteriorated.

A third approach could be to focus on solvent extraction methods. Preferentially, the use of organic solvents such as hexane or ether should be eliminated. An option would be to use supercritical fluids, which is already done on an industrial scale, or work on optimisation of aqueous extractions. These aqueous extractions might be improved by combining them with processes that enhance oil availability such as thermal treatment or use of enzymes.

Regarding the different oil recovery methods it is necessary to differentiate between the initial oil content of the oil-bearing material: *e.g.* oilseeds are treated differently compared to olive fruit pulp that contains much less oil and much more water. Subsequently, each material has its own optimal oil recovery parameters.

3.3.8 Conclusions

Recovery of oil from fruit or seed is mostly done by pressing. The most suitable process depends on the oil content of the seed or fruit. In general, seeds contain more oil compared to fruit. For materials having relatively high oil contents (>35%), which is the case for rapeseed and sunflower seed, a three-step process is often applied consisting of a pre-treatment, a pressing stage and (organic) solvent extraction. Olive pulp contains much more water and less oil and is treated in a totally different manner.

The main purpose of oilseed pre-treatment is to damage the oil cells and adjust the oil viscosity and moisture content to facilitate oil extraction. The most common pre-treatment method is thermal pre-treatment; other methods reported include microwave/radio frequency, pulsed electric field and enzymatic pre-treatment.

Primary Processing

After the pre-treatment stage, the material is pressed to remove most of the oil from the oilseed by expellers or extruders. Subsequently, the press cake is extracted with organic solvents such as hexane or ether, but preferentially with other media given the hazardous nature of organic solvents. Examples of other extraction media are supercritical fluid extraction and aqueous extraction processing.

When supercritical fluids are combined with mechanical expression the residual oil content of the press cake can be reduced to such a level that a subsequent solvent extraction step is no longer necessary. The process (gas-assisted mechanical expression or GAME) has potential on an industrial scale if it becomes a continuous process.

3.4 Protein and Amino Acid Isolation

The oil production from olives, rapeseed and sunflower results in waste streams that still contain valuable compounds such as proteins. In Table 3.7 the composition of rapeseed, sunflower and olive pulp is shown.[17] Rapeseed and sunflower seeds contain 40–50% protein, whereas the amount of protein in olive pulp only is 1.6%. Therefore, the extraction of peptides and amino acids is only beneficial with respect to sunflower and rapeseed.

Traditionally, based on their nutritional value, taste, physiological and chemical characteristics, amino acids are used in human food and animal feed.[62] Hydrolysis of proteins, mainly vegetable, into peptide and amino acids is performed on an industrial scale to produce food ingredients with a savoury, meat-like flavour. Monosodium glutamate has been used for a long time as a flavour enhancer and is the amino acid with the largest production capacity. In food, also glycine, cysteine and oligopeptides are used. Essential amino acids, such as lysine, methionine, tryptophan and threonine, are added to animal feed. In the last few decades, amino acids gained interest in the pharmaceutical, cosmetic and chemical industries.

Amino acids can be produced chemically, *via* enzymatic catalysis, by fermentation and by extraction from protein-rich materials.[63] The two biotechnological processes, fermentation and enzymatic catalysis, are most exploited. Extraction of amino acids is at present not suitable for large-scale production. However, due to the increasing amounts of biomass waste streams,

Table 3.7 Composition [%] on dry matter of rapeseed, sunflower and olive pulp.

	Rapeseed	*Sunflower*	*Olive pulp*
Water	5.0	6.7	50.0
Oil	41.6	53.3	22.0
Protein	26.2	22.9	1.6
Carbohydrates + fibres	23.4	14.0	24.9
Ash	3.8	3.1	1.5

the extraction from these materials will become more and more important in the future. Therefore, the search for other applications like chemicals will also become more relevant.

3.4.1 Protein Hydrolysis

To extract peptides in the form of amino acids from rapeseed and sunflower waste streams, the protein has to be hydrolysed. In principle, proteins can be hydrolysed under acidic or alkaline conditions (Figure 3.5). Proteins in solution or dispersion can already start hydrolysing under mild conditions, for example at pH 8 at 50 °C. However, for complete hydrolysis into amino acids more severe conditions are necessary. Complete hydrolysis of proteins into amino acids can be achieved by heating proteins for 24 hours at 110 °C in 6 N HCl. But even under such severe conditions, some peptide bonds, such as valine–valine and isoleucine–isoleucine, will not be cleaved. Besides this, some amino acids are chemically modified under these circumstances. To prevent oxidation of cysteine into cysteic acid and of methionine into methionine sulfoxide, and the formation of chlorotyrosine, acid hydrolysis has to be carried out under vacuum or an inert atmosphere. In contrast, tryptophan is completely destroyed and threonine and serine are partially (5–15%) destroyed.

Nearly complete hydrolysis under alkaline conditions is possible at 100 °C, 5 N NaOH over 18–24 hours. Under these circumstances, tryptophan stays intact, but other amino acids like cystine, cysteine, serine and threonine are destroyed or damaged due to racemisation. In addition, lysine can be transformed in lysoalanine, which is not allowed for food applications.

Besides chemical hydrolysis, the molecular weight of proteins can also be decreased by the use of proteases. For example, in the food industry proteolytic hydrolysis is performed to increase the solubility and emulsifying or foaming properties. Basically, enzymes can hydrolyse a protein in two ways; exo enzymes can sequentially remove single amino acids from the end of the protein chain, whereas endo enzymes can rupture the internal bonds in a random manner at any point along the chains. To achieve high degrees of hydrolysis often combinations of exo and endo proteases are used.

Apart from being an endo or exo type, proteases have an additional level of specificity towards certain amino acids. For example, trypsin is an endoprotease that cleaves only peptide bonds in which an arginine or lysine amino acid residue is involved (Figure 3.6). This enzyme hydrolyses proteins only at a limited number of sites and therefore has a narrow specificity. However, most of the proteases are less particular about the amino acids encountered and have a broad specificity.

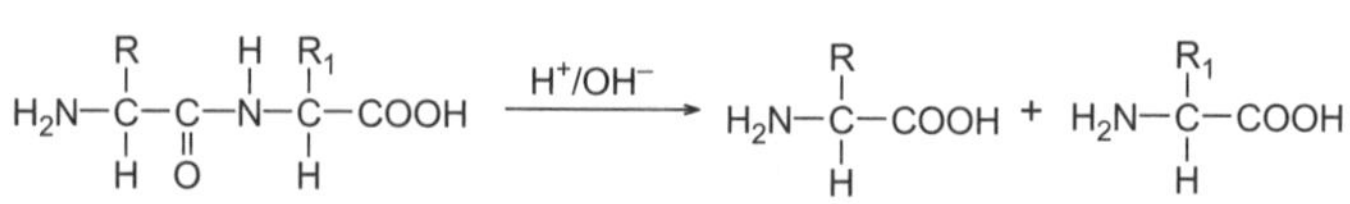

$$H_2N-\underset{\underset{H}{|}}{\overset{\overset{R}{|}}{C}}-\underset{\underset{O}{\|}}{C}-\underset{\underset{H}{|}}{\overset{\overset{H}{|}}{N}}-\underset{\underset{H}{|}}{\overset{\overset{R_1}{|}}{C}}-COOH \xrightarrow{H^+/OH^-} H_2N-\underset{\underset{H}{|}}{\overset{\overset{R}{|}}{C}}-COOH + H_2N-\underset{\underset{H}{|}}{\overset{\overset{R_1}{|}}{C}}-COOH$$

Figure 3.5 Cleavage of peptide bonds under alkaline or acidic conditions.

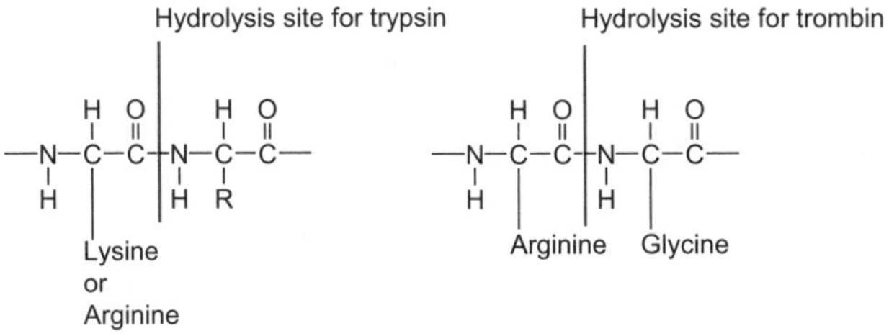

Primary Processing

Figure 3.6 Specificity of trypsin and trombin.

To produce hydrolysates with a high degree of hydrolysis, as mentioned, mixtures of different enzymes are used, but also cascade processes in which a series of proteases differing in specificity and optimal conditions are added. In a patent of Novo Nordisk[64] it is described that hydrolysates can be prepared having a degree of hydrolysis of up to 90%. However, in this case the dosage of the enzymes is rather high.

With respect to the hydrolysis of proteins to produce amino acids, the chemical pathway is cheaper and economically more feasible. Enzymatic methods may become more attractive in case of breakthroughs, like cheaper production methods or the development of less specific enzymes. Immobilisation of enzymes might also enhance the economic feasibility. The drawback of hydrolysis under alkaline or acidic conditions is that a lot of salt is being trapped in the amino acids/peptides after recovery. In addition, the severe conditions of the chemical hydrolysis can cause damage to certain amino acids. A non-conventional technique to hydrolyse proteins, by the use of solid acid catalysts is a promising option as it will not require the use of corrosive mineral acids. In this way, less salt is being generated and isolation of the amino acids is facilitated. In order to use solid bases or acids, protein solubility is required. This can be achieved by mild acid, base or enzymatic treatment.

3.4.2 Extraction Process of Peptides and Amino Acids

The recovery of amino acids and peptides from by-products of the oil production process can be performed by different techniques and scenarios:

- Extraction of proteins before oil recovery during pre-treatment;
- Extraction of amino acids and peptides from the press cake after oil recovery;
- Simultaneous extraction of amino acids, peptides and oil.

3.4.2.1 *Extraction before Oil Recovery*

When the main goal is to isolate proteins from a biomass source, proteins can also be extracted before oil recovery. During such a process, the pre-treated rapeseed or sunflower oilseeds are enzymatically treated with proteases, or chemically with acids or alkaline solutions in order to hydrolyse the proteins. To obtain optimal hydrolysis the seeds are milled to obtain non-defatted meals. After hydrolysis, both the amino acids/peptides and oil have to be isolated. A complicating factor in this can be the fact that low-molecular-weight fractions of proteins in general stabilise oil emulsions, which would negatively affect the recovery of the oil. Apart from this, it is also difficult to obtain peptide fractions that are free from oil.

An alternative way of recovering protein directly from oilseeds is by the use of ultra-filtration membranes in combination with reverse osmosis. This process uses less effluent and an increase in isolate yield was shown.

3.4.2.2 Extraction after Oil Recovery

Historically, the three most common processes for recovering oil from seeds are hydraulic pressing, the earliest processing method, and the currently used (combination of) expeller pressing and solvent extraction. As a solvent, hexane has been proven to be the most efficient and cost-effective means of recovering oil from oilseeds. Through these methods press cakes are formed from which peptide and amino acids can be isolated. Since the oil production is usually carried out at elevated temperature and/or in combination with hexane, the proteins will be partially damaged. The proteins will be denatured and chemical transformations will have taken place. The hydrolysis of the proteins will be negatively affected by this and it will be difficult to obtain free amino acids. Therefore, severe enzymatic or chemical hydrolysis processes will be necessary. Another possibility to recover the protein fractions from the dry press cake is by means of dry fractionation techniques (milling combined with air classification). By this, a fraction is obtained that is relatively rich in protein. From this fraction protein can be recovered by selectively dissolving the protein followed by hydrolysis into amino acids.

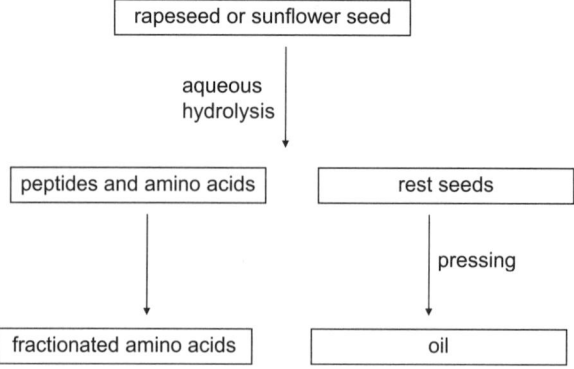

Figure 3.7 Production of amino acids from oilseeds.

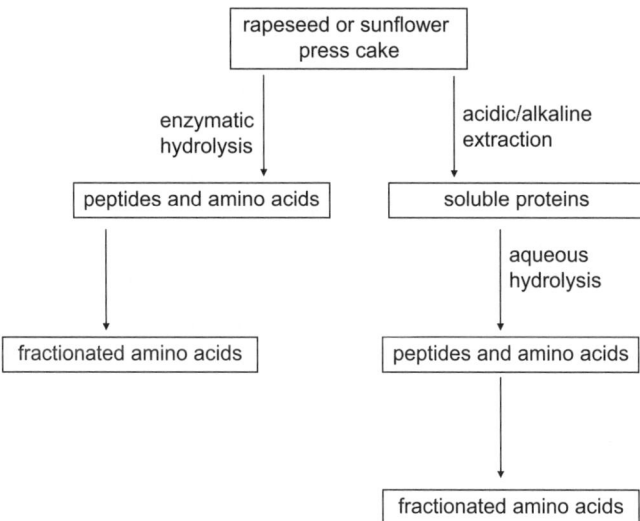

Figure 3.8 Production of amino acids from press cakes.

3.4.2.3 Simultaneous Extraction

Due to safety matters and environmental issues the industry is actively looking for alternative processes for hexane extraction. Aqueous extraction processing has gained considerable interest during the past decade as an environmentally friendly method for oil extraction.[33] Using this method the oil recovery is 65–75% whereas oil recovery using hexane is higher than 95%. Enzymes can be used to increase the oil recovery. Aqueous enzymatic extraction is a novel technology to isolate proteins and oil simultaneously. The combination of carbohydrases and proteases facilitates the separation of free oil. The general process to obtain protein from oilseeds is depicted in Figure 3.9.

A critical step that affects oil and protein yield in aqueous extraction is grinding of the starting material. Efficient extraction is possible when cell walls are broken down and small particles are formed. Regarding the extraction step itself, the ratios between the solid particles and liquid, pH, time, temperature, energy of agitation and the number of stages are important factors. In general, both high oil and protein yields are obtained when the extraction is performed away from the isoelectric point. Under acidic conditions protein concentrate is produced whereas under alkaline conditions isolates are obtained. Optimum pH conditions for oil and protein extraction of sunflower and rapeseed are respectively about pH 10 and pH 6.6. Protein yields in these cases are about 90%. Use of enzymes in general increases the oil and peptides/amino acids recovery. In respect to this, enzyme mixtures (pectinases, cellulases, proteases) with combined activity give the best results. Aqueous extraction in comparison to solvent extraction has an advantage in the way that higher quality of the protein is obtained.

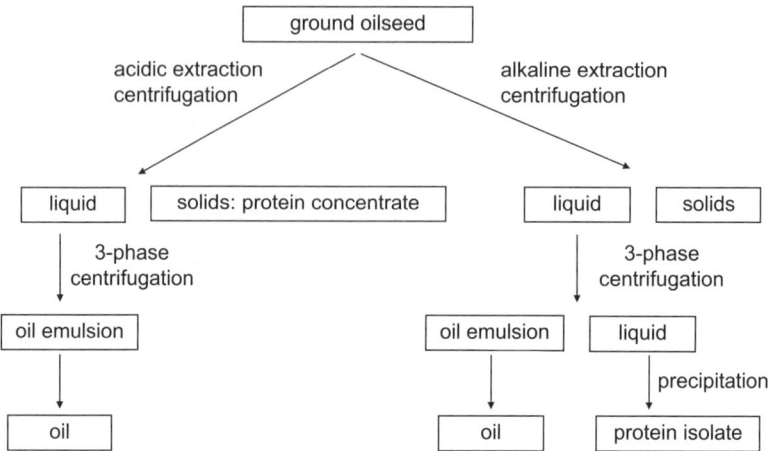

Figure 3.9 Simplified process for protein recovery.

To isolate protein after the extraction usually demulsification is needed and centrifuges are used to obtain the highest oil and recovery. After centrifugation usually three layers are formed consisting of oil, water and a solid phase.

Protein can be recovered from the aqueous phase and/or the solids. After the separation step, most of the proteins may be recovered as concentrate in the solid phase or as an isolate in the aqueous phase, depending on the pH of the extraction. When the extraction took place under alkaline conditions, protein was obtained by iso-electric precipitation. Despite the fact that the use of hazardous solvents is avoided, this method also has several drawbacks. Oil yields are still low compared to hexane extraction and therefore the residue and protein fraction can still contain a considerable amount of oil.

In several papers Zhang et al. investigated the aqueous extraction of rapeseed.[65,66] After wet milling, the slurry was treated with a combination of cell-wall degrading enzymes pectinase, cellulase and β-glucanase (4:1:1, v/v/v) at a concentration of 2.5% (v/w) for 4 hours. Alkaline extraction was performed at pH 10 for 30 minutes, followed by hydrolysis of the protein using Alcalase (1.5%) at pH 9 for 3 hours. After centrifugation and downstream processing the free oil and the aqueous phase, containing the hydrolysates, were obtained; yields for oil and protein were 75% and 80% respectively. Protein hydrolysates were composed of about 96% of peptides with a molecular weight less than 1500 Da, of which approximately 87% were less than 600 Da. The protein hydrolysates were purified using a macro-porous adsorption resin.

Protein isolate from defatted sunflower was subjected to hydrolysis using an endo protease, Alcalase, and an exo protease, Flavourzyme, resulting in a hydrolysate with a degree of hydrolysis of about 50%.[67] Another study was performed on the defatted oilseed sunflower whole meal. In this case a combination of Alcalase and Flavourzyme was again used for protein hydrolysis. The highest degree of hydrolysis that was achieved was 40%. Apart from the degree of hydrolysis, the amount of free amino acids was measured. It was revealed that the free amino acid found in the highest proportion in the hydrolysate was aspartic acid, which accounted for over 50% of the free amino acids present.

Hydrolysis of proteins from by-products of oil production results in complex mixtures containing many different compounds. To extract peptides and amino acids from these mixtures, reactive extraction, electrodialysis and ion exchange chromatography can be used. The separation of a mixture of amino acids into pure fractions can be achieved by selectively modifying one (or more) of the amino acids. Based on the modification process the isolation technique can be selected. For example, when the charge of certain amino acids is changed, electrodialysis could be an option.

3.4.2.4 Chemicals Derived from Amino Acids

The potential of biomass as a feedstock for energy supply or as a feedstock for chemical building blocks has recently obtained much attention, which is reflected by the publication of various roadmaps describing the (general) potential of biomass as a feedstock in the coming decades. Biomass consists of lignocellulosic and related materials (70–75% of biomass), oils and fatty acids (15–20%) and proteins (about 5%). These compounds comprise carbon, hydrogen, oxygen and nitrogen as elements. This implies that in principle almost all of today's important organic bulk chemicals can be completely or to a very significant extent based on biomass. In 2004 in the USA, the Pacific Northwest National Laboratory and the National Renewable Energy Laboratory published a roadmap study 'Top Value Added Chemicals from Biomass'.[68] From an initial list of more than 300 candidates, based on criteria such as current state of the art, known market data, properties and performance, this list was narrowed down to 12 chemical building blocks that can be derived from sugars and derivatives. As these are all derived from easily accessible carbohydrate resources (*e.g.* wheat, maize, potato, sugar beet) the perspectives for these building blocks will be similar in Europe compared to the USA. These platform chemicals are used to produce all the major bulk chemicals. The majority of the bulk chemicals can be produced based on six platform chemicals (ethylene, propylene, C4-olefins, benzene, toluene and xylene). On the short to medium term it may be expected that the chemical industry is more prone to adapt existing bulk chemicals based on biomass than new bulk chemicals based on biomass, having a unique structure, different from today's chemical building blocks.

Traditionally, the chemical industry has used fossil feedstocks for the production of chemicals. Due to the decrease in availability of fossil feedstocks and environmental issues, biomass is becoming an important raw material for the production of chemical compounds. Apart form these aspects, the use of biomass can also have advantages in synthetic pathways. The petrochemical industry uses simple molecules, like ethylene, to produce functionalised chemicals for which co-reagents, catalysts and many processing steps are used. Therefore, it is more efficient to make functionalised chemicals starting from materials such as amino acids, which already possesses functional groups.[69] Naphtha has a caloric value of 45 GJ per tonne. Process energy like heat and electricity are necessary to produce chemical products with a much lower caloric value than naphtha. Biomass-related materials, such as proteins and amino acids, have a caloric value of 20–30 GJ per tonne. This value is in general similar to chemical products. Thus, in comparison with the petrochemical approach, less energy is wasted. Additionally, fewer processing steps (less energy) are needed to synthesise functionalised chemicals.

In the literature chemical reactions in which amino acids are involved are diverse. Most of the reactions focus on the transformation of amino and carboxylic acid groups, for example the decarboxylation of α-amino acids.[70–73] In Table 3.8 chemical compounds that are derived from amino acids are given.

The conversion of lysine and phenylalanine in respectively ε-caprolactam and styrene is of much relevance for the chemical industry since these compounds are industrially produced on a bulk scale. The specific synthesis of N-methylpyrrolodone from glutamic acid is described by Lammens et al.[74] Notre[75] describes the synthesis of specific conversions to terminal alkenes starting from aliphatic carboxylic acids. This reaction can also be applied for the synthesis of acrylonitril from glutamic acid. A general method uses cross metathesis of olefins derived from amino acids e.g. by the enzymatic reactions of ammonia lyases.[76] At present, the processing of different biomasses, such as rapeseed and sunflower seeds, is mainly related to the feed and food area. However, it is expected that within a couple of years several biomass-related raw materials will be manufactured. The chemical industry can use these products as starting materials for the production of their polymers etc. This scenario assumes that different companies will be involved with the production of biomass-related products. For example, there are promising specific routes for Shell Chemical,

Table 3.8 Chemicals derived from amino acids.

Amino acid ⟶	Chemical
Alanine	Ethylamine
Asparagine/aspartic acid	2-amino-1,4-butanediol, amine tetrahydrofuran
Glycine	Nitric acid, oxalic acid
Leucine	Isoprene
Lysine	ε-caprolactam
Phenylalanine	Cinnamic acid, styrene
Proline	Pyrrolidone
Serine	Ethanolamine
Threonine	Isopropanol amine
Tryptophan	Catechol, muconic acid
Tyrosine	p-hydroxy-cinnamic acid
Valine	Isobutyraldehyde

Primary Processing 141

BASF and Lyondell. Biobased alternatives are possible by converting biomass into sugars, which subsequently can be modified to chemical products. Based on this approach, for Shell Chemical an alternative manufacture is possible for ethylene and butadiene. For BASF a biobased alternative can be set up for the production of hydroxyl-propionic acid and acrylic acid. For Lyondell a biobased route for the production of 1,4-butanediol, as well as a process for the isolation of amino acids from biomass, is possible. Brehmer[77] studied 16 different crops that were biorefined by various technologies to deliver chemicals in order to substitute the highest amount of fossil resources as possible. Very much depending on the composition of the crops, there are major differences in effective use of the agricultural land.

3.4.3 Conclusions

Traditionally, the production of amino acids involves the use of carbohydrates and nitrogen sources like ammonia. Therefore, it is interesting to investigate the possibilities of extracting peptides and amino acids from inexpensive protein sources like press cakes of oilseeds. The current processes are developed and optimised for oil recovery. When, besides oil, the extraction of protein fractions is also taken into account, changes in the processes are most probably necessary. After hydrolysis of the proteins and extraction of the different fractions, the fractionation into separate amino acids is crucial. When a large number and complex mixture of amino acids are generated chromatographic methods are required. It is clear that the chemical industry is more interested in the production of chemicals based on renewable resources and amino acids can play an important role in this.

3.5 Production of Levulinic Acid from Straw

3.5.1 Introduction

A major role as a platform chemical, obtained from oil extract residue straw, is levulinic acid or 4-oxopentanoic acid with the molecular formula $C_5H_8O_3$, molar mass 116.11 g/mol, density 1.1447 g/cm^3, melting point of 33–35 °C, boiling point of 245–246 °C, and solubility in water, ethanol and diethyl ether. It can also be obtained from cellulose and lignocellulose as discussed earlier.

A combine harvester can recover 60% of the rape straw from the field. Assuming a corn-straw ratio of rapeseed of about 1 : 2.9 at an average corn yield of about 3.5 tons per ha and year, about 10 tons of crop wastes remain on the field in the form of rape straw. After chaffing, a part of the rape straw is left on the field as humus forming substrate and nutrient supplier. The maximum recovery rate amounts to about 50–80% of the whole crop residues, *i.e.* to about 5–8 t/ha rape straw per year. In addition there is the possibility of extraction of residues from the press oilseed cake. Crude fibres extracted from press cake are an additional raw material for platform chemicals, such as levulinic acid. The crude fibres from rapeseed amount to 10.3% of the dry matter and from sunflower seed amount to 17.3% of dry matter.

During the conversion of cellulose-rich material, 5-hydroxymetylfurfural (HMF) is formed *via* acid-catalysed dehydration, which can be split *via* dehydration into levulinic acid and metanoic acid (formic acid). Even with raw materials that strongly vary in their quality from batch to batch, the yields that can be obtained are very high. The formic acid can be removed *via* distillation for further use. Today, hydrolysis processes of lignocellulosic material are carried out as a thermochemical (acid catalysed/steam assisted) of fungal (enzymatic) catalysed reactions. Acid catalysis is problematic due to the corrosive nature of the reaction media, which must be recovered before further steps and enzymatic reactions require longer residence times.

Levulinic acid is a versatile chemical intermediate. Levulinic acid is a potential precursor to nylon-like polymers, synthetic rubbers and plastics. It is a versatile synthetic intermediate, *e.g.* in the synthesis of pharmaceuticals. It is also a precursor in the industrial production of other chemical commodities such as methyltetrahydrofuran, valerolactone and ethyl levulinate (Figure 3.10).

3.5.2 A Short Survey on the Development of Levulinic Acid Chemistry

Since 1875, the basic chemistry and properties as well as the various synthesis routes of levulinic acid (4-oxopentanoic acid) have been investigated extensively. Although the potential of levulinic acid as an industrial chemical

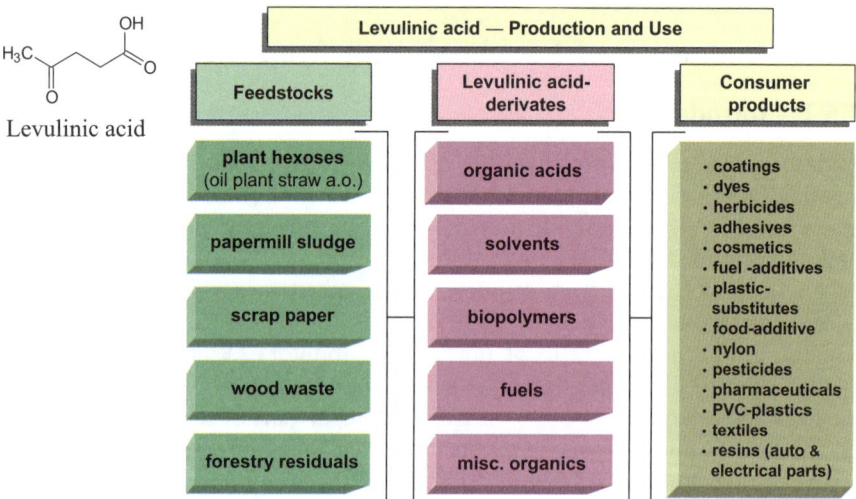

Figure 3.10 Production and use of levulinic acid.

Primary Processing

intermediate has been recognised due to the exceptional reactivity of the keto and carbonyl groups as well as of lactones, it has never been in commercial use in significant volumes.

In the 1940s levulinic acid was produced for the first time in the USA using starch and HCl in an autoclave. During this period several authors suggested the use of low-cost polysaccharide products including starch, cellulose and wood pulp as well as chitin, rice or soybean residues as hexose sources. Thomas and Barile[78] gave a comprehensive review of the various biomass feedstocks used until the middle of the 1980s.

During the 1970s the chemical community centred its attention again on levulinic acid as a chemical raw material[79,80] but only during the end of the 1990s did it seem that the alternative and cost-efficient production directly from biomass as demonstrated by the Biofine process[81,82] could overcome the encountered problems of expensive raw materials and low yields, excessive equipment costs and physical properties detrimental to easy recovery and handling. Ever since, levulinic acid production,[83-88] chemistry[89] and its derivatives[90] have been the centre of attention for researchers worldwide.

By 2005 the annual production of levulinic acid was estimated to be 454 tons at a sale price of about \$8.8–13/kg.[91] Since the price went down to \$3.2/kg in 2009,[92] this could have had an impact on production volumes.

The US Department of Energy selected levulinic acid from around 300 substances to be one of the 12 potential platform chemicals in the biorefinery concept.[93] The broad range of possible levulinic acid high-value derivatives has led to intense research efforts during the last decade. New synthesis routes that deliver chemical compounds of industrial relevance,[81,94,95] especially for applications as solvents,[90] monomers,[96-98] fuels[99] and fuel additives,[100-103] are being developed continuously in research groups and important industrial companies all over the world.

3.5.3 Levulinic Acid Production

Levulinic acid is a linear C5-alkyl carbon chain containing one carboxylic acid group in position 1 and one carbonyl group in position 4. It was first synthesised by A. Freiherrn von Grote and B. Tollens in 1875 by heating sugar candy in equal amounts with concentrated acid in water for several days. Formation of formic acid and water as well as large amounts of humin was observed during the reaction. The authors gave the name levulinic acid, since the levorotary fructose, called levulose, is the reactant for the acid generation.

Today the following four preparation routes for levulinic acid are of major interest:

i) Ring opening of furfuryl alcohol;
ii) Acid treatment of hexoses such as glucose, fructose, mannose or galactose from polymeric carbohydrates such as cellulose, hemicellulose or starch or mono- and disaccharides *via* formation of D-fructose and 5-hydroxymethylfurfural (5-HMF);

iii) Acidic treatment of pentoses such as xylose and arabinose from hemicellulose followed by a reduction step;
iv) Ozonolysis of unsaturated hydrocarbons from petrochemical raw material.

The preparation of levulinic acid *via* the ring opening of furfuryl alcohol was patented by Bernd and Guy and is currently applied in industry. The process proceeds by heating in the presence of water and a strong oxidising protonic acid and presents substantial cost and environmental problems. Furfuryl alcohol is obtained by the catalytic reduction of furfural that is produced industrially from sugarcane bagasse and corn cobs at a large scale.

In this review emphasis is put on the acid treatment of hexoses and its polymeric carbohydrates that occur in large quantities in nature. It seems to be the most relevant and promising reaction pathway for the future large-scale industrial production of levulinic acid.

3.5.4 Levulinic Acid from Hexoses *via* Formation of Fructose and 5-BHF

Principally, levulinic acid can be obtained by reacting fructose with any kind of acid at an increased temperature following the reaction mechanism *via* the formation of 5-hydroxymethylfurfural shown in Figure 3.12. Fructose is present in sucrose and can be obtained by the isomerisation of glucose as shown in Figure 3.11. Glucose is the second building block of sucrose as well as the hexose of other disaccharides such as maltose and cellobiose, which are cleaving products of amylose or starch and cellulose. Fructose can also be obtained by the isomerisation of mannose *via* glucose; however, this process is not commercially used. Mannose is both a structural and reserve carbohydrate

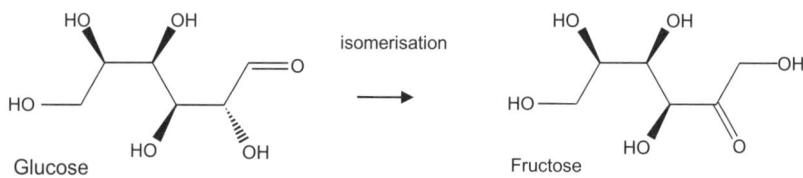

Figure 3.11 Isomerisation of glucose into fructose.

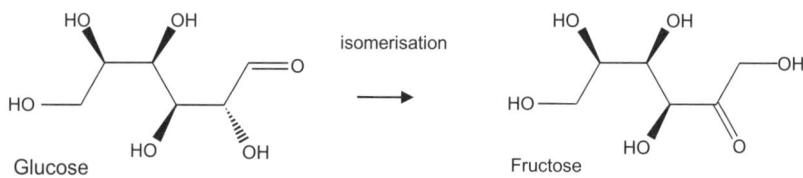

Figure 3.12 Formation of levulinic acid from fructose.

and is found in the form of mannan in plant bulbs, in the endosperm of seeds like ivory nut and guar gum as well as in seaweed and alga.

The isomerisation of glucose into fructose proceeds *via* the endiol form by the use of immobilised glucose isomerase. Today, the conversion of glucose for the production of high-fructose corn-syrups has become the largest immobilised biocatalytic process worldwide with annual production exceeding 8 MM tons/year.[104]

Heterogeneous isomerisation catalysts for glucose-fructose isomerisation that could integrate the acid catalysis of fructose into the valuable chemical intermediates 5-hydroxymethylfurfural (5-HMF) and levulinic acid are also the subject of study and could be a new breakthrough in the carbohydrate chemistry for chemical raw material production from biomass.

Levulinic acid production from fructose proceeds *via* the formation of 5-HMF, whereby yield and reaction velocity depend on the nature of the acid, acid concentration, temperature, pressure and the kind of solid-state acid catalyst.

Several kinetic studies starting from glucose for both treatments have been published. Also if the reaction mechanism in not yet fully clear and several reaction intermediates have been found, the authors[105] are in agreement that i) the carbohydrate is first hydrolysed by acid catalysis to form glucose, ii) 5-HMF is the intermediate formed from D-fructose by an overall first-order reaction *via* successive dehydration steps and iii) 5-HMF is finally hydrated and cleaved into levulinic and formic acid in equal molar ratio. Grethlein[106] presented a complicated reaction scheme for the conversion of lignocellulosic feedstock to levulinic acid on a molecular level. Timokhin *et al.*,[89] differently from other authors,[107,108] report detailed reaction schemes for the formation and degradation of the intermediate 5-HMF.

Recent kinetic studies performed by the team of Girisuta, Janssen and Heeres[109] at moderate temperatures confirmed the previous investigations and delivered a kinetic model for the acid-catalysed decomposition of glucose in a broad operating window (c[H_2SO_4]: 0.05–1 M, c[glucose]: 0.1–1 M, temperature: 140–200 °C. They found that higher temperatures favour humin formation. At lower temperatures the reaction becomes more selective for levulinic acid formation. The highest yield was obtained at higher acid concentration c[H_2SO_4] = 1 M, lower glucose concentration c[glucose] = 0.1–1 M and at 140 °C. Glucose was found to decompose in a consecutive reaction mode to give levulinic acid as the final product through 5-HMF as intermediate whereby both glucose and 5-HMF decompose in parallel mode to give insoluble humins as by-product. Their model implies that the highest yield of levulinic acid may be obtained in continuous reactor configurations by applying a dilute solution of glucose, high concentrations of sulfuric acid as the catalyst and by using a reactor configuration with high extent of back-mixing.

C. Chun *et al.* found that also if the formation of 5-HMF by neglecting the formation of intermediates can be modelled as consecutive reactions of overall first order the relative rate of 5-HMF formation with respect to glucose decomposition does not comply with the rate constant k_1 for the overall 5-HMF formation that increased with temperature and acid concentration.

They further concluded from the experimental results that the overall glucose decomposition rate increases more quickly than the overall 5-HMF formation rate and higher temperatures will promote glucose decomposition. They report that with glucose as a starting material at 170 °C and H_2SO_4 concentration of 5%, levulinic acid yields can reach up to 80.7% of theoretical yield.

Kinetic data for the levulinic acid formation from glucose at lower temperatures, namely at 98 °C, were reported by Tarabanko et al.[110] The authors report that at lower temperature the levulinic acid formation from glucose decomposition depends only slightly on the acid concentration, but the selectivity of the process strongly increases with rising acidity. Also large differences were found in the activity of the three acids investigated: HCl was found to be 12 times more active than H_2SO_4 at nearly equal selectivity, while H_3PO_4 was too weak for application as a catalyst. The authors attribute the higher conversion rate with HCl to the greater activity of protons in this acid.

Principally two different reaction regimes for the conversion of carbohydrate containing biomass into levulinic acid that differ in the reaction conditions but not significantly in the reaction pathway can be distinguished: i) the dilute acid treatment at high temperature and pressurised atmosphere and ii) the treatment with highly concentrated acid at lower temperatures and normal pressure. The low-temperature treatment of polymeric hexoses currently delivers maximum levulinic acid yields of 62.5% of theoretical values while the high-temperature treatment of carbohydrates like cellulose and starch delivers maximum yields of about 70–80% of theoretical values. A short description of the most significant processes for both treatments is given in the following sections.

3.5.4.1 Low Temperature and Concentrated Acid Dehydration Process

Hydrolysis of cellulose to hexoses at atmospheric pressure is usually performed with strong acids (HCl, H_2SO_4) at about 100 °C. In a second step the resulting hydrolysate is heated up to 110 °C with 20% HCl and kept at this temperature for 24–48 h. Free halogens, transition metals and anion-exchange resins can accelerate the reaction.[19] The reaction mixture is filtered to separate humin compounds and concentrates. Levulinic acid is isolated by distillation at reduced pressure or by extraction with ether, ethyl acetate or ethyl methyl ketone. The yield of levulinic acid is about 40% with respect to hexose content.

Dahlmann reacted the polymeric hexoses in a ratio 1:10 with 20% HCl at 108 °C for 6–8 h. Humin compounds were filtered off and the reaction mixture was concentrated to one-fifth of the initial volume by distillation. An appropriate solvent was added to the reaction mixture for levulinic acid extraction. After extraction the solvent was distilled off and the levulinic acid was distilled at reduced pressure. Dahlmann reported a yield based on hexose content of about 50–60% based on hexose from lignocellulosic feedstock and of about 57–60% from tapioca flour. These yields are by far the highest ever reported for the low-temperature treatment.

Ion-exchange resins were also tested as acid catalysts but the reaction rates were low.

3.5.4.2 High Temperature and Diluted Acid Hydration Process

The reaction when performed at higher temperatures delivers a considerable gain in time and permits the use of lower acid concentration. The effectiveness of the tested acids was HBr > HCl > H_2SO_4. Zeolites were found to manifest high catalytic activity in the preparation of levulinic acid from glucose and fructose during the high-temperature conversion.[60] Recently developed processes reach levulinic acid yields of about 60–70% based on the hexose content of the starting material. A brief summary on the most significant processes developed during the last two decades is discussed in the following paragraphs.

A breakthrough in levulinic acid production from lignocellulosic feedstock, called the Biofine process, was made possible in 1988 by Fitzpatrick.[82,111] In the process a novel reactor configuration is used that promotes levulinic acid production at high temperature through acid catalysed hydrolytic breakdown of cellulose to form levulinic acid while reducing char formation. Yields of levulinic acid from cellulose with H_2SO_4 as acid catalyst of over 70% of theoretical yield have been attained. The reactor system consists of a plug flow reactor followed by a lower temperature, completely mixed reactor. The conditions, namely 210–220 °C, 25 bar pressure, 12 s residence time, acid concentration 1.5–3%, in the first state favour the dominant first-order, high-temperature, acid-catalysed hydrolysis of cellulose and hemicellulose to soluble hexoses and pentoses intermediates. The completely mixed conditions in the second-stage reactor favour the first-order reaction sequence leading to levulinic acid at the expense of the higher order condensation reactions leading to tar. Typical conditions are CSTR mixing configuration, 190–200 °C, 1.5–3% acid concentration, 14 bar pressure and residence time 20 minutes. Additionally, the reaction conditions in the first stage followed by vapour separation in the second lower pressure stage favour high yields of furfural from the hemicellulosic fraction of the feed. Advantages of the Biofine process over other technologies are: i) short residence time, thus small reactor volume at high throughput; ii) high feedstock flexibility within a wide range of low-grade variable composition cellulosic feedstock, iii) continuous process control and iv) ease of scale-up. Moreover the process uses low-cost acid catalyst that is recycled within the process. By the use of only dilute mineral acid catalysed hydrolysis the process is unaffected by contaminants often found in waste feedstock and is thus very robust. Finally the process is reported to be energy self-sufficient.

A summary of the various pilot plants of different size that have demonstrated the feasibility of the process as well as an outlook on future commercial development are given in the following paragraphs.

Ghorpade and Hanna found that processing starch acid mixtures in a heated extruder increased levulinic acid yields due to increased acid hydrolysis of the starch by the extrusion process whereby lower screw speeds, namely a higher residence time, increased both glucose and levulinic acid yields. The preparation of levulinic acid from starch containing 20–70% amylose by a continuous reactive extrusion process was patented. The extrusion takes place in a twin-screw extruder having a plurality of temperature zones in the range of 80–160 °C. At a first stage the starch slurry is formed from 57–64 parts starch, 30–40 parts water and 3–6 parts mineral acid (HBr, HCl or H_2SO_4, H_2SO_4 being the most preferred) and pre-conditioned for 3–4 hours. The pH of the slurry is 1 and the temperature reaches 100 °C at the end of the pre-conditioning step. Successively the slurry is fed to the extruder where the temperature at the die is 150–160 °C. The residence time inside the extruder is about 100 s. The extrudate is filter pressed and vacuum distilled. By this technology the authors obtained levulinic acid yields of about 70% of theoretical.

Fang and Hanna proposed a method for preparing levulinic acid in a pressurised reactor using abundant and low-cost whole kernel grain flour as raw material. They observed that yields increased with increasing sulfuric acid concentration (2–8%) and temperature while the flour loading has to be kept quite low. At 200 °C, acid concentration of 8% and flour loading of 10% they obtained a maximum yield of 32.6% of levulinic acid based on raw material. Data on levulinic acid yield based on theoretical yield are unfortunately not given.

L. Yan *et al.* used bagasse and paddy straw for the production of levulinic acid. They reached maximum yields of up to 80% at 220 °C, 45 minutes reaction time and an HCl concentration of 4.45%, whereby the amount of water plays an important role. Too little and too high amounts of water had an adverse effect on levulinic acid concentration. They report to have found three by-products of great potential in considerable quantities, one being for sure formic acid, but without revealing the nature thereof.

3.5.5 The Bofine Demonstration Plants and Outlook on Future Industrial Scale Facilities

The Biofine process has been refined and expanded for over two decades, with initial test work being conducted during 1986–1987 at the National Renewable Energy Laboratory in Golden, Colorado, and funded jointly by Biofine and the New York State Energy Research and Development. At that time, Biofine developed a system that produced high yields of levulinic acid and furfural from biomass, and demonstrated the process over a wide range of operating conditions.

Successively a reactor system was installed at Dartmouth College in New Hampshire under a grant from Biofine and NYSERDA and operated in the period 1988–1996.

In 1997, the reactor system was scaled up to processing 1 dry ton per day of biomass in a pilot plant built at Epic Ventures, Inc., in South Glens Falls, New York. Paper sludges from nearby paper mills were initially used as pilot plant feedstocks and gave levulinic acid yields ranging from 0.42 to 0.595 kg per kilo of cellulose (between 59 and 83% of the theoretical maximum yield). The work

Primary Processing 149

was supported under a grant from the US Department of Energy and NYSERDA and co-funded by Biofine. This project demonstrated that high yields of levulinic acid of about 70% were obtainable in a multi-day operation at an increased scale.

In 2006, the 1 ton-per-day plant was moved from New York to its current location in Gorham, Maine, where the plant is capable of sustained multi-day operation at 2 ton-per-day capacity. The Biofine pilot plant in Gorham, Maine, is designed for development, process testing and feedstock evaluation, not for production. The outputs are used for the validation of yields of products and by-products from the feedstocks being tested.

A 50 ton-per-day facility is currently being planned for Old Town, Maine, which will use forest biomass from the same sources that supply large biomass-fired power plants throughout the region. That facility will be focused on large-scale production of ethyllevulinate for use as home heating oil. For its realisation Biofine has recently applied for a grant from the Department of Energy that focuses on integrated biorefineries.

Economic projections indicate that the levulinic acid production cost could fall to as low as $0.08–$0.22/kg depending on the scale of operation.

A further commercial plant for levulinic acid production is built in Caserta, Italy, and converts 3000 t/day of raw material (waste sludge from tobacco and paper industry).

3.5.6 Technology Draft for a Low-temperature Conversion Process of LCF to Levulinic Acid

Taking into account the surprisingly high yields of about 50–60% of theoretical of levulinic acid obtained from the low-temperature concentrated acid conversion of lignocellulose and starch feedstock obtained by Dahlmann and the lack of recent studies, we think it should be worthwhile to reconsider this technology from the point of view of industrial processing of LCF.

The technology will be studied for the conversion of straw from different crops (wheat, barley, rye) and residues from rape oil production that has a raw material price of about two-thirds relative to crop straw.

Figure 3.13 reports the technology draft derived from the experimental work of Dahlmann for the production of levulinic acid at about 110 °C and normal pressure through the hydrolytic breaking of the cellulosic material contained in wheat straw by highly concentrated HCl solution.

A 20-litre reactor is fed with 1 kg chopped wheat straw of about 36.7% cellulose content and 10 kg hydrochloric acid at a concentration of 20 wt%. The highly acidic straw slurry is kept at 108 °C for 6–8 hours under reflux and successively cooled down to 30–35 °C. The insoluble humin compounds are separated by filtration and washed with a little warm water. The washing water is added to the filtrate and the mixture is concentrated to 2 litres by distilling off acidic water containing formic acid as a by-product and HCl. Formic acid will be recovered by solvent extraction and successive distillation while hydrochloric acid will be recycled to the process.

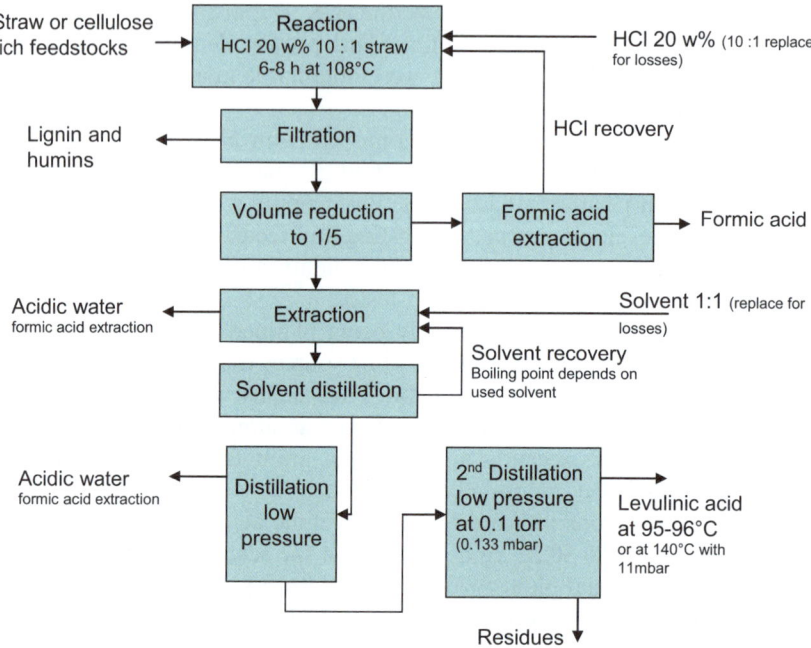

Figure 3.13 Process scheme for low temperature concentrated acid conversion of carbohydrate biomass to levulinic acid as main product and formic acid as valuable by-product.

The levulinic acid will be recovered from the concentrated filtrate by solvent extraction, whereby an appropriate solvent is added at a solvent/filtrate ratio of 1:1. After 15–20 hours of extraction, the two phases are separated. Levulinic acid is obtained by distilling off the organic solvent and successive two-fold vacuum distillation at 94–115 °C and 95–96 °C at 0.133 mbar. According to Dahlmann, a yield of 54.6–60.3% of levulinic acid based on the theoretical yield due to the initial cellulose content should be achieved in this way.

3.5.7 Outlook

Further research is necessary to study new methods of pre-treatment of biomass for production of levulinic acid within the LCF biorefinery system. Sub- and supercritical water have been gaining increasing attention as attractive solvents and reaction media for a variety of applications. It is cheap, non-toxic and non-flammable and offers some other advantages in the field of 'green chemistry'. At high temperatures water develops acidic characteristics and accelerates cleavage of chemical bonds and lignocellulose decomposition. One of the major motivations in hydrolysis of polymeric sugars in hemicellulose and cellulose is to implement sub- and supercritical water as environmentally friendly solvent in the pre-treatment stage.

3.6 Integrated Biorefinery

This section focuses on changes that could be made in existing plant oil mills to improve the recovery of valuable components of oilseeds while conserving the main unitary operations of the processing. Each technology will be considered to assess the opportunity of its adoption according to a short SWOT analysis.

3.6.1 Dehulling

Dehulling involves removing the fibrous envelopes of the seeds. The most common techniques involve controlled hitting to break the seeds and then proceeding to separation by size or in airflow. Without pre-treatment, these techniques are not very efficient since modern seeds have very high oil content and their meats (kernel) are fragile. In the case of sunflower, lots of cultivars have small seeds and adherence between kernels and hulls.

Dehulling can be considered for two purposes:

- Supply the mill with biomass for thermal energy production;
- Purification of the kernels for high-protein meals, with fewer fibres and less lignin and avoiding the presence of polyphenols responsible for colour problems and other disadvantages when meals are used to process protein concentrates or isolates.

In the first case, the percentage of hulls removed can be adjusted according to the amount of energy that can be used by the mill. When this amount is limited to the need of the thermal power for internal steam production only a few percent of the available mass is required. Using a larger percentage of lignocellulosic material it is possible in a high-pressure boiler to generate electricity and low-pressure steam for the process.

Considering the development of sunflower proteins concentrates or isolates would require near to 100% dehulling, this performance could only be achieved in a plant where two lines could function in parallel, the first one carrying out dehulling and returning partially dehulled material to the second one, which would crush whole seeds and partially dehulled seeds. The first line would produce only a small fraction of the overall. Since the protein market is much narrower than the feed market, this discrepancy won't be a real problem.

In the case of rapeseed, the main problem is the high oil content of the hulls fraction. Hulls contain naturally up to 10% of oil and imperfect dehulling allows more oil to be kept because of poor separation.

3.6.1.1 SWOT Analysis for Different Dehulling Strategies

Scenario 1: Minimal Dehulling Sunflower Seeds (Steam Production) Table 3.9
This technique is operational and doesn't require specific research to be implemented. From 4 to 8% of the mass of the seeds extracted is to be burned

Table 3.9 SWOT analysis minimal dehulling of sunflower seeds.

Strengths	Weaknesses	Opportunities	Threats
Techniques are available. Capital cost.	Technique for SF only. Lack of room in existing plants. Production of wastes (ashes).	Significant impact on GHG emissions. Improvement of meal quality.	Ability of SF seeds to dehulling (breeding for oil content has negative impact on hullability).

Table 3.10 SWOT analysis minimal dehulling of rapeseeds.

Strengths	Weaknesses	Opportunities	Threats
Partial oil recovery of the hulls is possible.	Lack of room in existing plants. Oil losses. Quality of oil recovered.	Significant impact on GHG emissions. Improvement of meal quality.	Sulfur in the hulls (emissions in flue gas).

in a fitted boiler, which supplies the mill with 10–12 bars of steam. Practical considerations prevent its generalisation: the technique works only for sunflower when most of the mill shifts from rapeseed to sunflower according to the market and the dehulling equipment necessitates much more room than pre-pressing and extraction plants. On the other hand, it could enable biodiesel production to strongly reduce its greenhouse gas emissions.

Proposed strategy: R&D to improve dehuller efficiency and reduce the size of the separators.

Scenario 2: Minimal Dehulling Rapeseeds (Steam Production) Table 3.10
In theory, rapeseed could be dehulled and its hulls could be used in the same fashion as sunflower hulls as fuel for steam production. Unfortunately, rapeseed hulls contain at least 10% of oil and its separation is not very efficient. Partial recovery of this oil by an expeller could limit the losses and also ease handling problems.

Proposed strategy: R&D to improve dehuller efficiency and reduce the size of the separators.

Scenario 3: Optimal Dehulling Sunflower Seeds and Rapeseeds for Energy Supply (Cogeneration) Table 3.11
Since the energy contained in the hulls exceeds the needs of oil mills for their steam consumption, it is possible to use this extra energy to co-generate steam and electricity. In this case the oil mill could become an energy producer. In the case of OSR maximisation of dehulling could lead to 80–100% dehulling. For SF, only 60 to 80% of hull removal seems feasible. In the state of the art, these performances are not realistic because of excessive oil losses during hull

Table 3.11 SWOT analysis optimal dehulling of sunflower seeds and rapeseeds.

Strengths	Weaknesses	Opportunities	Threats
Acceptable oil losses and high dehulling yield.	Lack of room in existing plants. Energy consumption for pre-treatment.	Significant impact on GHG emissions. Large improvement of meal quality. Demand for electricity from renewable sources.	Sulfur in the hulls (with OSR). Failure to reach this ambitious objective. Reversal of policy concerning bonus for renewable electricity.

Table 3.12 SWOT analysis maximal dehulling of sunflower seeds and rapeseeds.

Strengths	Weaknesses	Opportunities	Threats
Use of available technologies. Low losses of oil. High purity of the kernels.	Availability of surface to build secondary processing unit. Technology able to desolventise the meal without damaging the proteins. Polyphenols presence and colour problems	Reduction of GHG emission. Development of new products with cheap raw material. International demand for protein of food grade. Alternative proposal to soy protein.	Volatility of the prices. Competition with soy protein.

separation. Pre-treatments like drying could improve the yields but at a cost in energy. Further research is required to overcome these technical barriers.

Proposed strategy: study of pre-treatments aiming at improving hullability and research for improved decorticators and separators.

Scenario 4: Maximal Dehulling Sunflower Seeds and Rapeseeds for Protein Quality and Energy Production Table 3.12
In this scenario, the purity of the kernels fraction needs to be maximised and the protein quality must remain intact. Pre-treatment of the seeds targeting improved hullability could damage the proteins and the size of the market for high-value protein is much smaller than the one for feed. In consequence, we assume that proteins would be produced on a secondary plant contiguous to the existing one. This plant would process pure kernels extracted with poor yield from the main stream of seeds dehulled at large scale. After hull removal, kernels are isolated by slotted sieves in the case of rapeseed and by a combination of sieves and densimetric table in the case of sunflower. The amount of hulls actually extracted would be used as in scenarios 1 and 2.

Proposed strategy: R&D for cold pressing of pure kernels in order to maximise mechanical extraction of oil. By reducing the oil content of the material entering the extractor, it will be possible to reduce the amount of solvent kept in the marc and, therefore, the energy required to evaporate this hexane. Significant improvement of the desolventisation is expected. Further development of new processes to bring the residues of phenolic compounds to an acceptable level is required.

3.6.2 Cold Pressing

Cold pressing Table 3.13 could help to reduce the consumption of energy of oil mills and can be combined with dehulling. The technique was proposed by the Krupp Company in the 1980s but failed to match the requirements for cake extraction. Cold pressed cake contains all the water of the seeds so it is too wet to be sufficiently structured and to allow acceptable solvent percolation. More recently, cold pressing was considered again in order to supply crude oil with low phosphorus content for direct use in diesel motors. These presses were developed in order to maximise the oil yield. With reduced oil content, granulation could help to overcome the structure problem and cold pressed cake could be used with current extractors. Cooking prior to pre-pressing consumes approximately one-third of the energy employed for oilseed extraction. Given the efficiency requirements in the biofuel sector, cold pressing could enable the crusher to reach its objectives by introducing only a minor change in their processing scheme. An interesting spillover of the technique is a strong change in the phospholipids content of the crude oil: very low in oil stemming from pre-pressing and very high in extraction oil.

Proposed strategy: study the solvent extraction of cold pressed cake at pilot plant level in order to demonstrate the feasibility of the technique and study the consequences of higher water content of the cake on desolventisation.

3.6.3 Improvement of Meal Quality by Significant Reduction of Hexane Retention in Marcs

Marcs retain about 25–30% of solvent when they go out of extractors. Most of this solvent has taken the place of the oil that has been extracted and its concentration depends upon the oil content of the cake. By reducing this oil content, it would be possible to decrease the charge of the desolventiser and, in

Table 3.13 SWOT analysis cold pressing.

Strengths	Weaknesses	Opportunities	Threats
Simplicity of the technique. Impact on GHG. Quality of cold pressing oil.	Capital cost (reduction of throughput of the presses). High moisture of meals.	Exigencies of policy makers about energy efficiency. Possibility to benefit from grants for investment.	Possible problems of meal quality because of high level of water content of the cake.

Primary Processing

Table 3.14 SWOT analysis reduction of hexane retention in marcs.

Strengths	Weaknesses	Opportunities	Threats
Increased capacity of existing extractors and desolventisers. Increased use of RSM for monogastrics. Possibility to recover proteins for food applications. Improved energy efficiency. Possible gains on oil quality (phospholipids degradation).	Capital costs (reduced capacity of the presses). Not available. No strong market incentives. Presence of glucosinolates residues.	New markets for RSM. Global demand for high-quality proteins.	Market prices for improved RSM. Competition with soy meals. Insufficient quality gain. Possible ban of hexane.

Table 3.15 SWOT analysis supercritical CO_2 extraction.

Strengths	Weaknesses	Opportunities	Threats
Low cost of the solvent. Absence of toxicity and regulation hurdles. Renewable solvent. Low-temperature extraction.	Not available at affordable cost. Capital cost. Electricity consumption?	Possible ban of hexane. Compatible with 'organic' labelling.	Competition with other extraction techniques.

consequence, the heat damage on proteins Table 3.14. Available presses are able to produce cakes with less than 12% of oil provided that the seeds were correctly pre-treated. As the goal is to preserve the protein quality, it is not possible to increase the cooking intensity before the pressing.

By working simultaneously on pre-treatment and press technology, it should be possible to achieve better deoiling at the pre-pressing stage without damaging the proteins and without degrading energy efficiency. The oil content of the cake could be adjusted so that the intensity of desolventisation could achieve an optimum between glucosinolates destruction and protein degradation.

Proposed strategy: development of new equipment for pre-pressing and definition of the optimum desolventisation conditions.

3.6.4 Supercritical CO_2 Extraction

Supercritical CO_2 extraction does not function continuously and has large costs to be useful for extraction of common vegetable oils Table 3.15. New developments suggest that this technique could be used continuously with significantly lower implementation costs. Due to the high diffusivity and low viscosity of supercritical CO_2, extraction could be carried out with shorter residence time than with hexane. A major difficulty to overcome is the continuous passage of the material through high pressure and simultaneous release of this material from high pressure to atmospheric pressure. Thanks to the screw press, it is possible to reach the required pressure in order to have permanent plugs at each extremity of the extractor. The design of the extractor must be able to cope with the necessary residence time, the solvent recovery and safety requirements due to the high pressure.

The advantages of SFC are well known: absence of toxicity and explosion risks, low operational temperature, cheapness and the renewable character of the solvent. However, the technical constraint to maintain pressures over 100 bars is not very easy to cope with. A major effort in design of machinery is needed to achieve this goal.

Proposed strategy: costing an operational supercritical CO_2 extraction facility to estimate its viability in race condition against techniques like double pressing. If that economical study is favourable, we may consider working on equipment design.

3.6.5 Gas-assisted Oil Pressing

German researchers have recently published works on gas-assisted oilseed pressing Table 3.16. The principle of the invention is to help the deoiling by injection of CO_2 in the press barrel at high pressure. The gas at this pressure behaves as both a liquid and a gas. It diffuses in oil and lowers its viscosity. The proportion of liquid in the cake exiting the press doesn't change strongly but this 'liquid' is a mixture of oil and CO_2, which under atmospheric pressure evaporates resulting in decreased oil content in the cake. According to available information, the concept was never developed.

Harburg Freudenberger, a German press builder, in conjunction with Crown Iron Works, is developing a system named HIPLEX for gas-assisted mechanical extraction at an industrial scale.

3.6.6 Use of Alcohols as an Alternative for Hexane

Concerns about health and environment in relation to hexane could lead to its replacement by less harmful solvents like ethanol or isopropanol Table 3.17.

Table 3.16 SWOT analysis gas-assisted oil pressing.

Strengths	Weakness	Opportunities	Threats
Low cost of the solvent. Absence of toxicity and regulation hurdles. Renewable solvent. Good protein solubility in the meal.	Performances not available. Recovery of CO_2?	Possible ban of hexane. Compatible with 'organic' labelling.	Competition with other extraction techniques.

Table 3.17 SWOT analysis use of alcohols.

Strengths	Weaknesses	Opportunities	Threats
Solvent less harmful than hexane.	Unproved efficiency at pilot plant scale.	Possible ban of hexane.	Safety issues about flammability and explosivity.
Possibility to improve the feed quality of the meal.	Capital cost of equipment for solvent recovery.	Requirement to reduce the use of chemical pesticides.	Lack of renewable character of IPA.
Possibility to recover glucosinolates.	Unproved interest of glucosinolates for plant protection.	Development of membranes technologies.	
Possible retrofitting of existing plants.			

Isopropanol seems to be the best choice because it doesn't require low water content in pre-press cake as ethanol. Contrarily to hexane, alcohols are not highly miscible in oils so that extraction must be carried out at near to boiling point temperature and oil/solvent separation can be made by cooling the miscella. Due to this property, the higher vaporisation enthalpy should not result in high energy consumption for miscella evaporation.

However, the solvent/cake ratio must be increased to face the lower oil content of the miscella and the desolventisation of the marc will require more energy than with hexane. In conclusion, the overall energy requirement should not be decreased with the use of isopropanol.

An interesting spillover of this substitution is the extraction of polar compounds with the oil extraction. This effect is only significant if the lean miscella resulting from the cold decantation gets rid of its polar extracts, otherwise the solution becomes saturated and extraction stops. Separation by membranes is an energy-efficient way to recover these substances.

The rapeseed meal resulting from isopropanol extraction has higher protein content (33.5 to 37.5%), lower glucosinolates residues (from 25 μmol/g down to 10 μmol/g before toasting) and lower oligosaccharides. From a nutritional point of view this meal is significantly improved.

In the extract, valuable molecules such as glucosinolates could be recovered and used as biopesticides. Sugars could be fermented to produce alcohol or burned to supply energy for the processing.

Proposed strategy: further investigations on the feasibility of recovering glucosinolates from the non-lipidic extract, verification of the animal feed interest of IPA extracted meals, pilot plant scale in continuous setting up of the technique.

3.6.7 Simultaneous Extraction and Transesterification

In theory, it would be possible to skip the step of extraction of vegetable oil before carrying out the transesterification by immersing flaked oilseeds in ethanol and catalyst and then by using ethanol for extraction of ethyl esters of fatty acids Table 3.18.

Ethyl esters would be more environmentally friendly because methanol stems from petrochemistry when ethanol is produced by fermentation of biomass. Moreover, this method is hexane free.

A pilot plant with a capacity of 10 000 t/annum is being set up in France near La Rochelle. This pilot should solve important problems such as managing moisture of the seeds which must be low and the difficulty to percolate the solvent in oil-rich flakes. No information is available about the success of the project, which, given the announced schedule, suggests that these difficulties may have given more trouble than expected.

In Brazil, Petrobras has patented a method for simultaneous extraction of castor oil with ethanol, but the technique has not been industrialised.

Proposed strategy: wait until the final results of La Rochelle experience.

3.6.8 Isolation of Oil Bodies (Oleosomes) Table 3.19

In oilseeds, triglycerides are stored in specialised organites called oleosomes and characterised by a mono-molecular membrane including phospholipids and specific proteins: the oleosines. These proteins have an amphiphilic constitution, which gives them emulsifier properties. Practically, their presence explains why water extraction of oilseed oils has not yet proved its feasibility since a significant amount of oil can be kept in a very stable emulsion and

Table 3.18 SWOT analysis simultaneous extraction and transesterification.

Strengths	Weaknesses	Opportunities	Threats
Substitution of methanol by ethanol. No primary extraction step. Hexane free.	Requires very low water content in oilseeds. Difficult to make continuously because of poor percolation of oil-rich flakes.	Obligation to improve the carbon footprint of biofuels production. Better public image.	Technical hurdles. Competition with other techniques (ethanol extraction of pre-press cake?).

Table 3.19 SWOT analysis isolation oil bodies.

Strengths	Weaknesses	Opportunities	Threats
Creation of new products. Limited number of technical steps (cheap). Possibility to recover some soluble proteins?	Non-existent market. Recovery of insoluble material necessitates drying or fermentation.	High-value niche market (cosmetics). Environmentally friendly.	Insufficient market. Competition with soybean.

Primary Processing 159

requires complex additional separation steps. Rather than develop complex demulsification techniques, some thought they could sell these emulsions.

If marketing these emulsions as a food ingredient seems possible, as with some applications in lubricants and cosmetics, it seems unlikely that their market share could go beyond a niche.

At the same time, water extraction requires entirely different processing and has little similarity to existing facilities. Small-scale water extraction plants could be developed beside existing mills in order to supply new products such as proteins isolates, emulsions and intermediates for further transformations.

Proposed strategy: marketing approach and R&D to develop uses of emulsions in foods and beverages.

3.6.9 Water Extraction

Water extraction Table 3.20 is the standard processing of fruit oils (olive, palm) but doesn't function for oilseeds because of emulsion formation during wet milling. Several methods are proposed in a scientific paper to solve this problem. In Germany, GEA has developed the Friolex process, in which an alcohol in a mixture with water is used to help demulsification. The nature of the alcohol used is not published. American researchers have proposed to use proteases to digest the oleosines and several alternatives have been published to address the emulsions issue.

From a practical point of view, the Friolex technique seems the one that could function with fewer new products to manage since alcohol avoids protein solubilisation. Resulting meals are likely to have the same composition than in regular processing if the water/alcohol mixture is not retreated and with a purification step if the liquid, glucosinolates and oligosaccharides are to be removed.

Compared to current processes, such a method would avoid hexane but requires more energy for drying.

Proposed strategy: economical analysis of the Friolex technology and verification of its actual efficacy.

3.6.10 Anaerobic Digestion of Residues

Methanation of biomass is a way to recover energy from waste residues, especially when the moisture content of these products is high and direct combustion is not possible Table 3.21.

Table 3.20 SWOT analysis water extraction.

Strengths	Weaknesses	Opportunities	Threats
No hexane. No new products to deal.	Energy required for drying meal. Possible degradation of protein quality.	Possibility to recover soluble material to improve meal quality and recover glucosinolates.	Patent cost.

Table 3.21 SWOT analysis anaerobic digestion of residues.

Strengths	Weaknesses	Opportunities	Threats
Better energy efficiency than aerobic digestion.	Feeble quantity. Relatively higher capital cost than classical water treatment.	On-site use of the gas.	

Compared to direct combustion, methanation requires digesters with a relatively low concentration of dry matter (around 20%) and duration of treatment of at least 20 days. In a 600 000 t/year facility, the methanation of 10% of the entering seeds would require the treatment of 180 t/day of hulls at 10% DM, *i.e.* filling a near to 1000 m^3 digester and at the end of cycle recovering and treating digestion residues. By comparison, combustion just requires a burner and generates only ashes.

Methanation seems better suited to process moisture-rich wastes than dry ones. Nevertheless, at a lesser scale, an oil mill produces liquid effluents that could be used in an anaerobic digester to produce methane. Most of the organic matter comes from scrubbers, which remove dust in the vapours of the desolventiser before the condensers. The order of magnitude of this matter represents only 0.1% of entering seeds.

In the frame of water extraction, methanation of aqueous residues could become a way to valorise organic mater of lower value.

3.6.11 Recovery of Gums from Water Degumming

Water degumming consists of removing hydratable phospholipids from crude plant oils before storage in order to avoid deposit of these gums. Currently, gums are reincorporated in meals because of their water content, which makes them unstable, and because the market for lecithin cannot absorb all the potential production, therefore their recovery is not worth the cost of drying.

These gums contain phospholipids and triglycerides that could be valorised for fatty acids supply. Acidulation, which is the normal way to produce acid oils, cannot be carried out on gums because of the tension-active properties of the product. New technologies could help to improve the valorisation of this material for biofuels.

3.6.12 Integrated Scheme Biorefinery

Based on the information that is available from the literature or from oil-producing companies, several changes were proposed to adapt the traditional oil-recovery process in order to increase the value and optimise the utilisation of the oil-bearing crops. In Figure 3.14 a possible integration of such a biorefinery scheme in the existing industry is depicted. This is a simplified

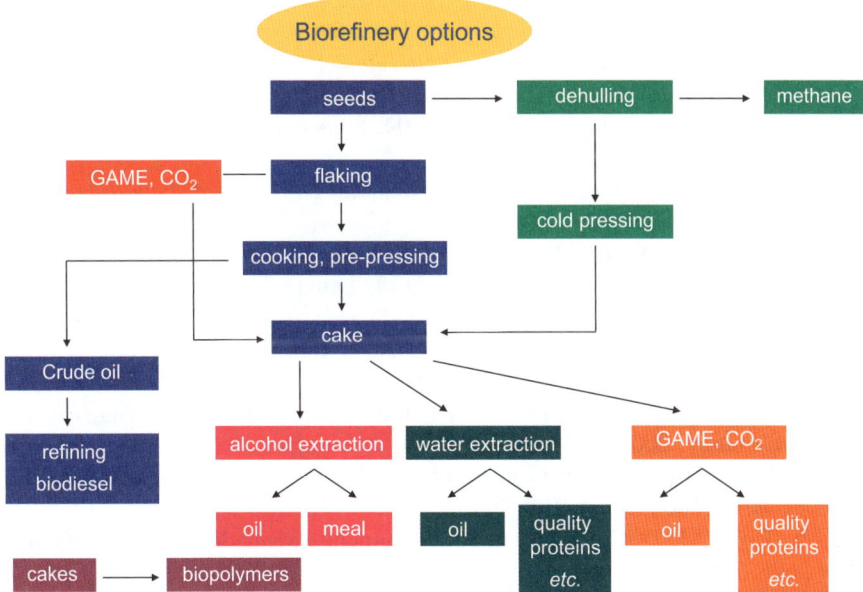

Figure 3.14 Option for a an integrated biorefinery process.

production process and other biorefinery schemes can also be created. The choice for changing the process is dependent on, amongst others, the benefits that can be obtained.

References

1. J. A. Ikebudu, S. Sokhansanj, R. T. Tyler, B. J. Milne and N. S. Thakor, *Can. Agr. Eng.*, 2000, **42**(1), 4.1–4.13.
2. F. H. Schneider, *Fette, Seifen, Anstrichmittel*, 1979, **81**(1), 11–16.
3. A. F. Mustafa, D. A. Christensen and J. J. McKinnon, *Can. J. Anim. Sci.*, 1996,, **76**(4), 579–586.
4. H. Liu, R. Przybylski, K. Dawson, N. A. M. Eskin and C. G. Biliaderis, *J. Am. Oil Chem. Soc.*, 1996, **73**(4), 493–498.
5. J. Evrard, F. Labalette and X. Pinochet, *Compte rendu du 12e congrÃš international sur le colza 26-30 mars 2007, Wuhan, Chine*, 2007, **14**(6), 326–331.
6. J. Denise, *Le raffinage des corps cras*, 1983, Westhoek editions, 153.
7. L. Denis and F. Vear, *Euphytica*, 1996, **87**(3), 177–187.
8. A. Bazus, L. Rigal and A. Gaset, *Revenue Francaise des corps gras*, 1992, **39**(11–12), 345–350.
9. M. M. Conghos, M. E. Aguirre and R. M. Santamaria, *Environ. Tech.*, 2006, **27**(9), 969–978.

10. B. Wolfgang, *Fett/Lipid*, 1997, **99**(2), 46–51.
11. R. Ohlson, *J. Am. Oil Chem. Soc.*, 1992, **69**(3), 195–198.
12. C. Oberndorfer and W. Lucke, *Fett/Lipid*, 1999, **101**(5), 164–167.
13. I. Irfan and E. Pawelzik, *Fett/Lipid*, 1999, **101**(5), 168–171.
14. M. Guderjan, P. Elez-Martinez and D. Knorr, *Innovat. Food Sci. Emerg. Tech.*, 2007, **8**(1), 55–62.
15. M. Guderjan, S. Toİpfl, A. Angersbach and D. Knorr, *J. Food Eng.*, 2005, **67**(3), 281–287.
16. H. M. Obergfoll, *Ocl-Oleagineux Corps Gras Lipides*, 1997, **4**(1), 35–37.
17. H. Dominguez, M. J. Nunez and J. M. Lema, *Food Chem.*, 1994, **49**(3), 271–286.
18. G. Bianchi, *Eur. J. Lipid Sci. Tech.*, 2003, **105**(5), 229–242.
19. G. Bianchi, *Ocl-Oleagineux Corps Gras Lipides*, 1999, **6**(1), 49–55.
20. E. Molina-Alcaide and D. R. Yanez-Ruiz, *Animal Feed Science and Technology*, 2008, **147**(1–3), 247–264.
21. J. A. Alburquerque, J. Gonzalvez, D. Garcia and J. Cegarra, *Bioresour. Technol.*, 2004, **91**(2), 195–200.
22. A. Roig, M. L. Cayuela and M. A. Sánchez-Monedero, *Waste Manag.*, 2006, **26**(9), 960–969.
23. G. Rodriguez, A. Lama, R. Rodriguez, A. Jimenez, R. Guillen and J. Fernandez-Bolanos, *Bioresour. Technol.*, 2008, **99**(13), 5261–5269.
24. M. S. Haddadin, S. M. Abdulrahim, G. Y. Al-Khawaldeh and R. K. Robinson, *J. Chem. Tech. Biotechnol.*, 1999, **74**(7), 613–618.
25. J. Fernandez-Bolanos, G. Rodriguez, R. Rodriguez, A. Heredia, R. Guillen and A. Jiminez, *J. Agr. Food Chem.*, 2002, **50**(23), 6804–6811.
26. J. Fernandez-Bolanos, G. Rodriguez, R. Rodriguez, R. Guillen and A. Jimenez, *Grasas Y Aceites*, 2006, **57**(1), 95–106.
27. H. Niewiadomski, in *Rapeseed – Chemistry and Technology*, 1990.
28. P. D. Fullbrook, *J. Am. Oil Chem. Soc.*, 1983, **60**(2), 476–478.
29. B. Srivastava, Y. C. Agrawal, B. C. Sarker, Y. P. S. Kushwaha and B. P. N. Singh, *Journal of Food Science and Technology – Mysore*, 2004, **41**(1), 88–91.
30. O. Valentova, Z. Novotna, Z. Svoboda, P. Pejchar and J. Kas, *J. Am. Oil Chem. Soc.*, 2002, **79**(12), 1271–1272.
31. G. Andrich, S. Balzini, A. Zinnai, V. De Vitis, S. Silvestri, F. Venturi and R. Fiorentini, *Eur. J. Lipid Sci. Tech.*, 2001, **103**(3), 151–157.
32. S. Isobe, F. Zuber, K. Uemura and A. Nogushi, *J. Am. Oil Chem. Soc.*, 1992, **69**(9), 884–889.
33. A. Rosenthal, D. L. Pyle and K. Niranjan, *Enzym. Microb. Tech.*, 1996, **19**(6), 402–420.
34. J. A. Ward, *J. Am. Oil Chem. Soc.*, 1984, **61**(8), 1358–1361.
35. P. Evon, V. Vandenbossche, P. Y. Pontalier and L. Rigal, *Ind. Crop. Prod.*, 2009, **29**(2-3), 455–465.
36. U. Salgin, O. Doker and A. Calimli, *J. Supercrit. Fluids*, 2006, **38**(3), 326–331.
37. I. A. Kartika, P. Y. Pontalier and L. Rigal, *Ind. Crop. Prod.*, 2005, **22**(3), 207–222.

38. I. A. Kartika, P. Y. Pontalier and L. Rigal, *Bioresour. Technol.*, 2006, **97**(18), 2302–2310.
39. C. Dufaure, J. Leyris, L. Rigal and Z. Mouloungui, *J. Am. Oil Chem. Soc.*, 1999, **76**(9), 1073–1079.
40. C. Dufaure, Z. Mouloungui and L. Rigal, *J. Am. Oil Chem. Soc.*, 1999, **76**(9), 1081–1086.
41. P. Evon, V. Vandenbossche, P. Y. Pontalier and L. Rigal, *Ind. Crop. Prod.*, 2007, **26**(3), 351–359.
42. H. Dominguez, J. Sineiro, M. J. Nunez and J. M. Lema, *Food Res. Int.*, 1995, **28**(6), 537–545.
43. H. G. Schwartzberg, *Sep. Purif. Meth.*, 1997, **26**(1), 1–213.
44. A. A. Clifford, *Gas assisted press extraction of oil, UK Patent application*, 2000.
45. W. K. Rice, US 4744926, 1988, **45**.
46. N. Foidl, 1998, EP 0822893.
47. T. Homann, J. Schulz and R. Zmudzinski, 2006, EP 1717014.
48. R. Eggers and E. G. Schade, 1988, DE 3322968.
49. M. J. Venter, R. Hink, N. J. M. Kuipers and A. B. de Haan, *Innovat. Food Sci. Emerg. Tech.*, 2007, **8**(2), 172–179.
50. M. J. Venter, N. J. M. Kuipers and A. B. de Haan, *Chem. Ing. Tech.*, 2004, **76**(9), 1403–1404.
51. M. J. Venter, N. J. M. Kuipers and A. B. de Haan, *J. Food Eng.*, 2007, **80**(4), 1157–1170.
52. M. J. Venter, P. Willems, S. Kareth, E. Weidner, N. J. M. Kuipers and A. B. de Haan, *J. Supercrit. Fluids*, 2007, **41**(2), 195–203.
53. M. J. Venter, P. Willems, N. J. M. Kuipers and A. B. de Haan, *J. Supercrit. Fluids*, 2006, **37**(3), 350–358.
54. P. Willems, N. J. M. Kuipers and A. B. de Haan, *J. Food Eng.*, 2008, **89**(1), 8–16.
55. P. Willems, N. J. M. Kuipers and A. B. de Haan, *J. Supercrit. Fluids*, 2008, **45**(3), 298–305.
56. P. Willems, N. J. M. Kuipers and A. B. de Haan, *J. Food Eng.*, 2009, **90**(2), 238–245.
57. S. Voges, R. Eggers and A. Pietsch, *Separ. Purif. Tech.*, 2008, **63**(1), 1–14.
58. J. M. Del Valle and J. C. De La Fuente, *Crit. Rev. Food Sci. Nutr.*, 2006, **46**(2), 131–160.
59. L. A. Johnson, P. J. Wan and W. Farr, *Introduction to Fats and Oils*, 2000, pp. 108–135.
60. R. Sharma and P. C. Sharma, *J. Sci. Ind. Res.*, 2007, **66**(1), 52–55.
61. S. Pereira-Freitas, L. Hartman, S. Couri, F. H. Jablonka and C. W. P. de Carvalho, *Fett/Lipid*, 1997, **99**(9), 333–337.
62. W. Leuchtenberger in *Biotechnology*, VCH, Weinheim, 2nd edn, 1996, vol. 6, pp. 465–502.
63. W. Leuchtenberger, K. Huthmacher and K. Drauz, *Appl. Macrobiol. Biotechnol.*, 2005, **69**, 1–8.

64. Applicant: Novo Nordisk Inventor: Kofoed, L WO 98/18343, 1997.
65. S. B. Zhang, *J. Am. Oil Chem. Soc.*, 2007, **84**, 693–700.
66. S. B. Zhang, Z. Wang and S. Y. Xu, *J. Am. Oil Chem. Soc.*, 2007, **84**(1), 97–105.
67. C. Ordonez, C. Benitez and J. L. Gonzalez, *Bioresour. Technol.*, 2008, **99**, 4749–4754.
68. T. Werpy and G. Petersen, *Top Value Added Chemicals from Biomass*, 2004.
69. E. Scott, F. Peter and J. Sanders, *Appl. Microbiol. Biotechnol.*, 2007, **75**(4), 751–762.
70. P. M. Könst, P. M. C. C. D. Turras, M. C. R. Franssen, E. L. Scott and J. P. M. Sanders, *Adv. Synth. Catal.*, 2010, **352**(9), 1493–1502.
71. E. L. Scott, J. v. Haveren and J. P. M. Sanders, *The Biobased Economy: Biofuels, Materials and Chemicals in the Post-oil Era*, ISBN: 9781844077700, 2010.
72. P. M. Könst, M. C. R. Franssen, E. L. Scott and J. P. M. Sanders, *Green Chem.*, 2009, **11**(10), 1646–1652.
73. T. M. Lammens, D. De Biase, M. C. R. Franssen, E. L. Scott and J. P. M. Sanders, *Green Chem.*, 2009, **11**(10), 1562–1567.
74. T. M. Lammens, M. C. R. Franssen, E. L. Scott and J. P. M. Sanders, *Green Chem.*, 2010, **12**(8), 1430–1436.
75. J. L. Notre, E. L. Scott, M. C. R. Franssen and J. P. M. Sanders, *Tetrahedron Lett.*, 2010, **51**(29), 3712–3715.
76. J. P. M. Sanders, J. v. Haveren, E. L. Scott, D. S. v. Es and J. L. Notre, PLT/NL 2010/050286, 2010.
77. B. Brehmer, R. M. Boom and J. Sanders, *Chem. Eng. Res. Des.*, 2009, **87**(9), 1103–1119.
78. J. J. Thomas and R. G. Barile, in *Proc. Energy from Biomass and Wastes VIII*, 1461–1494, 1984.
79. M. Kitano, F. Tanimoto and M. Okabayashi, *Chem. Econ. Eng. Rev.*, 1975, **7**(7), 25–29.
80. R. A. Schraufnagel and H. F. Rase, *Ind. Eng. Chem. Prod. Res. Dev.*, 1975, **14**(1), 40–44.
81. J. J. Bozell, L. Moens, D. C. Elliott, Y. Wang, G. G. Neuenscwander, S. W. Fitzpatrick, R. J. Bilski and J. L. Jarnefeld, *Resour. Conservat. Recycl.*, 2000, **28**(3–4), 227–239.
82. W. Fitzpatrick Stephen, US 4897497, 1990.
83. V. M. Ghorpade and M. A. Hanna, in US 5859263, 1999.
84. Q. Fang and M. A. Hanna, *Bioresour. Technol.*, 2002, **81**(3), 187–192.
85. J. Y. Cha and M. A. Hanna, *Ind. Crop. Prod.*, 2002, **16**(2), 109–118.
86. K. I. Seri, T. Sakaki, M. Shibata, Y. Inoue and H. Ishida, *Bioresour. Technol.*, 2002, **81**(3), 257–260.
87. C. Chang, P. Cen and X. Ma, *Bioresour. Technol.*, 2007, **98**(7), 1448–1453.
88. L. Yan, N. Yang, H. Pang and B. Liao, *Clean – Soil, Air, Water*, 2008, **36**(2), 158–163.
89. B. V. Timokhin, *Russ. Chem. Rev.*, 1999, **68**(1), 73–84.

90. L. E. Manzer, in *ACS Symposium Series*, 2006, pp. 40–51.
91. S. W. Fitzpatrick, in *ACS Symposium Series*, 2006, pp. 271–287.
92. A. D. Patel, J. C. Serrano-Ruiz, J. A. Dumesic and R. P. Anex, *Chem. Eng. J.*, 2010, **160**(1), 311–321.
93. T. Werpy and G. Petersen, in *Top Value Added Chemicals from Biomass, Results of Screening for Potential Candidates from Sugars and Synthesis Gas*, 2004, **1**.
94. D. J. Hayes, S. W. Fitzpatrick, M. H. B. Hayes and J. R. H. Ross, *Biorefineries, Principles and Fundamentals*, 2005, **1**, 139–164.
95. F. M. A. Geilen, B. Engendahl, A. Harwardt, W. Marquardt, J. Klankermayer and W. Leitner, *Angew. Chem. Int. Ed.*, 2010, **49**(32), 5510–5514.
96. L. E. Manzer, *Appl. Catal. Gen.*, 2004, **272**(1–2), 249–256.
97. J. P. Lange, J. Z. Vestering and R. J. Haan, *Chem. Comm.*, 2007, **33**, 3488–3490.
98. Y. Isoda and M. Azuma, in JP 08053390, 1996.
99. D. M. Alonso, J. Q. Bond, J. C. Serrano-Ruiz and J. A. Dumesic, *Green Chem.*, 2010, **12**(6), 992–999.
100. G. W. Huber, S. Iborra and A. Corma, *Chem. Rev.*, 2006, **106**(9), 4044–4098.
101. I. T. Horváth, H. Mehdi, V. Fábos, L. Boda and L. T. Mika, *Green Chem.*, 2008, **10**(2), 238–242.
102. J. P. Lange, R. Price, P. M. Ayoub, J. Louis, L. Petrus, L. Clarke and H. Gosselink, *Angew. Chem. Int. Ed.*, 2010, **49**(26), 4479–4483.
103. J. C. Serrano-Ruiz, D. Wang and J. A. Dumesic, *Green Chem.*, 2010, **12**(4), 574–577.
104. M. Moliner, Y. Román-Leshkov and M. E. Davis, *Proc. Natl Acad. Sci. USA*, 2010, **107**(4), 6164–6168.
105. M. S. Feather and J. F. Harris, *Adv. Carbohydr. Chem. Biochem.*, 1973, **28**, 212–217.
106. H. E. Grethlein, *J. Appl. Chem. Biotechnol.*, 1978, **28**(4), 296–308.
107. H. E. Van Dam, A. P. G. Kieboom and H. Van Bekkum, *Starch/Stärke*, 1986, **38**(3), 95–101.
108. J. Horvat, B. Klaić, B. Metelko and V. Šunjić, *Tetrahedron Lett.*, 1985, **26**(17), 2111–2114.
109. B. Girisuta, L. P. B. M. Janssen and H. J. Heeres, *Chem. Eng. Res. Des.*, 2006, **84**(5A), 339–349.
110. V. E. Tarabanko, M. Y. Chernyak, S. V. Aralova and B. N. Kuznetsov, *React. Kinet. Catal. Lett.*, 2002, **75**(1), 117–126.
111. S. W. Fitzpatrick, in US 5608105, 1997.

CHAPTER 4
Secondary Processing of Plant Oils

ZSANETT HERSECZKI,[a] ABBAS KAZMI,[b]
RAFAEL LUQUE[c] AND DIEGO LUNA[c]

[a] University of Pannonia, Cooperative Research Centre for Environmental and Information Technology, H-8200 Veszprem, POB 158, Hungary; [b] Green Chemistry Centre of Excellence, University of York, York, YO10 5DD, UK; [c] Departamento de Química Orgánica, Universidad de Córdoba, Campus de Rabanales, Edificio, Marie Curie, E-14014 Córdoba, Spain; Seneca Green Catalyst S.L., Campus de Rabanales, E14014, Cordoba, Spain

4.1 Applications of Glycerol

Glycerol, or propan-1,2,3-triol, is an important by-product of biodiesel production generated from the transesterification reaction of triglycerides (Scheme 4.1) from virgin vegetable oils or fats as well as waste oils, with alcohols including methanol and ethanol, in the presence of a homogeneous base catalyst such as NaOH or KOH,[1,2,3] and acid catalyst.[4] In general, the production of 10 kg of biodiesel yields approximately 1 kg of crude glycerol (10% (w/w)),[5] and currently the world's capacity for biodiesel production is dramatically increasing. Further increases in biodiesel production rates will significantly raise the quantity and surplus of crude glycerol and partially purified glycerol in the environment. In contrast to the surplus of impure glycerol, high-purity glycerol is an important industrial feedstock that finds applications in the food, cosmetic and pharmaceutical industries, as well as

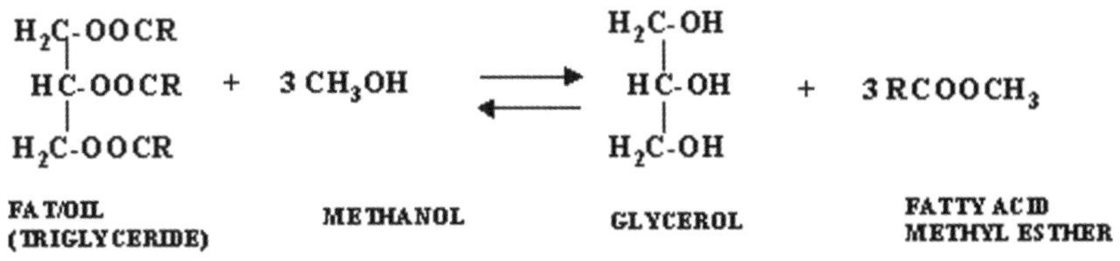

$$\begin{array}{c}\text{H}_2\text{C-OOCR}\\ |\\ \text{HC-OOCR}\\ |\\ \text{H}_2\text{C-OOCR}\end{array} + 3\,\text{CH}_3\text{OH} \;\rightleftarrows\; \begin{array}{c}\text{H}_2\text{C-OH}\\ |\\ \text{HC-OH}\\ |\\ \text{H}_2\text{C-OH}\end{array} + 3\,\text{RCOOCH}_3$$

FAT/OIL (TRIGLYCERIDE)   METHANOL   GLYCEROL   FATTY ACID METHYL ESTER

Scheme 4.1

other more minor uses. However, its refining is generally costly, especially for medium- and small-sized plants.[6]

Two phases are produced after transesterification and distillation of the excess alcohol during biodiesel production. The upper ester phase (EP) contains the main product (biodiesel) while the lower glycerol phase (GP) consists of glycerol and many other chemical substances including water, organic and inorganic salts, a small amount of esters and alcohol, traces of glycerides and vegetable pigments. The exact composition of the raw GP depends on the transesterification methodology and the separation conditions employed in the biodiesel production; in any case, glycerol concentration is usually between 30 and 60 wt% with larger biodiesel plants having the highest purities (75–90%).[7] The remainder of the crude glycerol consists primarily of unconverted triglycerides, unconverted methanol, biodiesel, soaps and contamination. This dilution means that the actual amount of glycerol formed is much larger, between 100/90 (1.1) and 100/55 (1.8) times as much. Table 4.1 shows typical composition data for biodiesel-derived glycerol. Most of the contaminants can be traced back to the biodiesel synthesis process (*e.g.* unreacted methanol not completely evaporated and/or reacted). Furthermore the concentrations of Na and K can tell whether sodium or potassium hydroxide were utilised as catalysts in the transesterification reaction. Alkali metals including Na, K, Ca and Mg are naturally present in vegetable oils. Sulfates and phosphates may be also present from the neutralisation of the mixture with sulfuric or phosphoric acid.

4.1.1 Existing and Novel Glycerol Purification Technologies

The existing biodiesel-derived crude glycerol is of poor quality and requires expensive refining technologies before it can be rendered suitable for new product technologies. Scaling up of such methodologies is considered to be essential for these technologies to become economically feasible.

Different processes have been implemented to refine glycerol. However, all of them involve soap splitting followed by two main separation steps: salt and methanol removal. Some of the separation techniques require vacuum due to the sensitivity of glycerol under heat, which normally splits into water and decomposes at 180 °C. Generally speaking, the following technologies may be used to further purify glycerol (after the soap-splitting step): fractional distillation, ion-exchange, adsorption, precipitation, extraction, crystallisation and dialysis.

Table 4.1 Composition of G-phase.

Property	Value	Unit
Genetically modified origin	Possible	
Glycerol content	77–90%	wt% A.R.
Ash content	3.5–7%	wt% A.R.
Moisture content	0.1–13.5%	wt% A.R.
Lower calorific value	14.9–17.5	MJ/kg A.R.
Kinematic viscosity	120	mm^2/s
3-monopropylenediol	200–13,500	ppm
Methanol	0.01–3.0%	wt%
MONG*	1.6–7.5%	wt%
pH	4.5–7.4	
Sulfate	0.01–1.04	wt%
Phosphate	0.02–1.45	wt%
Acetate	0.01–6.0	wt%
Na	0.4–20	g/kg
K	0.03–40	g/kg
Ca	0.1–65	mg/kg
Mg	0.02–55	mg/kg
Fe	0.1–30	mg/kg
Mn	<0.5	mg/kg

*MONG = non-glycerol organic matter.

A glycerol soap-splitting followed by a combination of methanol recovery/drying, fractional distillation, ion-exchange (by using zeolite or resins) and adsorption (using active carbon powder) seems to be the most common purification pathway. Well-known companies involved in crude glycerol purification plants are Desmet Ballestra and Buss-SMS-Canzler (ion exchange equipment). Chemical companies like Rohm & Haas and Lanxess supply ion-exchange granulates while a company such as Norit supplies powder and granulated activated carbon.[8]

4.1.1.1 Soap Splitting as a Glycerol Pre-treatment Step

Three steps can be distinguished in the purification process. The first step involves neutralisation using an acid to remove the catalyst and soaps. The reaction of an acid with the soaps will give free fatty acids (FFAs) and salts while its reaction with the base catalyst generates salts and water. Since FFAs are insoluble in glycerol, these components will rise to the top so that they can be skimmed off. Some glycerol insoluble salts will also precipitate out. Ooi et al. attempted to recover crude glycerol, fatty acids and salts from a glycerol residue waste derived from a palm kernel oil methyl ester plant by using 6% (v/v) H_2SO_4.[9] They reported that such chemical treatment at low pH was beneficial in increasing glycerol content as well as reducing the ash content in the recovered crude glycerol. However, the non-glycerol organic matter (MONG) content was slightly superior. The treatment also increased the recovered salts and reduced the crude glycerol proportions, and did not affect the recovery of crude fatty acids but, despite these properties, it was relatively suitable to recover crude glycerol from a dilute glycerol source (10–20% (w/w) glycerol) with a high NaCl (60–70% (w/w)) presence, such as that found in glycerol residues.

The second step involves methanol removal. The methanol contained in the glycerol phase can be removed *via* flash evaporation or using falling film evaporators. Falling film evaporators have an advantage of keeping the contact time short and are best suited for these processes owing to the temperature susceptibility of glycerol, which can result in its decomposition. The purity of glycerol will be approximately 85% after methanol removal. In the third step, glycerol can be further purified to 99.5% by using a combination of adsorption, vacuum distillation and ion exchange processes.[10] Yong *et al.* reported the recovery of crude glycerol from glycerol residues by a simple distillation at 120–126 °C and 0.04–0.4 mbar to yield around 141.8 g glycerol/kg glycerol residue (~14% yield) at an acceptable purity of 96.6% (w/w) glycerol, with 0.03% (w/w) ash, 1% (w/w) H_2O and 2.4% (w/w) MONG as contaminants.[11] Hazimah *et al.* used the combined processes of chemical and physical treatment (acid protonation, ether and ethanolic extractions, filtration and distillation) to recover glycerol and diglycerol from glycerol pitch, recovering a high-purity glycerol (~99.1–99.8% (w/w)) with low content of contaminants (0.11–0.80% (w/w) H_2O, 0.054% (w/w) ash and 0.56% (w/w) soap).[12]

4.1.1.2 Conventional Processes for Glycerol Purification

The conventional process for glycerol purification comprises various steps including pre-treatment, concentration, purification and refining.

The pre-treatment step is used to remove colour and odour matters as well as any remaining fat components from crude glycerol. In the pre-treatment step, sodium hydroxide is employed to remove fat components *via* saponification reaction and activated carbon is utilised for bleaching purposes.

The concentration step involves the removal of ionic substances using ion exclusion chromatography. In this process, a bed filled with strongly acidic exchange resins is charged with a glycerol stream. The principle used for the separation is the Donnan exclusion. Ionic substances are repelled from the resin surface and remain in the liquid volume due to their charge, while non-ionic substances can be accommodated in the pores of the resins. The column is subsequently rinsed with water, which firstly removes the ionic substances in the liquid and then the non-ionic ones. In some cases, ion exchangers (both cationic and anionic) are employed instead of water, particularly when the concentration of ionic substances in the glycerol stream is very high. The following step is the purification of the glycerol phase by means of ion-exchangers, normally used in pairs (cationic and anionic). In the case of cationic exchangers, positive ions are exchanged for protons (H^+), while in anionic exchangers negative ions are exchanged for hydroxide ions (OH^-). This purification step is able to remove inorganic salts, fat and soap components as well as colour- and odour-causing matters. The crude glycerol phase can be purified by ion exchange on Amberlite-252 (strong acid resin) and it has also been suggested that the macroporous Amberlite could be useful for the removal of sodium ions from glycerol/water solutions with high salt concentrations.[13] Another technology to purify glycerol with high salt content is by using an ion-exchange Ambersep BD50.[14]

The subsequent step will be based on the treatment of glycerol in multiple vacuum flash evaporators (10–15 kPa vacuum), which results in 90–95% concentration (Figure 4.1). An alternative way to do the same job is to use thin film distillation (Figure 4.2). In thin film distillation, the glycerol stream is distributed as a thin film on the wall of the evaporator and heated externally. Glycerol will fall down to the bottom of the evaporator as a residue while highly volatile components including methanol and water are evaporated and collected at the top. The final concentration of glycerol to 99.5% is carried out under vacuum (0.5–1 kPa) in forced circulation evaporators.[15]

4.1.1.3 Recent Development in Glycerol Purification Processes

John E. Aiken reported various improvements in glycerol purification processes. He proposed five separation steps, which can be conducted under either batch or continuous conditions (Figure 4.3). This process is claimed to be able to produce glycerol of higher than 99.5% purity from typical crude glycerol containing a mixture of mono-, di- and triglycerides, excess methanol, water, fatty acid alkyl esters, quantities of residual catalyst and salts.[16]

(i) First reactor
Crude glycerol, whose purity is typically 86–92%, is preheated and then fed to the first reactor, which is generally used to recover triglycerides by reacting entrained methyl esters and glycerol to produce glycerides and methanol (reversed biodiesel production reaction). Nitrogen is sparged to provide agitation and to remove methanol and water; thus, the reaction is shifted to glycerides formation. The temperature inside the reactor is maintained at 120–160 °C. A gas effluent stream is then passed through a condenser. Upon separation from condensed methanol and water in a condenser, nitrogen is then recycled to the reactor.

(ii) Second reactor
The liquid effluent stream from the first reactor is heated to maintain the second reactor at 120–160 °C. In this reactor, unreacted methyl esters are reacted to produce methanol and triglycerides. Wash water (containing glycerol) is also added to the second reactor. Similarly, sparging nitrogen is used to stir the mixture inside the reactor and to remove methanol and water. Entrained methanol and water are condensed. After being separated from nitrogen, wash water is recycled. The operating conditions are adjusted in such a way that the glycerol effluent stream contains a maximum of 0.5 wt% of methanol and approximately 5 wt% of water.

Secondary Processing of Plant Oils 171

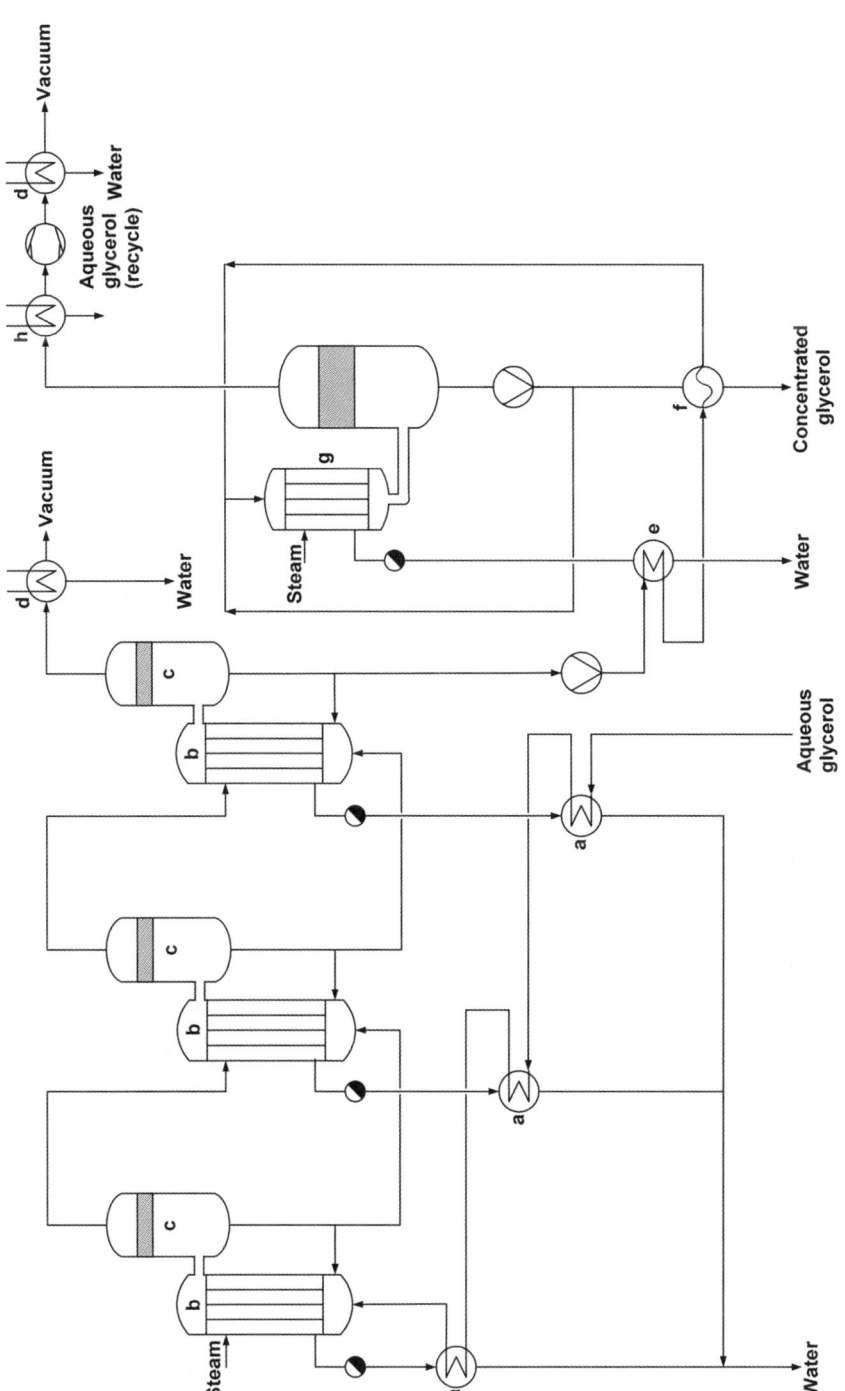

Figure 4.1 Continuous glycerol concentration: (a) Feed heater: (b) Evaporator: (c) Separator with demister: (d) Water condenser: (e) Glycerol heater: (f) Glycerol heater/final product cooler: (g) Falling film evaporator: (h) Glycerol condenser.

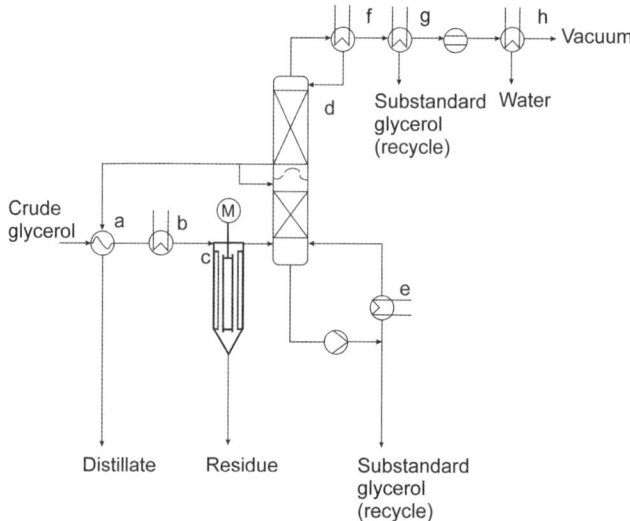

Figure 4.2 Continuous glycerol distillation (Cognis): (a) Economizer: (b) End heater: (c) Thinfilm distillation: (d) Fractionating column: (e) Reboiler: (f) Reflux condenser: (g) Glycerol condenser (h) Water condenser.

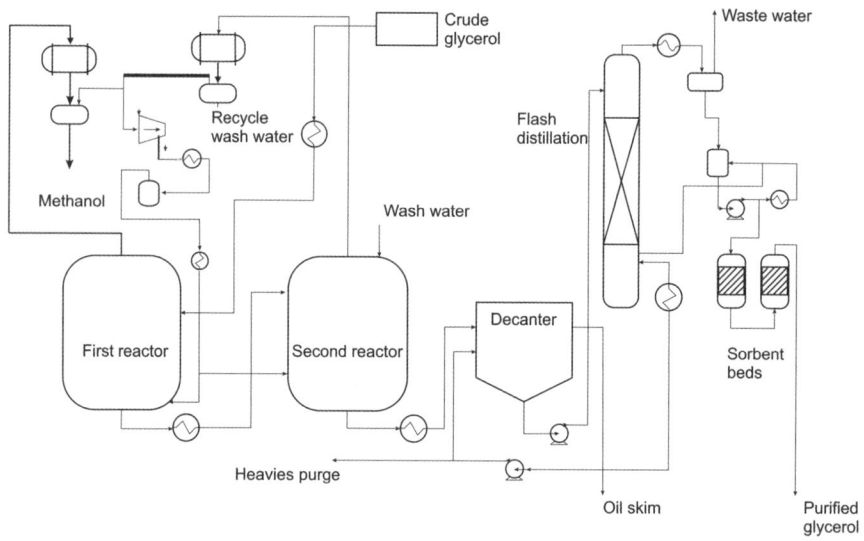

Figure 4.3 Simplified flow sheet of the recent development process, based on US 7 126 032 B1.

(iii) Decanter
A decanter is placed after the second reactor. It serves as a feed tank for the flash distillation column and a separator to remove the oil layer of the glycerol stream by lowering the pH below 7 and skimming it

from the glycerol layer. The recycle stream from the bottom of the flash distillation column is mixed with the glycerol stream in this tank.

(iv) Flash distillation column/stripper

The flash distillation column consists of a packed-bed column with a steam-heated reboiler. This column operates at a temperature of 185 °C and a pressure of 5–20 mmHg. There is no reflux returned to the top of the column. About 80–90% of glycerol in the feed stream is drawn as overhead product, which is then condensed in two condensers in series. The first condenser is utilised to condense glycerol, while the second condenses water, which will be sent to a wastewater stream.

The bottom product of the column, which contains glycerol and heavy compounds, is pumped back to the decanter. Some of it is purged continuously or intermittently to prevent salts and glycerol build-up in the decanter.

(v) Adsorption columns

The last step of glycerol refining is the removal of colour and trace impurities. Many different materials may be used as adsorbents, including activated carbons, ion exchange resins and molecular sieves. The purified glycerol is then pumped to a storage tank.

4.1.1.4 Chromatography and Regenerative Column Adsorption

Different adsorption techniques used in the separation of glycerol are proven technologies. Some biodiesel equipment vendors purify glycerol using activated carbon powder from suppliers including Norit. The main components to separate are:

- Glycerol (key component);
- Water;
- Ions (like K^+);
- Saponification residues;
- Methanol traces.

Activated carbon has surface areas between 500 and 1500 m^2/g and sizes <150 micron, being an especially suitable adsorption medium for organic molecules. However, it is rather expensive to regenerate the carbon. Operational costs will be high when using a column adsorption, because of the high viscosity of the crude glycerol and the high pressure drop. Activated carbon is normally employed due to its good properties in wastewater purification. New developments in adsorption techniques are mostly based on chromatography separation. Originally, these techniques were applied to the separation of small quantities of samples in a laboratory. Currently, capacities and applications have been widely expanded. Table 4.2 shows some chromatography techniques and their properties.

Companies such as Rohm & Haas and Lanxess sell granular ion exchange resins that are also used for glycerol purification (a.o. salts, colour and odour removal). But most important is the separation of water and glycerol molecules

Table 4.2 Summary of chromatography separation processes.

Method	Separation parameters	Important parameters
Gel permeation	Particle size	Column length
Ion exchange chromatography	Charge	pH, ionic strength
Hydrophobic interaction	Hydrophobicity	Polarity, ionic strength
Reversed phase	Hydrophobicity	Polarity, ionic strength
Affinity chromatography	Biospecific interaction	Ligand, eluent

Table 4.3 Summary of chromatography separation processes.

Chromatographic technique	Resolution	Capacity
Gel permeation	Moderate	Moderate
Ion exchange	Low/moderate	Very high
Hydrophobic interaction	High	High
Reversed phase	Very high	High
Affinity	Very high	High

based on affinity and particle size. Water molecules that are bound to glycerol are difficult to separate. It is therefore important to find a suitable type of adsorbent with respect to high separation efficiency (resolution) at a high volume flow capacity and low pressure drops. Table 4.3 summarises some chromatographic techniques with their resolutions and volume flow capacities.

It should be possible to design new types of adsorbent media specific for glycerol that meet these process criteria. The first process set-up can be an ion exchange column with a second column added. This column could be based on the affinity properties of the adsorbent and the (gel) permeation principle. This principle is shown in Figure 4.4.

Typical properties of gel permeation are:

- Particle size: 0.1–0.2 mm;
- Column length: up to 1 metre;
- Liquid flux: up to 5×10^{-5} m/s.

These characteristics are difficult to apply at large scale, but new developments show potential for higher flows at stable pressure drops; however, these are far from economically feasible applications.

4.1.2 Transformation of Glycerol into High-quality Products through Green Chemistry and Biotechnology

Glycerol can be used as a building block for many chemicals such as 1,3-propanediol, lactate and succinate. In fact many companies have initiated commercial plans to manufacture high-value chemicals such as epichlorohydrin (Solvay SA) and proplylene diol (Ashland/Cargill) from glycerol feedstocks.

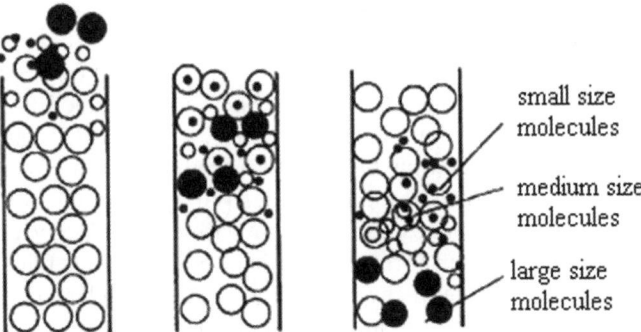

Figure 4.4 Gel permeation principle.

The market volatility in the price of glycerol has caused concern for these projects; however, the long-term fundamentals remain strong.

Glycerol has been known since 2800 BC mainly as a by-product of soap production.[17] Currently glycerol has numerous applications in personal care, food, tobacco, detergents, cellophane, explosives and pharmaceuticals.[18] Leffingwell and Lesser identified 1582 applications for glycerol in 1945.[19] However, in recent times, many glycerol production plants are closing and new plants utilising glycerol as a raw material are starting up.[20] Global glycerol production increased from 60,000 tons in 2001 to 800,000 tons in 2005, partly due to biodiesel production. The amount of glycerol being used in technical applications is around 160,000 tons per year, and this is expected to grow at a rate of 2.8% per year.[21]

Glycerol is a raw material for the production of flexible foams and rigid polyurethane foams. It is known to provide properties including flexibility, pliability and toughness in surface coatings and paint-regenerated cellulose films, meat casings and special quality papers.[22] Glycerol has the ability to absorb moisture from the atmosphere and is therefore used in many adhesives and glues to prevent early drying. In food applications, non-toxic glycerol is used as solvent, sweetener and preservative. Many polyols such as sorbitol, manitol and maltitol are used as sugar-free sweeteners; however, they are facing fierce competition from glycerol. Glycerol has similar sweetness to sucrose and has the same energy as sugar. Furthermore, it does not raise blood sugar levels and does not feed plaque bacteria. Glycerol is also employed as an emollient, humectant and lubricant in many products in the personal care industry including toothpaste, mouthwashes, shaving cream and soaps.[23]

A detailed revision of glycerol transforming processes and applications can be found in '*The Future of Glycerol – New Usages for a Versatile Raw Material*' by Mario Pagliaro and Michele Rossi.[23] The book was published in 2008 and focuses on key chemical and biochemical transformations with detailed processing conditions. In this book, relevant information on the sustainability and economics of glycerol and biofuels production is discussed. The detailed synthetic chemistry involved in the transforming processes has also been reviewed by Behr et al. in his paper entitled 'Improved utilisation of renewable resources: New important derivatives of glycerol'.[24]

A more detailed revision of chemicals that can be derived from glycerol was conducted in 2008 by Zheng, Chen and Shen in 'Commodity Chemicals Derived from Glycerol, an Important Biorefinery Feedstock'.[25] Many important chemicals have been identified that can be produced from glycerol-derived platform chemicals and their respective industrial applications are discussed. Furthermore, this review maps the reaction pathways of a glycerol-derived platform chemical which can form many other commodity chemicals that are not easily identifiable. Some of the important commodity chemicals identified include acrolein, dichloropropanol, epichlorohydrin, dihydroxyacetone, 1,3-propanediol, 1,2-propanediol, glycerol carbonate, diacylglycerol (DAG), monoglyceride (MG), oxygenate fuels, glyceric acid, tartronic acid and mesoxalic acid.

Biochemical methods can likewise be employed to transform glycerol into commodity chemicals as highlighted by Yazdani et al.[26] For some transformations a detailed description of the processes involved are shown in Figure 4.5, including the overall production costs.

Some of the important commodity chemicals produced using anaerobic fermentation include succinic acid, 1,3-propanediol, propionic acid, formic acid, butanol and ethanol. A recent paper by Silva et al. reviews glycerol as a source for industrial microbiology.[27] This report identifies various microbial reaction pathways to produce many chemicals from glycerol-derived platform chemicals and an example is shown in Scheme 4.2.

4.1.2.1 Glycerol Transforming Processes

Aqueous Phase Reforming
One of the major achievements in glycerol chemistry is the development of aqueous phase reforming processes (APRs), which involve the conversion of glycerol to hydrogen and carbon monoxide (synthesis gas). The reported process conditions require 250 °C and the utilisation of a Pt-Re catalyst in a single reactor.[28] This process can also theoretically produce high yields of hydrogen from glycerol at low CO concentrations due to favourable water-gas shift (WGS) thermodynamics. This requires significantly lower energy consumption than traditional methane reforming.

Fischer–Tropsch
The synthesis gas can be used as a building block for chemicals and fuels *via* Fischer–Tropsch synthesis (FTS). Through FTS, syngas can be converted to a range of useful liquid hydrocarbons (mainly linear alkanes, although alkenes and alcohols can also be produced under certain conditions) using iron and cobalt catalysts. The temperatures used in the process typically range from 150 to 300 °C and pressures of one to a few atmospheres are common. High temperatures lead to gasolines and linear low molecular mass hydrocarbons whereas lower temperatures and high pressures favour the formation of longer chain hydrocarbons (*e.g.* waxes).

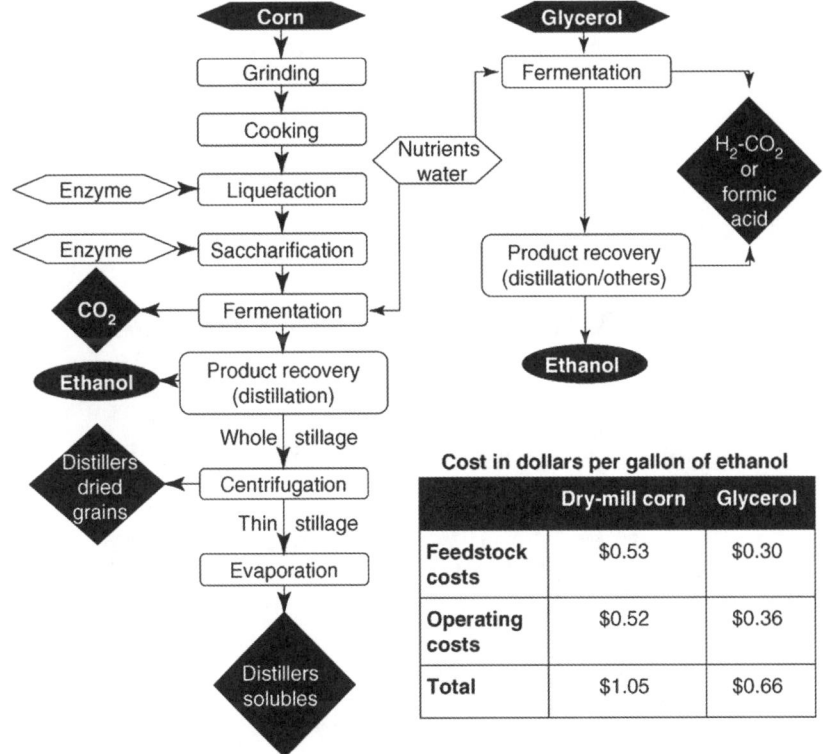

Figure 4.5 Comparison of ethanol production from corn-derived sugars (dry grind ethanol) with ethanol production from glycerol.

Selective Reductions

The main processes utilised to reduce glycerol to glycols are hydrogenolysis, dehydroxylation and biotechnology *via* bacteria.

Propylene glycol is commercially produced *via* hydrogenolysis using a copper chromite catalyst at 200 °C at pressures below 10 bar. Wang and co-workers showed that it was possible to produce 1,3-propanediol *via* selective dehydroxylation.[29] The central hydroxyl group of glycerol is selectively converted to a tosyloxy-group, which is removed using hydrogenolysis. The biological reduction to 1,3-propanediol involves the use of bacterial strains from groups such as *Citrobacter, Enterobacter, Ilyobacter, Klebsiella, Lactobacillus, Pelobacter* and *Clostridium*. Freund showed in 1881 that PDO could be produced using *Clostridium*, a widely available microorganism found in nature.[30] The process involves a two-step enzyme-catalysed reaction sequence in which a dehydratase catalyses the conversion of glycerol to 3-hydroxypropionaldehyde, which is subsequently reduced to PDO by a NAD^+-linked oxidoreductase.

Halogenations

The chlorination of glycerol *via* a 1,3-dichloro-2-propanol intermediate yields epichlorohydrin, an important and valuable chemical. 1,3-Dichloro-2-propanol

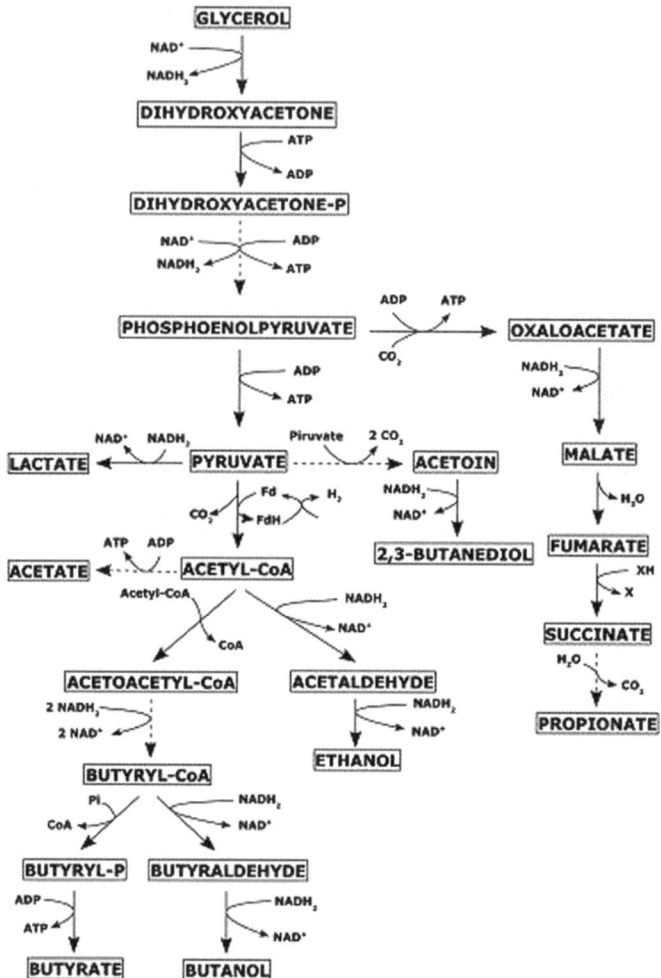

Scheme 4.2

can be produced directly from glycerol using HCl as catalyst. Its subsequent dehydrochlorination using NaOH generates epichlorohydrin and NaCl.

Dehydrations
Glycerol dehydration can also produce relevant chemicals including acrolein, 3-hydroxypropionaldehyde and acrylic acid. Protonated glycerol is more susceptible to dehydration due to reduction in the energy barrier of the intermediate state. Therefore, acrolein can only be produced in acidic conditions. The reaction can be conducted in either the liquid or the gas phase at high temperatures and/or vacuum that are normally used to drive the dehydration. In the presence of molecular oxygen, acrylic acid can be produced *via* a one-step oxy-dehydration step.

Etherifications

Glycerol alkyl ethers can be synthesised by etherification of alkenes including isobutylene in the presence of an acid catalyst at temperatures from 50 °C to 150 °C. The typical molar ratios used in the reaction are 1:2 (glycerol: isobutylene) and the yield can be improved by optimising the reaction conditions. Glycerol can be etherified to form polyglycerol *via* anionic polymerisation of glycidol through a cation exchange equilibrium initiated by partially deprotonated 1,1,1-tris(hydroxymethyl) propane. The resulting polymer usually has a polydispersity of below 1.5 and a molecular weight ranging from 1000 to 3000 gmol^{-1}.

Esterifications

Glycerol can be esterified with carboxylic acids or *via* carboxylation and nitration.[31] Reaction with carboxylic acids results in the formation of monoacylglycerols and diacylglycerol. Monoacylglycerols are produced at commercial scale by either continuous chemical glycerolysis of fats and oils (250 °C, alkaline, N_2 atmosphere) or by direct esterification with fatty acids.[32] The reaction of glycerol with dimethyl carbonate can also produce a high yield of glycerol carbonate in the presence of a biocatalyst (*e.g.* lipases). Glycerol can be converted to glycidyl nitrate by nitration, which can be subsequently polymerised to form a valuable polymer.

Selective Oxidations

The oxidation of glycerol can be catalysed using highly active aerobic catalysts such as platinum and palladium. Supported gold catalysts are well known for catalytic stability, resistance to oxygen and tolerance against inhibition by aliphatic and aromatic amines. Organocatalysts such as 2,2,6,6-tetramethylpiperidine-1-oxyl (TEMPO) can be used for the selective oxidation of glycerol to mesoxalic acid. TEMPO has also been used in electrochemical oxidation where glycerol is converted to 1,3-dihydroxyacetone (DHA). The reaction proceeds by applying a small electric potential to a solution containing glycerol, water and 15 mol% TEMPO using a glassy carbon anode. DHA can also be produced using biological oxidation *via* microorganisms or enzymes. Other oxidation products include glyceraldehydes, glyceric acid, glycolic acid, hydroxypyruvic acid, oxalic acid and tartronic acid.

Pyrolysis

Glycerol was identified as a feedstock for pyrolysis in 1985,[33] well before the growth in the biodiesel market. Recent research by Valliyapan and co-workers has focused on optimising conditions for hydrogen or syngas production.[34] Pyrolysis carried out in a continuous down-flow fixed-bed microreactor can take place with flowrates of nitrogen from 30 to 70 ml/min, temperatures of 650 to 800 °C and at atmospheric pressure. It was shown that the type and size of packing material in the tubular reactor can affect the conversion of glycerol and subsequent product distribution. Typical products include carbon monoxide, hydrogen, carbon dioxide, methane and ethane. At lower temperatures under steam or supercritical water conditions, longer molecules such as acrolein, formaldehyde and acetaldehyde are observed.[33,35]

Biotransformations
Glycerol can be converted to a very large number of chemicals using microorganisms and enzymes. The aerobic conversion of glycerol to 3-hydroxypropionaldehyde (3-HPA) was reported in 1985 by Slininger and Bothast.[36] The cells of *klebsiella pneumonia* can be grown on a rich glycerol medium and production of 3-HPA starts when these microorganisms are added to a buffer containing semi-carbazide and glycerol. It was shown that a yield of up to 84% could be obtained; however, this yield is sensitive to cell age and cultivation medium. The optimal processing conditions for this experiment were 32 °C, pH 7–8 and glycerol concentrations of 20–50 g/l.

4.1.2.2 Valuable Chemicals from Glycerol

Hydrogen
Hydrogen is the simplest element known to mankind and the most abundant element (75%), but is very rare in its molecular form (1%). Most hydrogen is found in water and hydrocarbons and one of the cleanest methods to produce hydrogen from water is by using light, photovoltaic cells and water electrolysis.[37] The current demand for hydrogen (5×10^{10} kg/year) worldwide is met by the reaction of crude-oil-derived methane with water. The main application for hydrogen is as a reactant in the hydrogenation process. It is also used for internal-combustion engines, gas-fired turbines and fuel cells.[23]

Succinic Acid
Succinic acid is a valuable chemical that has numerous uses in food additives, soldering fluxes, pharmaceutical products, surfactants, green solvents and biodegradable plastics. It is used as an intermediate for the synthesis of 1,4-butanediol, tetrahydrofuran, γ-butyrolactone and linear aliphatic esters.[27] Novel biodegradable plastics can also be synthesised from it including polybutyrate succinate and poly-(1,3-propylene succinate). The US market for succinic acid is 4.5×10^8 kg/year with a market price of \$2.8/kg and succinates are currently produced from hydrocarbons for all non-food applications.[38] For food applications, succinate is produced from sugar fermentation,[39] as well as from glycerol using *A. Succiniciproducens*.[40] However, due to high costs associated with separation and purification of succinate from the broth and the formation of by-products, further optimisation of the process is required.[41]

Ethanol
Ethanol is the most widely used biofuel in the world and is mainly produced in the USA using corn as a feedstock and in Brazil from sugar cane. Conventional petrol motor engines can process low-ethanol-petrol blends without major

modification; however, some cars have been modified to process 85% ethanol blends. Although most bioethanol is currently produced from sugar and starch, research is being conducted on exploiting glycerol as a natural resource. Yazdani and Gonzalez have shown that it is possible to efficiently convert crude glycerol to ethanol using engineered *Escherichia coli* (*E. coli*).[42]

Propylene Glycol
1,3-propanediol is produced from crude-oil-derived ethylene oxide using chemical catalysts developed by Shell.[43] It is also produced biologically by genetically engineered *E. coli* feeding on glucose. Novel polyester fibres are formed, which are used in carpet and textile applications when polymerised with terephthalic acid. Furthermore 1,2-propanediol is a high-volume chemical with a volume of 500 000 tonnes/year, which is used to produce antifreeze compositions. Anti-freeze can be produced containing 70% propylene glycol and 30% glycerol directly from biodiesel facilities.[44] Mono-propylene glycol [E1520], diacetin [E1517] and triacetin [E1518] are approved solvents in the flavourings industry.[45] A joint venture by Ashland Inc. and Cargill will be initially producing 65,000 tonnes/year of high-grade propylene glycol from biodiesel glycerol. The initial capital investment is about 100 million dollars and the plant is expected to be constructed in Europe.[46] Selective reductions can be employed to produce 1,2- and 1,3-propanediol from glycerol *via* hydrogenation using metallic catalysts. Propylene glycol can also be produced *via* dehydroxylation and biological reduction as discussed earlier.

Dihydroxyacetone
Dihydroxyacetone (DHA) has a global market of only 2000 tonnes;[23] however, it is very valuable and used in the cosmetics industry as an artificial tanning agent and as a synthon in organic synthesis.[47] DHA affects the sensory quality of wine with a sweet/etherish property. It is also known to react with proline to produce a 'crust-like' aroma and has anti-microbial properties. DHA can be produced *via* electrochemical oxidation, microorganisms and enzymes.

Acrolein
Acrolein is an important intermediate in the chemical industry and is used to produce polyester resin, polyurethane, propylene glycol, acrylic acid and acrylonitrile. A direct application of acrolein is its use as a herbicide for managing aquatic plants. Methionine is produced from acrolein and is used for animal feed to ensure sufficient growth, health and reproduction. Several million tonnes of acrolein are produced each year; however, it has serious health risks. Due to its toxic and explosive nature, high safety standards are required. Acrolein is currently produced industrially by oxidation of crude-oil-derived propene. Recent developments have led to hybrid processes where glycerol and propylene are converted to acrolein simultaneously. The first facility of such kind is being developed by Arkema, Texas.[23] Glycerol can be dehydrated to acrolein and 3-hydroxypropionaldehyde as discussed earlier.

Glycerol Tertiary Butyl Ether (GTBE)
Glycerol tertiary butyl ethers can be added to diesel fuels, which will help meet European biofuel targets and reduce particulate matter, carbon monoxide and unregulated aldehydes in the emissions.[48,49] Furthermore, the addition of these ethers into biodiesel decreases the cloud point (normally around 0 °C) to a value closer to conventional diesel (−16 °C).[50] In 2007, Appleby and Spooner-Wyman have shown that glycerol ethers can significantly reduce the NO_x emissions generated from the burning of pulverised coal.[49,51] These ethers are produced by etherification of alkenes such as iso-butylene in the presence of an acid catalyst. Industrially these ethers are produced from by-products of the biodiesel and ethanol industries using olefins derived from crude oil.[52] A number of companies in the USA and Europe are involved in commercialising GTBE.[53]

Mono- and Di-acylglycerol (DAG)
Mono- (MAG) and di-acylglycerol (DAG) are amphiphilic molecules and are used as ingredient blending agents in dairy/bakery products, margarines and sauces.[54] They are used as texturing agents to improve the consistency of creams and lotions.[55] These molecules have excellent lubricating and plasticising properties and are used in oils for textile machines.[56] Currently, there are two industrial methods for manufacturing acylglycerols, which are continuous chemical glycerolysis (250 °C, alkaline, N_2) and direct esterification of glycerol with fatty acids.[57] It is also possible to produce these molecules using biological processes (*e.g.* enzymatic conversion) as shown by Vicente and co-workers,[58] which offers high selectivity and lower temperatures and pressures to those of conventional chemical methods. Furthermore, the costs of enzymatic conversion can be significantly reduced by scaling-up, making it a potentially attractive commercial method.

Citric Acid
Citric acid is industrially produced from sugars using cultures of *Aspergillus Niger* and is used as a food preservative, flavourant, metabolite, environmentally benign cleaning agent and an antioxidant. Furthermore, it is used as an additive in the pharmaceutical, cosmetic and toiletry industries.[59] Glycerol is being considered as a cheaper alternative to sugar feedstock as Papanikolaou and others have shown by producing citric acid from raw glycerol using *Yarrowia lipolytica*.[60] Novel biodegradable polyesters can be formed from the reaction of citric acid and glycerol, which find potential applications in packaging and similar products.[61,62] Rectangular slabs of a highly cross-linked citric acid-glycerol copolymer matrix have been shown to release *in vitro* drugs such as sulfadiazine, paracetamol, diazepam, quinine hydrochloride and doxycycline hydrochloride.[63] Jiugao *et al.* investigated the effects citric acid would have on the properties of glycerol-plasticised thermoplastic starch (GPTPS).[64] Results showed that the addition of citric acid increased adhesion between citric acid, glycerol, water and starch in

thermoplastic starch. Rheological investigations showed that citric acid could decrease the shear viscosity and increase fluidity of thermoplastic starch.

Table 4.4 shows a summary of many reported chemicals that can be derived from glycerol. The respective applications for each chemical have also been listed.

Table 4.4 An alphabetical list of chemicals that can be derived from glycerol *via* chemical (C) or biochemical (B) methods.

Chemical	C	B	Method	Applications
Acetaldehyde	C		Pyrolysis	Intermediate in the production of acetic acid, certain esters and a number of other chemicals
Acetate		B	Microorganism (enterobacteriaceae)	Synthetic chemistry
Acetol	C		Dehydration	Intermediate in propane diol (1,2) synthesis
Acrolein	C		Dehydration, pyrolysis	Used in the preparation of polyester resin, polyurethane, propylene glycol, acrylic acid and acrylonitrile
Acrylic acid	C		Oxydehydration	Plastics, coatings, adhesives, elastomers, floor polishes and paints
Alcohols	C		Fischer–Tropsch	Beverage (ethanol only), as fuel and for many scientific, medical and industrial utilities
Alkanes	C		Fischer–Tropsch, pyrolysis	Fuels, chemicals, waxes, bitumen and numerous other applications
Alkenes	C		Fischer–Tropsch, pyrolysis	Synthesis and numerous other applications
Butanediol (2,3)		B	Microorganism (klebsiella pneumoniae)	Used in the resolution of carbonyl compounds in gas chromatography
Butanol		B	Microorganism (C. Pasteurianum)	Used as a solvent, as an intermediate in chemical synthesis and as a fuel
Butyraldehyde		B	Microorganism (C. Pasteurianum)	Chemical intermediate and has many other applications
Carbon monoxide	C		Aqueous phase reforming, pyrolysis	Chemical building block
Citric acid		B	Microorganism (Yarrowia lipolytica)	Flavouring and preservative in food, water softener, detergents and other uses
Diacyl glycerol	C	B	Esterification, enzymes	Additive, emulsifier

Table 4.4 (*Continued*)

Chemical	C	B	Method	Applications
Dichloropropanol (1,3)	C		Halogenation	Intermediate in Epichlorohydrin synthesis
Dihydroxyacetone	C	B	Electrochemical oxidation, microorganisms, enzyme (glycerol dehydrogenase)	Tanning, wine-making
Epichlorohydrin	C		Halogenation, dehydrochlorination	Building block in the manufacture of plastics, epoxy resins, phenoxy resins and other polymers. Used as a solvent and a precursor to glycidyl nitrate
Ethanol		B	Bacteria (*Escherichia coli*)	Solvent, fuel, antiseptic, antidote and many others
Ethylene glycol	C		Reduction	Coolant, hydrate inhibition, manufacturing, chemistry, geothermal, laboratory and other uses
Formaldehyde	C		Pyrolysis	Chemical building block, textiles, disinfectant, photography and other uses
Formic acid	C	B	Oxidation by-product	Preservative, antibacterial agent, organic chemistry, fuel cells and others
Fumarate		B	Anaerobic (Anaerobiospirillum succiniciproducens)	Intermediate in succinate synthesis
Glyceraldehyde	C		Oxidation	Intermediate in carbohydrate metabolism
Glycerol carbonate	C		Esterification	Solvent, additive and chemical intermediate
Glyceroldimethacrylate	C		Esterification	Building block for polymers, medical applications
Glycerol *tert* butyl ethers	C		Etherification	Fuel additive
Glycidyl nitrate	C		Esterification	Energetic binder, smoke propellant, explosives
Hydrogen		B	Aqueous phase reforming, pyrolysis, bacteria (*Escherichia coli*)	Internal-combustion engines, gas-fired turbines, fuel cells, chemical building block
Hydroxypropionaldehyde	C	B	Dehydration, aerobic conversion (klebsiella pneumonia), enzyme (glycerol dehydratase)	Chemical intermediate and precursor to acrolein

Table 4.4 (*Continued*)

Chemical	C	B	Method	Applications
Hydroxybutanone		B	Microorganism	Food and cigarette additive
Hydroxyethanoic acid	C		Oxidation by-product	Skin care products, organic synthesis and other uses
Hydroxypyruvic acid	C		Oxidation	Intermediate in biological synthesis
Lactate		B	Microorganism (enterobacteriaceae)	Polymer precursor, plastics and other uses
Malate		B	Microorganism	Intermediate in fumarate synthesis
Mesoxalic Acid	C		Oxidation	Chemical intermediate, antidote to cyanide poisoning
Monoacyl glycerol	C	B	Esterification, enzymes	Additive, emulsifiers
Monobenzoyl glycerol	C		Esterification	Chemical intermediate
Oxalic acid	C		Oxidation	Chemical intermediate and many other uses
Oxaloacetate		B	Microorganism	Intermediate in malate synthesis
Phosphoenolpyruvate		B	Enzyme (dihydroxyacetone kinase)	Intermediate in pyruvate synthesis
Propanediol (1,2)	C	B	Hydrogenolysis, dehydroxylation, bacteria (Clostridium)	Moisturiser, lubricant, food additive, anti-freeze, coolant and many other uses
Propanediol (1,3)	C	B	Hydrogenolysis-dehydroxylation, acetalisation-detosylation, bacteria (Clostridium), enzyme (1,3 propane diol dehydrogenase)	Polyester fibres, carpets and textiles
Propanol	C		Hydrogenolysis	Solvent in the pharmaceutical industry, resins and cellulose esters
Propenol (2,1)	C		Dehydration, hydrogenation	Pesticide and chemical building block
Propionate		B	Bacteria (Propionibacterium acidipropionici)	Food preservative, chemical intermediate, pesticide, pharmaceutical, flavouring solvent
Pyruvate		B	Enzyme (Dihydroxyacetone Kinase)	Intermediate in biosynthesis
Succinic acid		B	Anaerobic (Anaerobiospirillum succiniciproducens)	Food additives, soldering fluxes, pharmaceutical products, surfactants, green solvents, biodegradable plastics and chemical intermediate
Tartronic acid	C		Oxidation	Chemical intermediate in mesoxalic acid formation

4.1.2.3 Employing Crude Glycerol from Biodiesel Production as an Alternative Green Reaction Medium

Glycerol is principally used as a highly refined and purified product in all the above-mentioned applications (as a reactant or as an additive). However, further purifications are needed including bleaching, deodorising and ion exchange to remove trace impurities if it is aimed to be employed in food, cosmetics and drugs. Purifying it to that stage, however, is very costly and generally out of the range of economic feasibility for biodiesel plants. Hence, as more and more crude glycerol is generated by the biodiesel industry, it is very important that economical methods of low-grade glycerol utilisation be explored to further defray the cost of biodiesel production in the growing global market.[65] Glycerol was recently reported to be a versatile and alternative green solvent in a variety of catalytic and non-catalytic organic reactions and synthesis methodologies, yielding high product conversions and selectivities.[66–68]

Solvents are utilised daily in numerous industrial processes as reaction media. Besides reactant and catalyst solubility, heat and momentum transfer, the chemical, physical and biological nature of the solvent also plays a key role from the environmental, economic, safety, handling and product isolation viewpoints.[69] In this respect, the use of glycerol as reaction medium tolerated a range of solubilities from various reactants and catalysts, and provided an easy separation of products and transition metal complex recycling, as well as the possibility to work under microwave irradiation conditions.

Crude glycerol from the alcoholysis of triglycerides has successfully been employed as a green solvent in base-catalysed aldol condensation (Scheme 4.3a) and palladium catalysed Heck carbon–carbon coupling (Scheme 4.3b) without any purification.[65]

The oil source did not affect reaction performance, yet the reactions in crude glycerol usually showed lower conversions than those conducted in pure glycerol. However, the residual base, which was used as catalyst in the alcoholysis reaction, was further used as catalyst in the investigated processes.[65]

4.1.2.4 Future Vision

The biodiesel market worldwide is at an early stage of its evolution. This is evident by the collapse of several large biodiesel manufacturers including D1

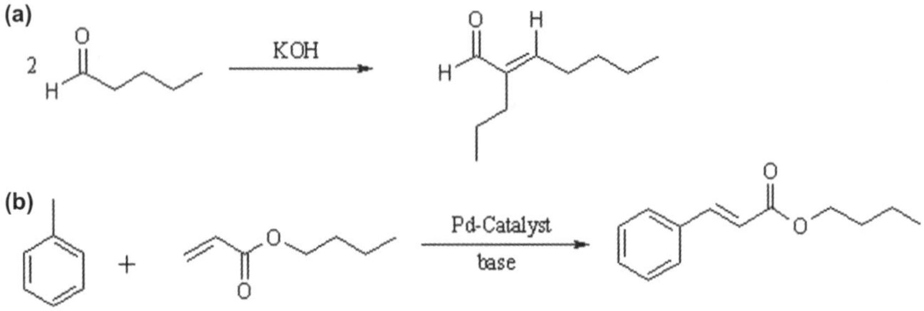

Scheme 4.3

oils and Biofuels Corporation in the UK. On the other hand, many producers from the USA are exporting diesel to Europe and the rest of the world. In the future, the world's largest plant oil producing countries such as Brazil, Indonesia and Malaysia are likely to produce biodiesel and export it rather than just exporting the oil. Recent developments in utilising India's hundreds of millions of hectares of brownfield land for jatropha production could further affect the biodiesel markets as well as having many economic, social and environmental implications for the country. The market for glycerol is therefore likely to remain volatile in the near future. Chemical industries need to be approached at a local, national and international level to determine their requirements and then research needs to be conducted on glycerol in association with biodiesel producers, chemists, biologists and engineers to provide a solution.

4.2 Novel Routes to Biodiesel Incorporating Glycerol into Their Composition

Triglycerides (TGs) of vegetable oils and fats are becoming increasingly important as alternative fuels for diesel engines due to the diminishing petroleum reserves. However, their high viscosities and low volatilities do not allow their direct use or use in oil/petrol blends,[70,71] in any diesel engine type.[72] Nowadays, the main process developed to overcome this drawback is the methanolysis reaction of such feedstocks to produce biodiesel, a biodegradable, non-toxic diesel fuel substitute that can be used in unmodified diesel engines.[1,73] Biodiesel has a significant added value compared to petro-diesel because of its higher lubricity, which extends engine life and reduces maintenance costs as well as contributing to fuel economy.[74] The conventional methodology in the production of biodiesel primarily involves the use of NaOH and KOH as homogeneous catalysts. Three molecules of fatty acid methyl esters (FAME) and one molecule of glycerol are generated for every molecule of TG.[75]

However, the process is far from being environmentally friendly as the final mixture needs to be separated, neutralised and thoroughly washed, generating a great amount of waste in terms of salt residues. Also, the catalyst cannot be recycled. These several additional steps inevitably put the total overall biodiesel production costs up, reducing at the same time the quality of the glycerol obtained as by-product.[76] Ethanol could potentially be used instead of methanol, but the rates of reaction are comparatively slower.

Several reports can be found on the production of biodiesel involving other chemicals,[74,77] or enzymatic catalytic protocols as greener alternatives.[78,79] The increasing environmental concerns have led to a growing interest in the use of enzyme catalysis as it usually produces a cleaner biodiesel under milder conditions.[80] It also generates less waste than the conventional chemical process. Many reports on the preparation of biodiesel using free[81] or immobilised lipases can also be found.[82–84] Despite various attempts in which the enzyme was tested for the efficient production of biodiesel, the FAME conversions were lower than 60%.[85–88] However, Paula *et al.* have very recently obtained biodiesel synthesis from babassu oil and ethanol, propanol or butanol was feasible and, regardless of the kind of alcohols, results revealed that the immobilised pig pancreatic lipase (PPL) could efficiently convert triglycerides to fatty acid alkyl esters attaining yields varying from 75 to 95%.[89]

The major drawback of the process is the high cost due to the various steps involved that can limit somehow the widespread use of enzymes. Moreover, the true limitation of the enzymatic method compared to the conventional base catalysed process deals with the alcoholysis of only two fatty acid esters of glycerol. Lipases have indeed a peculiar 1,3-regioselectivity which means that they selectively hydrolyse the more reactive 1 and 3 positions in the triglyceride.[90] In this regard, the production of biodiesel using lipases needs to take into account such regiospecific character.[91,92] In general, the challenging full alcoholysis of triglycerides involves long reaction times and gives conversions lower than 70 wt% in fatty acid methyl or ethyl esters.[93,94]

A series of improvements in conversion levels and/or the use of methanol as alcohol to mimic the results of the base-catalysed transesterification reaction are currently ongoing as a consequence of the present legal regulations for biodiesel (EN 14214). Reasonably good results are sometimes reported due to the 1,2-acyl migration in the monoglycerides.[95–97]

The current standard biodiesel production (under alkaline chemical conditions) is considered to be the most technically simple way to reduce the viscosity of vegetable oils from a range of 11–17 times.[98–100] Various fuel properties of pure soybean oil, three B100 biodiesel types (soybean methyl esters, rapeseed methyl esters and rapeseed ethyl esters) and high-grade petrodiesel are summarised in Table 4.5.

The viscosity is the only significant parameter that may affect the performance of the diesel engine, as the other parameters are very similar. Interestingly, diglycerides (DG) and triglycerides (TG) are mainly responsible for the increase in viscosity of pure vegetable oils. Therefore, a novel biofuel containing a FAME/MG or FAEE/MG blend (in which we exclude the presence of significant quantities of DG and TG) can be expected to have similar physical

Table 4.5 Physico-chemical properties of soybean oil, biodiesel (B100) obtained from soybean oil and rapeseed oil and No. 2 diesel (D2).[100]

Properties	Soybean oil	FAME[a]	FAME[b]	FAEE[c]	D2
Specific gravity (g cm^{-3})	0.920	0.86	0.8802	0.876	0.8495
Viscosity (40 °C)	46.68	6.2	5.65	6.11	2.98
Cloud point (°C)	2	−2.2	0	−2	−12
Pour point (°C)	0	−9.4	−15	−10	−18
Flash point (°C)	274	110	179	170	74
Boiling point (°C)	357	366	347	273	191
Cetane number	48.0	54.8	61.8	59.7	49.2
Sulfur (%wt)	0.022	0.031	0.012	0.012	0.036
Heat of combustion (kJ kg^{-1})	40.4	40.6	40.54	40.51	45.42

[a]FAME stands for fatty acid methyl esters from soybean oil.
[b]FAME stands for fatty acid methyl esters from rapeseed oil.
[c]FAEE stands for fatty acid ethyl esters from rapeseed oil.

properties to those of conventional biodiesel, eliminating the production of glycerol as by-product. Besides, the transformation of glycerol in an alternative product miscible with FAME can also avoid the formation of such glycerol and therefore suppress any of the previously devised strategies for glycerol valorisation. The achievement of a glycerol-free biofuel is most convenient and advantageous in a market flooded by the production of glycerol as by-product in the preparation of biodiesel.[18,24,26,101,102]

Furthermore, the biofuel obtained is cleaner and the efficiency of the production can be increased more than 10% when the glycerine is somehow integrated into the biofuel. The last step of washing and cleaning of the biodiesel in the conventional synthetic process [to clean the biodiesel and remove the traces of glycerol up to 0.02% glycerol (EN 14214)] can therefore be removed, reducing costs and generation of wastewater.[103]

High levels of glycerol in the fuel causes various problems including coking, an increase in the viscosity of the fuel and a potential dehydration to acrolein, which can be further polymerised. Coking can also generate deposits of carbonaceous compounds on the injector nozzles, pistons and valves in standard engines, reducing the efficiency of the engines.[104,105]

Recent investigations have also shown that minor components of biodiesel, usually considered contaminants under the biodiesel standard EN 14214, including free fatty acids and monoacyl glycerols, are essentially responsible for the lubricity of low-level blends of biodiesel and petrodiesel. Pure FAME exhibited a reduced lubricity compared to the biodiesel containing these compounds.[10,106–110] The presence of greater quantities of monoglycerides and/or free fatty acids enhances the lubricity of biodiesel, which is another key feature of this novel biofuel that incorporates high amounts of MG.

4.2.1 Novel Biofuels Integrating Glycerol into Their Composition

The production of a biofuel integrating glycerol into its composition is currently a target of high interest, given that the market is already virtually swamped by the production of glycerol, obtained precisely as a by-product in the manufacture of biodiesel.[24,101] Table 4.6 shows the evolution of the drop in the price of this product in recent years.

Glycerol cleaning from biofuels production cannot be avoided, as its presence in the fuel is limited by the standard EN 14214 to less than 0.02% due to its undesirable reactions with oxygen inside the engine at high temperatures. Under these conditions, glycerol undergoes a process of oxidation and

Table 4.6 Changes in the price of glycerine in recent years.[101]

Glycerol	Year 2000	Year 2005	Year 2010
Refined	€1.5	€0.9	€0.6
Unrefined	€1.1	€0.6	€0.3

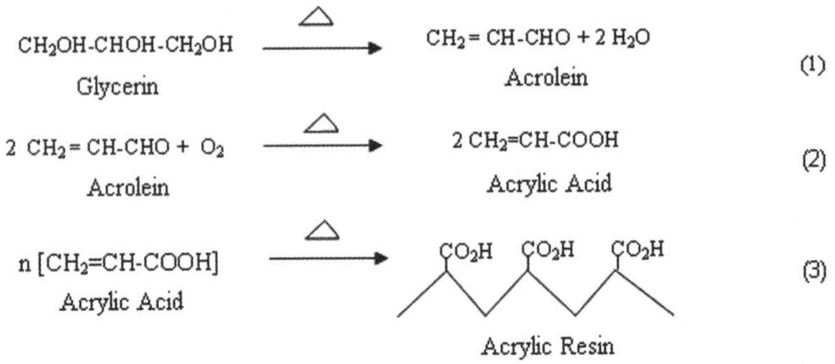

$$CH_2OH\text{-}CHOH\text{-}CH_2OH \xrightarrow{\Delta} CH_2=CH\text{-}CHO + 2\,H_2O \quad (1)$$
Glycerin → Acrolein

$$2\,CH_2=CH\text{-}CHO + O_2 \xrightarrow{\Delta} 2\,CH_2=CH\text{-}COOH \quad (2)$$
Acrolein → Acrylic Acid

$$n\,[CH_2=CH\text{-}COOH] \xrightarrow{\Delta} \text{Acrylic Resin} \quad (3)$$
Acrylic Acid

Scheme 4.4

polymerisation to acrolein, resulting in the formation of carbonaceous deposits on injectors, pistons and valves that reduce the engine efficiency (Scheme 4.4).[104,105]

Recent research has also shown that the minority components of biodiesel standard (which generally are considered pollutants under the standard EN 14214), including free fatty acids and monoacyl glycerol or monoglycerides (MG), are mainly responsible for the high lubricity that is obtained in the blends of biodiesel and petrodiesel, even at lower levels. FAMEs with very high purity showed a high reduction of lubricity as compared with the same FAMEs containing these compounds.[106–110] The presence of large quantities of monoglycerides and/or free fatty acids, or various derivatives of these, leads to a novel family of biofuels, which improves the lubricant power of biodiesel due to the incorporation of some derivatives of glycerol. Lubricity is a very important feature of biofuels, which improves performance and preserves the life of the engines.

In this respect, interesting results have been so far obtained using a pig pancreatic lipase (PPL),[111] in both free and immobilised forms. The use of such biocatalyst facilitates the 1,3 selective transesterification of TG to produce the corresponding 2-monoacil derivatives of glycerine (MG), and two moles of fatty acids ethyl esters (FAEE) with ethanol employed as alcohol in the reaction. An inherent advantage of enzymatic transesterification processes is that they are not restricted to the use of methanol as alcohol (as usual under conventional chemical reactions) but they can be conducted using different short-chain alcohol (ethanol, 1- and 2-propanol, 1- and 2-butanol, *etc.*) and their mixtures. The process is summarised in Scheme 4.5.

Other alternatives have also been reported in the aim to prepare related esters of fatty acids from TG that avoid the generation of glycerol as by-product in the reaction.

The transesterification of triglycerides with dimethyl carbonate (DMC) (Scheme 4.6) produces a mixture of three molecules of the FAME or FAEE and a molecule of glycerol carbonate (GC).[112–116] Such a FAME/GC mixture has

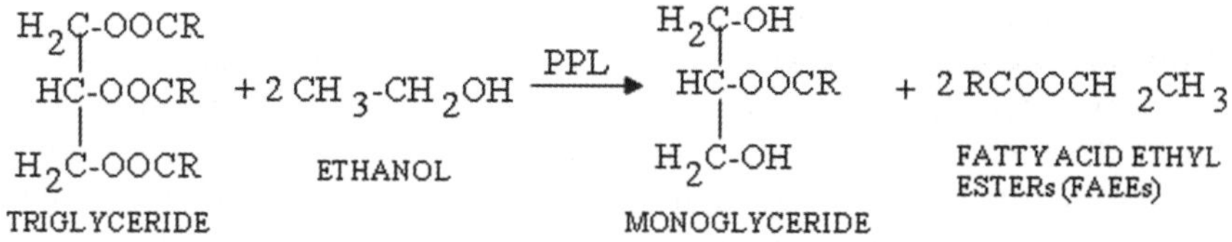

Scheme 4.5

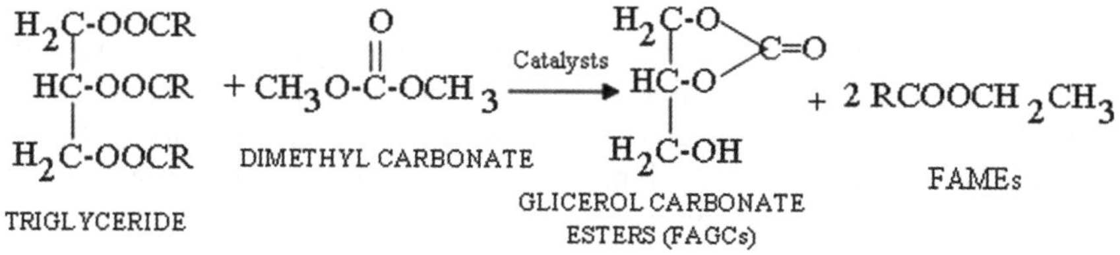

$$\begin{array}{l}\text{H}_2\text{C-OOCR}\\\text{HC-OOCR}\\\text{H}_2\text{C-OOCR}\end{array} + 2\,\text{CH}_3\text{-CH}_2\text{OH} \xrightarrow{\text{PPL}} \begin{array}{l}\text{H}_2\text{C-OH}\\\text{HC-OOCR}\\\text{H}_2\text{C-OH}\end{array} + 2\,\text{RCOOCH}_2\text{CH}_3$$

TRIGLYCERIDE — ETHANOL — MONOGLYCERIDE — FATTY ACID ETHYL ESTERs (FAEEs)

Scheme 4.6

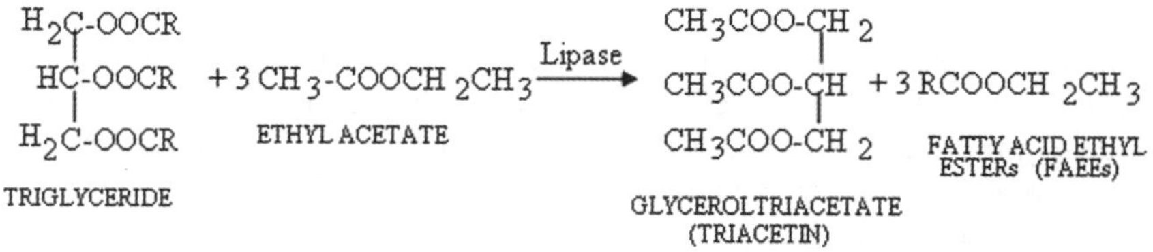

$$\begin{array}{l}\text{H}_2\text{C-OOCR}\\\text{HC-OOCR}\\\text{H}_2\text{C-OOCR}\end{array} + \text{CH}_3\text{O-}\underset{\text{O}}{\overset{\|}{\text{C}}}\text{-OCH}_3 \xrightarrow{\text{Catalysts}} \begin{array}{l}\text{H}_2\text{C-O}\\\text{HC-O}\\\text{H}_2\text{C-OH}\end{array}\!\!\!\!\!\!\!\!\!\!>\!\text{C=O} + 2\,\text{RCOOCH}_2\text{CH}_3$$

TRIGLYCERIDE — DIMETHYL CARBONATE — GLICEROL CARBONATE ESTERS (FAGCs) — FAMEs

Scheme 4.7

$$\begin{array}{l}\text{H}_2\text{C-OOCR}\\\text{HC-OOCR}\\\text{H}_2\text{C-OOCR}\end{array} + 3\,\text{CH}_3\text{-COOCH}_2\text{CH}_3 \xrightarrow{\text{Lipase}} \begin{array}{l}\text{CH}_3\text{COO-CH}_2\\\text{CH}_3\text{COO-CH}\\\text{CH}_3\text{COO-CH}_2\end{array} + 3\,\text{RCOOCH}_2\text{CH}_3$$

TRIGLYCERIDE — ETHYL ACETATE — GLYCEROLTRIACETATE (TRIACETIN) — FATTY ACID ETHYL ESTERs (FAEEs)

suitable physical properties to be employed as fuel, so it has been presented as a new biofuel denoted as DMC-BIOD.[11,113]

Similarly, Gliperol is another novel recently patented biofuel consisting of a mixture of three molecules of FAME and a molecule of glycerol triacetate (triacetin).[117] It can be obtained after the transesterification of a mole of TG with three moles of ethyl acetate using lipases as biocatalyst,[117–122] following the scheme showed in Scheme 4.7.

DMC-Biod, Gliperol and Ecodiesel are examples of novel biofuels that incorporate glycerol into their composition (as glycerol carbonate, triacetate or monoglyceride), thus avoiding the generation of waste or by-products in their preparation. A main difference with respect to conventional biodiesel (FAME) is that the production of such biofuels does not require any additional separation processes. MG, the DMC or triacetin may be perfectly incorporated (and thus burnt) with FAME in the diesel engines. In terms of green chemistry, the glycerol incorporation into the biofuel composition also increases the efficiency of the process (nominally from the current 90% to 100%), without causing substantial changes in the physical-chemical properties of the final products. The atomic efficiency also experiences the corresponding improvement.

The application of lipases, with their inherent ability to promote the 1.3-regiospecific alcoholysis of TG with short-chain alcohols, may constitute competitive procedures to those existing to biodiesel in the medium term, as compared to the conventional homogeneously catalysed method. The use of such biocatalysts for the production of these novel families of biofuels reduces the complexity of the process (avoiding washing steps to remove the residual glycerol), increases the process yield and minimises waste generation. In addition, biocatalysed pathways are conducted under comparatively milder reaction conditions (pH, temperature, pressure, *etc.*) as compared to the chemical routes for biodiesel production.

The actual existing limitations for the use of industrial lipases have been mainly associated with their high production costs, which can be overcome through the application of molecular technologies to achieve the production of enzymes purified in sufficiently high quantities as well as the long times of reaction required for the complete conversion of feedstocks to biodiesel. Indeed, another important constraint of the enzymatic method as compared with conventional basic catalysts relates to the ability of the enzymes to produce the 2-fatty acid esters of glycerol. This is due to the 1,3-regioselective character of many lipases, which selectively hydrolyse positions 1 and 3 of triglycerides.[90]

Therefore, the production of biodiesel with lipases must take into account that 1,3-stereoselective character.[91,92] Most of the described enzymatic processes to produce conventional biodiesel (exclusively a mixture of FAMEs) involve long reaction times and conversions less than 70% by weight of the methyl esters of fatty acids.[93,94]

4.2.2 Processing of Oils and Fats in the Actual Oil Refining Plants

An alternative directly accessible methodology to transform the triglycerides obtained from renewable sources is to carry out their processing in conventional crude oil refineries, together with the relevant portions of heavy crude oil of equivalent molecular weights. The production of high-quality diesel fuel from vegetable oils has been reported to take place *via* hydrocracking of triglycerides treated with high-molecular-weight hydrocarbons in conventional oil refineries.[123]

In this way, renewable liquid hydrocarbons (mainly alkanes) can be produced by treatment of mixtures of vegetable oils and fractions of heavy oil vacuum (HVO), in flows of hydrogen and conventional catalysts (sulfonated NiMo/Al_2O_3) under standard conditions of temperature (300–450 °C). The reaction involves the hydrogenolysis of C–C bonds from vegetable oils, which leads to a mixture of lower molecular weight alkanes by three different routes: decarbonylation, decarboxylation and hydrodesoxygenation. Figure 4.6 summarises the aforementioned processes.

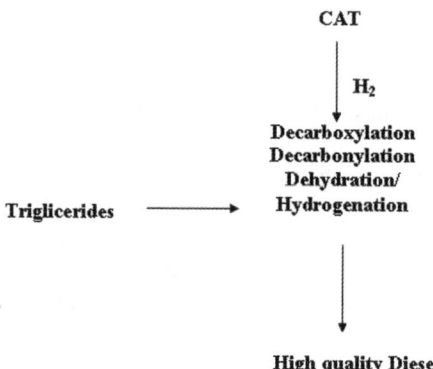

Figure 4.6 Production of high-quality biodiesel from vegetable oils, through overall hydrotreatment with petroleum hydrocarbons in conventional refineries.

4.2.3 Second-generation Technologies for the Production of Biodiesel-like Fuels

The main second-generation technologies that are currently being investigated intend to use engineered microorganisms for the production of biodiesel-type products or to convert cellulose waste into biodiesel *via* thermochemical processes.

The production of biofuels *via* microbial biotechnology is a very active field of research.[124] The biosynthesis of biodiesel-adequate FAEEs, referred to as microdiesel, in metabolically engineered *Escherichia coli* was recently reported.[125] This was achieved by heterologous expression in *E. coli* of the *Zymomonas mobilis* pyruvate decarboxylase and alcohol dehydrogenase and the unspecific acyltransferase from *Acinetobacter baylyi* strain ADP1. By this approach, ethanol formation was combined with subsequent esterification of the ethanol with the acyl moieties of coenzyme A thioesters of fatty acids, if the cells were cultivated under aerobic conditions in the presence of glucose and oleic acid. Ethyl oleate was the major constituent of these FAEEs, with minor amounts of ethyl palmitate and ethyl palmitoleate. FAEE concentrations of 1.28 g l^{-1} and an FAEE content of the cells of 26% (of the cellular dry mass) were achieved by fed-batch fermentation using renewable carbon sources. This novel approach might pave the way for the industrial production of biodiesel equivalents from renewable resources by employing engineered microorganisms, enabling a broader use of biodiesel-like fuels in the future.

On the other hand, thermochemical conversion processes can be subdivided into gasification, pyrolysis, supercritical fluid extraction and direct liquefaction.

Pyrolysis is the thermochemical process that converts biomass into liquid, charcoal and non-condensable gases, acetic acid, acetone and methanol by heating the biomass to about 450–500 °C in the absence of air. If the purpose is to maximise the yield of liquid products resulting from biomass pyrolysis, a low-temperature, high-heating-rate, short gas residence time process would be required. For high char production, a low-temperature, low-heating-rate process has to be chosen. If the purpose is to maximise the yield of fuel gas resulting from pyrolysis, a high-temperature, low-heating-rate, long gas residence time process would be preferred.[126]

Fischer–Tropsch synthesis (FTS) is one of the options available for utilising cellulosic biomass for fuel production and one of the most advanced second-generation processes that include gasification of biomass raw materials, cleaning and packaging of synthesis gas and subsequent synthesis of liquid (or gas) biofuel. This German process has been known since 1920, but in the past was used mainly for production of liquid fuels or natural gas from coal. However, the process, which uses biomass as a raw material, is still under development. Any type of biomass can in principle be utilised as a feedstock, including wood, grass, agricultural and forest residues.[127]

Fischer–Tropsch (FT) hydrocarbons can be produced by gasification of biomass, followed by downstream processing. In South Africa, FT-liquids have been produced from coal for many years and Malaysia has a plant producing FT-liquids from natural gas. Using biomass as feedstock for FT-synthesis is a relatively novel concept which brings with it several issues that still need to be overcome, most of them in the gas cleaning phase (Figure 4.7).

The biomass FT plant comprises:

- Biomass pre-treatment (chipping, drying);
- Gasification (resulting in syngas);

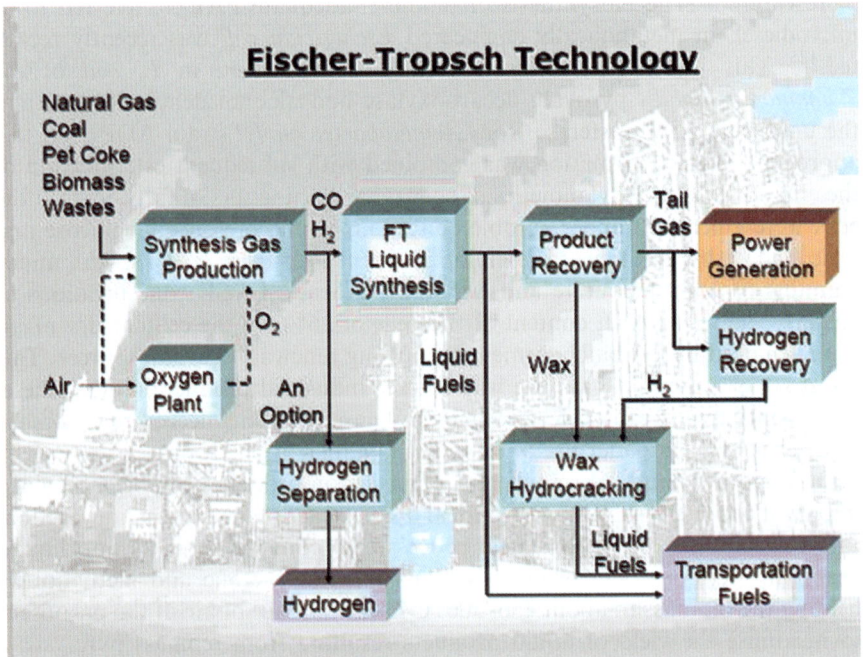

Figure 4.7 Fisher-Tropsch process.

- Gas cleaning and conditioning. The syngas needs to be purified to remove, *inter alia*, organic (BTX) and inorganic impurities and tars as well as impurities that can deactivate the FT and other catalysts;
- FT-reactor;
- Hydro-cracker.

As with the ethanol production process, different configurations are possible. Most configurations produce electricity and heat as by-products. Overall process efficiencies vary with plant design from 40% to 60–65%.[128]

Biomass is vaporised to produce synthesis gas, which is a mixture of carbon monoxide (CO) and hydrogen (H_2). Before the synthesis, this gas can be conditioned using steam *via* water gas shift processes to achieve a particular ratio of H_2/CO for the synthesis. The liquid products obtained from the synthesis gas, which include several fractions of hydrocarbons, are very pure (free of sulfides) and can be easily converted to fuel for cars. The Fischer–Tropsch biodiesel can be produced directly from FTS; however, a greater conversion to biodiesel can be achieved if the Fisher–Tropsch wax materials obtained as by-products are subsequently reformed *via* hydrocracking.

Fischer–Tropsch biodiesel properties (*e.g.* energy content, density and viscosity) are similar to diesel oil properties. This biodiesel can therefore be blended with petroleum diesel in any proportion without making any changes in car engines. However, the Fischer–Tropsch biodiesel has certain properties that make it a better fuel including a higher cetane number (improved car ignition) and it is a fuel with low aromatic content, which in turn results in lower NO_x emissions of particulates.

For the application of this technology, the main challenge is the production of synthesis gas. This synthesis gas is produced by gasification at a high temperature, similar to that employed for coal gasification. However, biomass has different properties from coal and several modifications in the conventional process are thus needed.

Firstly, a biomass pre-treatment step is needed and this requires different processes, because the grinding of the feedstock to small particles comparatively requires a lot of energy. Pyrolysis pre-treatment process can solve problems that arise when aggregates of small particles congregate in the power lines.

Secondly, the gasification temperature needs to be reduced due to the higher reactivity of the biomass compared with coal, thus requiring special designs for the gasification process and burners.

Thirdly, the composition of the ashes of the biomass is different from that of coal, which is manifested in different behaviours of ash and waste, which are important factors in the gasifier and requires further consideration. The behaviour of the ashes and waste is also important for the cooling of the synthesis gas.

Further research is therefore needed to develop such technologies, especially in the cleaning and packaging of the syngas, development of different catalysts and using by-products for the generation of electricity, heat and steam. FTS processes that employ biomass as a feedstock are currently being investigated in

Germany, Sweden and Austria. In Germany the company Choren, with the support of the German government, Volkswagen and DaimlerChrysler, is building a demonstration plant with an annual production capacity of 15 000 tons of FTS fuels.[129]

Another key technology for the production of biodiesel through thermochemical processes is Hydro Thermal Upgrading (HTU).[130] HTU aims to convert biomass, regardless of its origin, to different types of biofuels according to the general outline of Figure 4.8.

Hydrothermal technologies are broadly defined as chemical and physical transformations in high-temperature (200–600 °C), high-pressure (5–40 MPa) liquid or supercritical water. This thermochemical means of reforming biomass may have energetic advantages as a phase change to steam is avoided when water is heated at high pressures (also avoiding large enthalpic energy penalties). Biomass sources undergo a range of reactions, including dehydration and decarboxylation reactions, which are influenced by the temperature, pressure, concentration and presence of homogeneous or heterogeneous catalysts. Several hydrothermal conversion processes are currently under development or demonstration. Liquefaction processes generally involve lower temperature reactions (200–400 °C), which produce liquid products, often denoted as 'bio-oils' or 'bio-crudes'. Gasification processes generally take place at higher temperatures (400–700 °C) and can produce methane or hydrogen gases in high yields (Figure 4.9).[132]

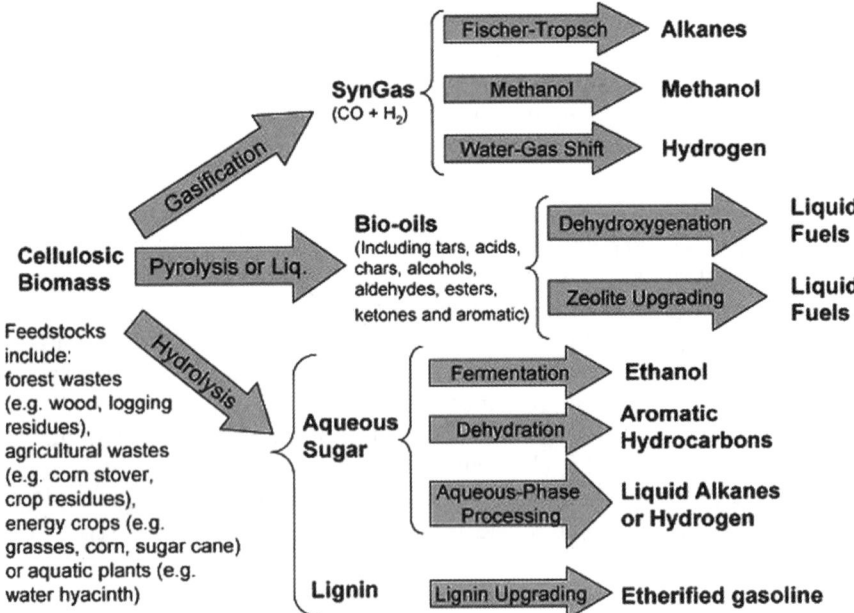

Figure 4.8 Strategies for production of fuels from lignocellulosic biomass adapted from Huber and Dumesic, 2006 (ref. 132).

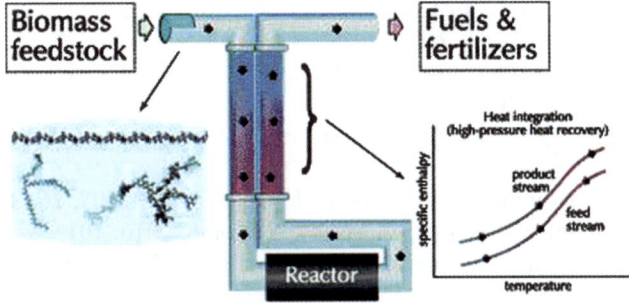

Figure 4.9 Strategies for biomass hydrothermal conversion processes (ref. 133).

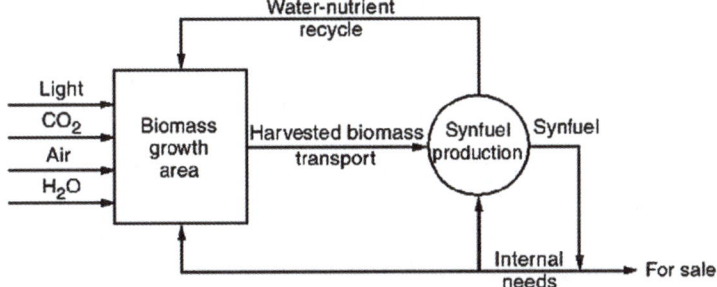

Figure 4.10 A conceptual model for integrated biomass production and conversion integration system.

Taking advantage of the oil productivity of many microalgae as compared to oil-producing crops, a great deal of research in the area of hydrothermal liquefaction technologies has been focused on the direct conversion of algal biomass to liquid fuels (Figure 4.10).

A problem associated with algal biomass is its relatively high water content that normally requires pre-treatment to remove it and increase the energy density. This requirement consequently increases the energy cost. However, direct hydrothermal liquefaction in sub-critical water conditions can be employed to convert the wet biomass to liquid fuel without the need to reduce the water content.

Overall, by adopting integrated approaches, including wastewater treatment, nutrients and heavy metals recovery by algae culture, whereby additional economic benefits are created, the obstacle of high cost of biodiesel production from algae may be overcome.[133]

References

1. H. Fukuda, A. Kondo and H. Noda, *J. Biosci. Bioeng.*, 2001, **92**, 405.
2. S. V. Ghadge and H. Raheman, *Bioresour. Technol.*, 2006, **97**, 379.

3. Z. M. Wang, J. S. Lee, J. Y. Park, C. Z. Wu and Z. H. Yuan, *Korean J. Chem. Eng.*, 2007, **24**(6), 1027.
4. Z. M. Wang, J. S. Lee, J. Y. Park, C. Z. Wu and Z. H. Yuan, *Korean J. Chem. Eng.*, 2008, **25**(4), 670.
5. Z. Chi, D. Pyle, Z. Wen, C. Frear and S. Chen, *Process. Biochem.*, 2007, **42**, 1537.
6. D. T. Johnson and K. A. Taconi, *Environ. Prog.*, 2007, **26**, 338.
7. M. Hájek and F. Skopal, *Bioresour. Technol.*, 2010, **101**, 3242–3245.
8. R. Brockmann, L. Jeromin, W. Johannisbauer, H. Meyer, O. Michel and J. Plachenka, *US Pat.* 4,655,879, 1987.
9. T. L. Ooi, K. L. Yong, K. Dzulkefly, W. M. Z. Wan Yunus and A. H. Hazimah, *J. Oil Palm Res.*, 2001, **13**(2), 16.
10. G. Knothe, J van Gerpen and J. Krahl, *The Biodiesel Handbook*, AOCS Press, Illinois, 2005.
11. K. C. Yong, T. L. Ooi, K. Dzulkefly, W. M. Z. Wan Yunus and A. H. Hazimah, *J. Oil Palm Res.*, 2001, **13**(2), 1.
12. A. H. Hazimah, T. L. Ooi and A. Salmiah, *J. Oil Palm Res.*, 2003, **15**(1), 1.
13. M. Carmona, J. Valverde and A. Pérez, *J. Chem. Tech. Biotech.*, 2008, **84**, 738–744.
14. X. Lancrenon and J. Fedders, *Biodiesel Mag.*, 2008, June, www.biodieselmagazine.com.
15. R. Christoph, B. Schmidt, U. Steinberner, W. Dilla and R. Karinen, *Ullmann's Encyclopedia of Industrial Chemistry*: electronic release, 6th edn, 2006.
16. J. E. Aiken, *US Pat.* 7,126,032 B1, 2006.
17. J. A. Hunt, *Pharm. J.*, 1999, **263**, 985.
18. M. Pagliaro, R. Ciriminna, H. Kimura, M. Rossi and C. D. Pina, *Angew. Chem. Int. Ed.*, 2007, **46**, 4434–4440.
19. G. Leffingwell and M. Lesser, *Merck Index*, Merck, 11th edn, 1945, **705**.
20. M. McCoy, *Chem. Eng. News*, 2006, **84**, 6–7.
21. J. Bonnardeaux, *Glycerin Overview*, Report for the Western Australia Department of Agriculture and Food, November, 2006.
22. K. D. Weiss, *Prog. Polym. Sci.*, 1991, **22**, 203–245.
23. M. Pagliaro and Michele Rossi, *The Future of Glycerol: New Usages for a Versatile Raw Material*, Royal Society of Chemistry, 2008.
24. A. Behr, J. Eilting, K. Irawadi, J. Leschinski and F. Lindner, *Green Chem.*, 2008, **10**, 13–30.
25. Y. Zheng, X. Chen and Y. Shen, *Chem. Rev.*, 2008, **108**, 5253–5277.
26. S. Yazdani and R. Gonzalez, *Curr. Opin. Biotechnol.*, 2007, **18**, 213–219.
27. G. Silva, M. Mack and J. Contiero, *Biotechnol. Adv.*, 2009, **27**, 30–39.
28. R. R. Soares, D. A. Simonetti and J. A. Dumesic, *Angew. Chem. Int. Ed.*, 2006, **45**, 3982.
29. K. Wang, M. C. Hawley and S. J. DeAthos, *Ind. Eng. Chem. Res.*, 2003, **42**, 2913.
30. H. Biebl, K. Menzel, A. P. Zeng and W. D. Deckwer, *Appl. Microbiol. Biotechnol.*, 1999, **52**, 289.

31. V. L. Budarin, J. H. Clark, R. Luque, D. Macquarrie, A. Koutinas and C. Webb, *Green Chem.*, 2007, **9**, 992–995.
32. N. O. V. Sonntag, *J. Am. Oil Chem. Soc.*, 1992, **59**, 795.
33. M. J. Antal Jr., W. S. L. Mok, J. C. Roy and T. A. Raissi, *J. Anal. Appl. Pyrol.*, 1985, **8**, 291.
34. T. Valliyappan, N. N. Bakhshi and A. K. Dalai, *Bioresour. Technol.*, 2008, **99**, 4476–4483.
35. W. Bühler, E. Dinjus, H. J. Ederer, A. Kruse and C. Mas, *J. Supercrit. Fluids*, 2002, **22**, 37.
36. P. J. Slininger and R. J. Bothast, *Appl. Environ. Microbiol.*, 1985, **50**, 1444–1450.
37. L. Schlapbach and A. Zuttel, *Nature*, 2001, **414**, 353–358.
38. *Biorefineries – Industrial Processes and Products*, ed. B. Kamm, P. Gruber and M. Kamm, Wiley-VCH, 2006, vol. 1.
39. J. G. Zeikus, M. K. Jain and P. Elankovan, *Appl. Microbiol. Biotechnol.*, 1999, **51**, 545–552.
40. P. C. Lee, W. G. Lee, S. Y. Lee and H. N. Chang, *Biotechnol. Bioeng.*, 2001, **72**, 41–48.
41. H. Song and S. Y. Lee, *Enzyme Microbiol. Technol.*, 2006, **39**, 352–361.
42. S. S. Yazdani and R. Gonzalez, *Metab. Eng.*, 2008, **10**, 340–351.
43. K. T. Lam, J. P. Powell and P. R. Wieder, WO Patent 9 716 250, 1997.
44. C. Boswell, *Chemical Marketing Reporter*, 24 January, 2005.
45. EU directive 2006/52/EC.
46. Forum: Chem. Eng. Technol. 6/2007. Chemical Engineering & Technology, 30, 681–682.
47. S. Wei, Q. Song and D. Wei, *Prep. Biochem. Biotechnol.*, 2007, **37**, 67.
48. F. J. Liotta, Jr., L. J. Karas and H. Kesling, *US Pat.* 5308365, 1994.
49. J. K. Spooner-Wyman, D. B. Appleby and D. M. Yost, *SAE Spec. Publ.*, 2003, **SP-1791**, 1–14.
50. H. Noureddini, *US Pat.* 6174501, 2001.
51. D. B. Appleby and J. K. Spooner-Wyman, *US Pat.* 7195656, 2007.
52. www.cpsbiofuels.com
53. GTBE: A renewable remedy for diesel soot emissions, project funded by the European Eureka Network.
54. M. A. Jackson and J. W. King, *J. Am. Oil Chem. Soc.*, 1997, **74**, 103.
55. D. E. Stevenson, R. A. Stanley and R. H. Fumeaux, *Biotechnol. Lett.*, 1993, **15**, 1043.
56. A. Coteron, M. Martinez and J. Aracil, *J. Am. Oil. Chem. Soc.*, 1998, **75**, 657.
57. N. O. V. Sonntag, *J. Am. Oil. Chem. Soc.*, 1992, **59**, 795.
58. M. Vicente, J. Aracil and M. Martinez, *Bioperspectives*, 11 May, 2005.
59. C. R. Soccol, L. P. S. Vandenberghe, C. Rodrigues and A. Pandey, *Food Technol. Biotechnol.*, 2006, **44**, 141–9.
60. S. Papanikolaou, L. Muniglia, I. Chevalot, G. Aggelis and I. Marc, *J. Appl. Microbiol.*, 2002, **92**, 737–744.
61. R. D'Aquino, *Chem. Eng. Prog.*, Oct. 2005.

62. D. Pramanick and T. T. Ray, *Polymer Bull.*, 1988, **19**, 365–370.
63. D. Pramanick and T. T. Ray, *J. Appl. Polymer Sci.*, 2003, **40**, 1511–1517.
64. Y. Jiugao, N. Wang and X. Ma, *Starch*, 2005, **57**, 494–504.
65. A. Wolfson, *Ind. Crop. Prod.*, 2009, **30**, 78–81.
66. A. Wolfson, C. Dlugy and Y. Shotland, *Environ. Chem. Lett.*, 2007, **5**, 67–71.
67. A. Wolfson and C. Dlugy, *Chem. Papers*, 2007, **61**, 228–232.
68. Y. Gu and F. Jerome, *Green Chem.*, 2010, **12**, 1127–1138.
69. B. Reichardt, *Solvent Effects in Organic Chemistry*, Verlag Chemie, Weinheim, 1979.
70. K. P. McDonnel, S. M. Ward and D. J. Timoney, *J. Agr. Eng. Res.*, 1995, **60**, 7–14.
71. M. P. Dorado, J. M. Arnal Gomez, A. Gil and F. J. Lopez, *Trans. ASAE*, 2002, **45**, 519–523.
72. A. Monyem, J. H. Van Gerpen and M. Canakci, *Trans. ASAE*, 2001, **44**, 35–42.
73. A. C. Pinto, L. L. N. Guarieiro, M. J. C. Rezende, N. M. Ribeiro, E. A. Torres, W. A. Lopes, P. A. P. Pereira and J. B. de Andrade, *J. Braz. Chem. Soc.*, 2005, **16**, 1313–1330.
74. M. G. Kulkarni and A. K. Dalai, *Ind. Eng. Chem. Res.*, 2006, **45**, 2901–2913.
75. F. Ma and M. A. Hanna, *Bioresour. Technol.*, 1999, **70**, 1–15.
76. M. Verziu, B. Cojocaru, J. Hu, R. Richards, C. Ciuculescu, P. Filip and V. I. Parvulescu, *Green Chem.*, 2008, **10**, 373–381.
77. M. H. Zong, Z. Q. Duan, W. Y. Lou, T. J. Smith and H. Wu, *Green Chem.*, 2007, **9**, 434–437.
78. M. Kaieda, T. Samukawa, A. Kondo and H. Fukuda, *J. Biosci. Bioeng.*, 2001, **91**, 12–15.
79. S. V. Ranganathan, S. L. Narasimhan and K. Muthukumar, *Bioresour. Technol.*, 2008, **99**, 3975–3981.
80. A. Salis, M. Monduzzi and V. Solinas, *Industrial Enzymes, Structure, Function and Applications*, Springer, The Netherlands, 2007, pp. 317–339.
81. D. Royon, M. Daz, G. Ellenrieder and S. Locatelli, *Bioresour. Technol.*, 2007, **98**, 648–653.
82. Y. Watanabe, Y. Shimada, A. Sugihara, H. Noda, H. Fukuda and Y. Tominaga, *J. Am. Oil Chem. Soc.*, 2000, **77**, 355–360.
83. A. F. Hsu, K. Jones, T. A. Foglia and W. N. Marmer, *Biotechnol. Appl. Biochem.*, 2002, **36**, 181–186.
84. A. Macario, G. Giordano, L. Setti, A. Parise, J. M. Campelo, J. M. Marinas and D. Luna, *Biocatal. Biotrans.*, 2007, **25**, 328–335.
85. S. Shah, S. Sharma and M. N. Gupta, *Energy Fuels*, 2004, **18**, 154–159.
86. Y. Yesiloglu, *J. Am. Oil Chem. Soc.*, 2004, **81**, 157–160.
87. G. D. Yadav and S. R. Jadhav, *Microporous Mesoporous Mater.*, 2005, **86**, 215–222.
88. P. D. Desai, A. M. Dave and S. Devi, *Food Chem.*, 2006, **95**, 193–199.

89. A. V. Paula, D. Urioste, J. C. Santos and H. F. de Castro, *J. Chem. Technol. Biotechnol.*, 2007, **82**, 281–288.
90. U. T. Bornscheuer, *Enzyme Microb. Technol.*, 1995, **17**, 578–586.
91. W. Li, W. Du and D. Liu, *Energy Fuels*, 2008, **22**, 155–158.
92. M. Tüter, B. Babali, Ö. Köse, S. Dural and H. A. Aksoy, *Biotech. Lett.*, 1999, **21**, 245–248.
93. V. Rathore and G. Madras, *Fuel*, 2007, **86**, 2650–2659.
94. E. Hernandez-Martın and C. Otero, *Bioresour. Technol.*, 2008, **99**, 277–286.
95. M. Oda, M. Kaieda, S. Hama, H. Yamaji, A. Kondo, E. Izumoto and H. Fukuda, *Biochem. Eng. J.*, 2005, **23**, 45–51.
96. W. Du, Y. Xu, D. Liu and J. Zeng, *J. Mol. Catal. B: Enzym.*, 2005, **37**, 68–71.
97. F. Camacho, A. Robles, P. A. González, B. Camacho, L. Esteban and E. Molina, *Appl. Catal. Gen.*, 2006, **301**, 158–168.
98. G. Vicente, M. Martinez and J. Aracil, *Bioresour. Technol.*, 2007, **98**, 1724–1733.
99. X. Lang, A. K. Dalai, N. N. Bakhshi, M. J. Reaney and P. B. Hertz, *Bioresour. Technol.*, 2001, **80**, 53–62.
100. C. Peterson and D. Reece, *Trans. ASAE*, 1996, **39**, 805–816.
101. A. Corma, G. W. Huber, L. Sauvanaud and P. O'Connor, *J. Catal.*, 2007, **247**, 307–327.
102. D. R. Dodds and R. A. Gross, *Science*, 2007, **318**, 1250–1251.
103. J. Van Gerpen, *Fuel Proc. Technol.*, 2005, **86**, 1097–1107.
104. M. Mittelbach, *Bioresour. Technol.*, 1996, **56**, 7–11.
105. M. Mittelbach and C. Remschmidt, *Biodiesel: The Comprehensive Handbook*, Boersedruck Ges. M. B. H., Vienna, Austria, 2nd edn, 2005.
106. J. Hu, Z. Du, C. Li and E. Min, *Fuel*, 2005, **84**, 1601–1606.
107. G. Knothe and K. R. Steidley, *Energy Fuels*, 2005, **19**, 1192–1200.
108. G. Knothe and K. R. Steidley, *Fuel*, 2005, **84**, 1059–1065.
109. G. Knothe, *Lipid Technol.*, 2006, **18**, 105–108.
110. G. Knothe and K. R. Steidley, *Fuel*, 2007, **86**, 2560–2567.
111. D. Luna, F. M. Bautista, V. Caballero, J. M. Campelo, J. M. Marinas and A. A. Romero, *Pat. No.* PCT/ES 2007/000450, 2007.
112. J. M. Renga and F. D. Coms, *Pat. No.* WO9309111, 1993.
113. M. Notari and F. Rivetti, *Pat. No.* WO2004/052874, 2004.
114. J. A. Kenar, G. Knothe, R. O. Dunn, T. W. Ryan III and A. Matheaus, *J. Am. Oil Chem. Soc.*, 2005, **82**, 201–205.
115. D. Fabbri, V. Bevoni, M. Notari and F. Rivetti, *Fuel*, 2007, **86**, 690–697.
116. E. Z. Su, M. J. Zhang, J. G. Zhang, J. F. Gao and D. Z. Wei, *Biochem. Eng. J.*, 2007, **36**, 167–173.
117. J. Kijeński, A. Lipkowski, W. Walisiewicz-Niedbalska, H. Gwardiak, K. Różyczki and I. Pawlak, *Eur. Pat.* EP1580255, 2004.
118. Y. Xu, W. Du, D. Liu and J. Zeng, *Biotech. Lett.*, 2003, **25**, 1573–6776.
119. Y. Xu, W. Du and D. Liu, *J. Mol. Catal. B: Enzym.*, 2005, **32**, 241–245.
120. W. Du, Y. Xu, D. Liu and J. Zeng, *J. Mol. Catal. B: Enzym.*, 2004, **30**, 125–129.

121. J. Kijeński, *Pol. J. Chem. Tech.*, 2007, **9**, 42–45.
122. M. K. Modi, J. R. C. Reddy, B. V. S. K. Rao and R. B. N. Prasad, *Biores. Technol.*, 2007, **98**, 1260–1264.
123. G. W. Huber, P. O'Connor and A. Corma, *Appl. Catal. A: General*, 2007, **329**, 120–129.
124. L. P. Wackett, *Microb. Biotechnol.*, 2008, **1**, 211–225.
125. R. Kalscheuer, T. Stölting and A. Steinbüchel, *Microbiology*, 2006, **152**, 2529–2536.
126. A. Demirba, *Energ. Convers. Manag.*, 2001, **42**, 1357–1378.
127. A. Klerk, *Green Chem.*, 2008, **10**, 1249–1279.
128. C. N. Hamelinck and A. P. C. Faaïj, *International Sugar Journal*, 2006, **108**, 168–175.
129. B. E. Kampman, L. C. Boer and H. Croezen, *Biofuels Under Development, An Analysis of Currently Available and Future Biofuels, and a Comparison with Biomass Application in Other Sectors*, Delft, CE, 2005.
130. L. Petrus, A. Minke and M. A. Noordermeer, *Green Chem.*, 2006, **8**, 861–867.
131. G. W. Huber and J. A. Dumesic, *Catal. Today*, 2006, **111**, 119–132.
132. A. A. Peterson, F. Vogel, R. P. Lachance, M. Fröling, M. J. Antal Jr. and J. W. Tester, *Energy Environ. Sci.*, 2008, **1**, 32–65.
133. V. Patil, K. Q. Tran and H. R. Giselrød, *Int. J. Mol. Sci.*, 2008, **9**, 1188–1195.

CHAPTER 5
Assessment of Economic and Environmental Cost-benefits of Developed Biorefinery Schemes

MICHAEL BINNS, ANESTIS VLYSIDIS AND
CONSTANTINOS THEODOROPOULOS*

School of Chemical Engineering and Analytical Science, University of Manchester, Manchester, M60 1QD, UK

5.1 Introduction

Biorefineries are facilities that can integrate a range of different processes depending on the feedstock and the desired products. In general a biorefinery will convert a specific biomass into energy (*e.g.* biofuels such as biodiesel or bioethanol), chemicals (*e.g.* levulinic acid and 1,3-propanediol) or materials that can be used as animal feed, fertiliser or construction materials.[1] Due to the renewable nature of biomass these refineries can provide sustainable routes for the production of both energy and chemicals. In addition to helping to meet these demands biorefineries are developing and implementing new technologies and new processes, which can make use of the untapped potential in some biomass feedstocks.[2,3]

The aim of much of the research into biorefineries is to enhance, to scale-up and to combine different processes into an integrated biorefinery that can make the most out of the available biomass, usually giving multiple different products. There has been great interest in using biorefineries to produce biofuels such as biodiesel and bioethanol either from traditional crops like rapeseed[4] or from more novel sources like algae,[3,5] jatropha[6] or coconut oil.[7]

Although most biorefineries produce a single 'main' product, which is the reason they are constructed, there are many cases where they can be modified or configured so they can produce additional valuable chemicals (or other products) by adding or combining different technologies to give an 'integrated biorefinery'. Integration of biorefineries is currently a hot topic as it allows biorefineries to generate additional revenue[8] or to implement green technology,[9,10] which makes them more sustainable. Sustainability is the reason for investigating the valorisation of side-streams through additional processing. This includes converting the glycerol, a by-product from biodiesel production, into valuable chemicals such as 1,3-propanediol[11] or succinic acid[12,13] or into biogas.[14] Also there are other options including converting fermenter residues into biogas.[2] A good summary of at least ten top value-added products can be found in the review by Bozell and Petersen.[15] These studies are complemented with holistic studies considering the complete life cycle of the biomass from cradle to grave.[16] This should include everything from growing of crops up to the final use of end products considering both economic and environmental issues. However, for the purpose of comparisons decisions made at the biorefinery could be considered the most significant as many different options for processing will have the same fixed supply of raw materials (farming and supply of biomass).

Optimisation of a biorefinery involves the construction of an objective function, which can include economic factors, environmental factors or both. Structural optimisation involves finding the best configuration of process units (the flowsheet) that can be achieved using either superstructure optimisation[17] or evolutionary optimisation[18] or by simply comparing several different configurations.[19] However, changing the structure of a biorefinery can require a large capital investment so it is also advisable to consider optimisation of the process variables such as pressures, temperatures and feed ratios. For this purpose local, deterministic nonlinear methods such as sequential quadratic programming[20] or stochastic methods (that can probabilistically converge to global optima) such as simulated annealing[21] or genetic algorithms[22] can be employed.

Multi-objective optimisation is necessary when there are economic and environmental goals that need to be satisfied. One method of combining the objectives is to assign a value to the environmental impact (for example applying a carbon tax to the emissions) so that both objectives can be considered in terms of economics. An alternative strategy is to find multiple different solutions, which gives the optimum profits with different constraints on the emissions. This is possible using a systematic method changing emission constraints or using multi-objective evolutionary algorithms[23] with the results giving a pareto curve[24] showing how the optimal profits change with the emissions.

This work analyses several biorefinery schemes developed using information generated within the FP7 EU project SUSTOIL. The economic and environmental impacts of different process options are considered with the aim to highlight the best processes and those with the greatest potential. To give weight to these conclusions the results from optimisation and multi-objective optimisation are used to show the optimal operating conditions for each biorefinery scheme. Comparisons are also made between the optimum configurations of each biorefinery scheme at the same scale in order to give a 'fair' analysis of which is best and which has greater potential.

5.2 Methodology

5.2.1 Simulation Software

For the modelling and simulation of different biorefinery schemes Aspen Plus has been the main software package used to perform simulations. This is supported by calculations performed in the mathematical language MatLab and data generated within the EU project SUSTOIL in addition to information from journals and databases. The main components of our simulation technology involved:

- Constructing realistic flowsheets of the various biorefinery processes using Aspen Plus;
- Creating an interface between the Aspen Plus simulator and MATLAB
  - to facilitate parametric analysis of the extracted results through automated modification of input files;
  - to read and analyse result files in order to evaluate objective functions in optimisation studies.

5.2.2 Optimisation Methods

5.2.2.1 Identifying Optimisation Parameters

Optimisation is the process of modifying a number of parameters (degrees of freedom) in an informed way in order to minimise an objective function. The identification of appropriate degrees of freedom is an important issue. The parameters identified must have an effect on the objective function (*e.g.* increase or decrease profits). It is also important that sufficient data must be available in order to simulate the system as parameters change so that the objective function can be calculated. For example it is not possible to use the capacity of a plant as a degree of freedom if we do not know how the operating costs change as the capacity is increased/decreased.

5.2.2.2 Constructing an Objective Function

An important objective is to minimise costs or to maximise the profits of a process. It should be mentioned here that proper construction of such objectives functions require knowledge of the costs of the raw materials and of the energy usage.

The costs of a system can be defined as:

$$\text{Costs} = CR + EC + CP + WC + EF$$

and the profits can be defined as:

Profit = VP − CR − EC − CP − WC − EF

where:

VP = Value of products

CR = Cost of raw materials

EC = Energy costs

CP = Capital costs (scaled to €/hr based on 10 years lifespan)

WC = Cost of liquid and solid waste disposal

EF = Error function (= 1×10^9 if an error in the simulation occurs otherwise = 0)

These objective functions can be modified slightly to account for different systems. For example some systems may not have any solid or liquid wastes because the by-products are sold, reused or converted into some other useful product.

The error function (EF) is added to account for situations where the simulator fails. Failures can sometimes be avoided by providing appropriate bounds for the parameters, which can be found through a combination of experience and knowledge of the system. However, in many cases when the parameters are being changed automatically these errors occur for many unknown reasons (missing libraries, computer running out of memory, unexpected behaviour of particular units *etc.*). An error function allows the system to reject simulations that fail and to move on.

5.2.2.3 Optimisation Methods: Deterministic and Stochastic

Since all the schemes considered in this work are nonlinear and include a number of constraints, an appropriate deterministic optimisation algorithm is the Sequential Quadratic Programming (SQP) built into the MatLab function 'fmincon', which can handle both equality and inequality constraints and will generally find a solution relatively fast. However, due to the existence of non-convexities in the parametric spaces of such nonlinear problems, this local optimisation method is likely to find the nearest minimum and to miss the global optimum in a system with multiple minima.

To address this issue we have developed our own version of a stochastic optimisation method, namely simulated annealing (SA), based on the original method of Kirkpatrick *et al.*[21] SA is an optimisation method that randomly changes the parameters in a system. One feature of simulated annealing is that the algorithms are designed to also allow some 'bad' parameter changes, which move the solution away from the optimum so that the algorithm can 'escape' from local minima. The details of how SA operates are shown in Figure 5.1.

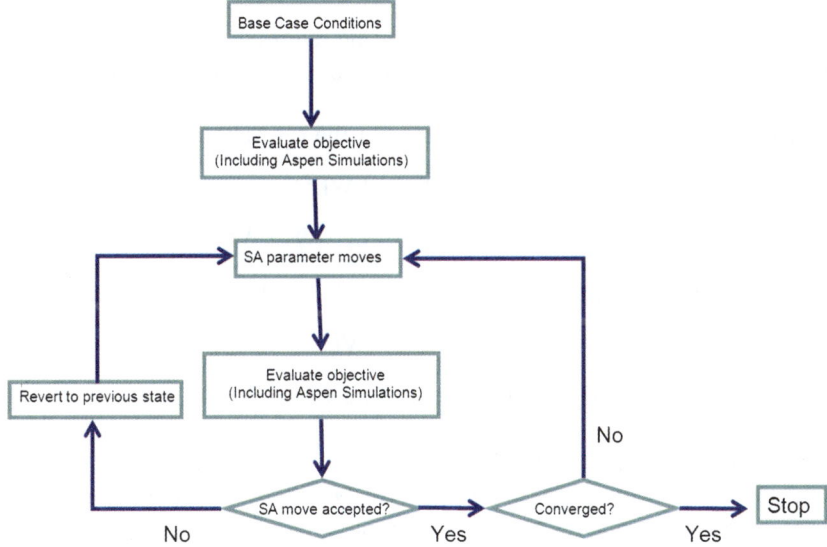

Figure 5.1 Simulated annealing algorithm.

5.2.3 Life Cycle Analysis

Life cycle analysis (LCA) is a useful tool to assess the environmental impact of a process and can include a number of different environmental or social factors. These are in addition to and sometimes in competition with the economic goals of a biorefinery. Some of these environmental factors include:

- Emissions;
- Liquid and solid wastes;
- Renewable *vs.* non-renewable energy consumption;
- Land usage (for growing crops and for the biorefineries).

In this work we have concentrated mainly on emissions as the main environmental factor. Liquid and solid wastes are considered by including a cost for disposal based on the amount produced as part of the economic analysis. Emissions can be considered in a similar way with a tax or penalty per kilogram of carbon dioxide emitted. However, instead we have considered the quantities of emissions and the effect any legislation or limitations might have on the different biorefinery schemes considered.

5.2.3.1 Calculating Emissions

The processes involved in a biorefinery can lead to emissions, directly or indirectly. Direct emissions are emitted on-site as by-products, *e.g.* when feedstock is burnt. Indirect emissions account for all other sources of emissions (emitted on-site and off-site), which are required for the biorefinery to operate. This includes emissions resulting from the transport of feedstock from the farm to the biorefinery (or for any intermediate transportation) as well as emissions generated producing the electricity and heating required, normally provided by gas or oil boilers. In the biorefinery schemes considered we have assumed that all heat energy is provided by a gas boiler.

The gas boiler properties were taken from the GEMIS database:

$$\text{Cost} = €0.0179/\text{kWh}^{25}$$

$$\text{Emission factor} = 0.201 \text{ kg } CO_2/\text{kWh}^{25}$$

For other process units that require electricity we have considered:

$$\text{Cost} = €0.0772/\text{kWh}^{26,27}$$

$$\text{Emission factor} = 0.537 \text{ kg } CO_2/\text{kWh}^{28}$$

These emission factors give the amount of CO_2 released for each kWh of energy provided for that type of energy. As CO_2 is the dominant gas emitted in all the biorefinery schemes considered we have based all our calculations on these emissions.

5.2.3.2 Calculating Costs and Profits

Costs and profits for each biorefinery scheme are calculated based on the flowrates into and out of the system, in addition to the specific costs associated with each process unit. Energy costs (shown in Table 5.1) can be computed using the same cost per unit of energy.

As can be seen in Table 5.1 electricity is the most expensive energy source; however, it is also the only type of energy that can be used to run some pieces of equipment. The pumps and the oilseed press, for example, both operate using electricity in the biodiesel case (fortunately the electricity requirement is relatively low). Heating is based on energy provided by gas boilers and cooling costs are based on the use of cooling water.

5.2.4 Multi-objective Optimisation

In most problems involving the optimisation of chemical plants the objective is either to maximise the profits or to minimise the costs of the system by changing certain parameters such as feed rates, temperatures or pressures. This is *single-objective* optimisation. For example, in a system with more objectives, *e.g.*

Table 5.1 Energy types and costs.

Energy type	Cost (€ per kWh)
Heating	0.0179^{25}
Cooling	0.0009^{29}
Electricity	$0.0772^{26,27}$

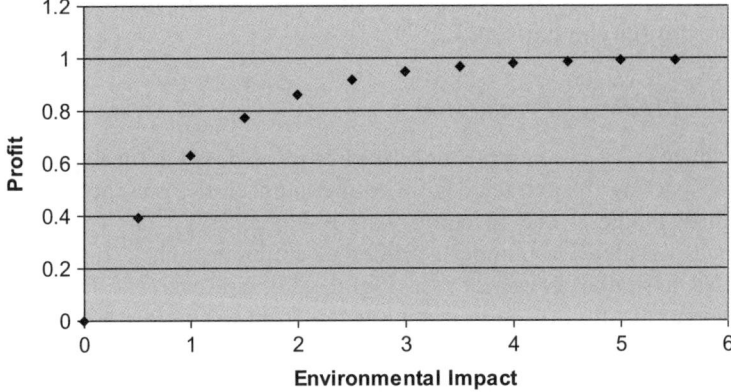

Figure 5.2 Diagram of a pareto curve.

maximising profits and simultaneously minimising emissions *multi-objective optimisation* is necessary.

There are two possible strategies for combining the two objectives, the first being to add the two objective functions together and optimise the overall objective function obtained:

Overall objective = economic objective + environmental objective

However this has the obvious problem that the units of the two objectives are entirely different, one might be in euros and the other in kilograms of CO_2. An alternative method for multi-objective optimisation, which we have employed here, involves optimising one objective while treating the other objective as a constraint. Performing multiple optimisations with different values for the constraints gives a set of different solutions. This is the method employed by Azapagic and Clift where the range of solutions obtained gives a pareto or 'non-inferior' curve.[24]

The pareto curve depicted in Figure 5.2 shows how profitability of a process changes with constraints on environmental factors. Government legislation can, for example, limit CO_2 emissions for plants of a certain size. The same methodology can be applied with multiple environmental factors resulting in multi-dimensional pareto surfaces. This also increases the amount of computations required. Hence, we have limited our multi-objective optimisation to consider only a single environmental factor.

5.2.5 Biorefinery Schemes

The techniques described above are used to model and simulate several biorefinery schemes in order to evaluate and optimise them where possible. For this purpose the structures of each biorefinery are proposed and the details, assumptions and parameter values are specified here. This is validated using information from a number of universities and companies which participated in the EU project SUSTOIL and through data from the literature. These models, simulations and calculations are used for optimisation, multi-objective optimisation and for comparisons.

5.2.5.1 Biodiesel Production

The production of biodiesel from oilseed crops is possible through transesterification, reacting the extracted oil with methanol in the presence of a catalyst to produce biodiesel and glycerol. This is modelled by Zhang *et al.*, where they have considered a number of different configurations.[19] In this work we have used a similar flowsheet (see Figure 5.3); however, we have modelled the transesterification reaction using kinetics from the literature[30] and we have added steps for extracting the oil from seeds and for purification or bioconversion of the crude glycerol to add value to the process.

Feed
The main feed for this process is rapeseed, which is assumed to contain three different components:

- Water – to account for the moisture in the seeds;
- Oil – the active component we wish to extract and convert to biodiesel;
- Solid residue – the solid matter that remains after all the liquids are removed.

Since it is not possible to perfectly separate these three components the extraction process will give rapeseed meal as a by-product that contains the solid residue with a small amount of oil. We have assumed the following composition for the seeds:

40% oil

10% moisture

50% solid residue

This is a valid assumption because oil content can vary from 40 to 46.5%[26,31] and moisture content is similar to the 9% given by FERA and Stephenson *et al.*[26,32] We have also assumed that the oil is 100% triolein, a triglyceride combining three molecules of oleic acid, which is a valid assumption because

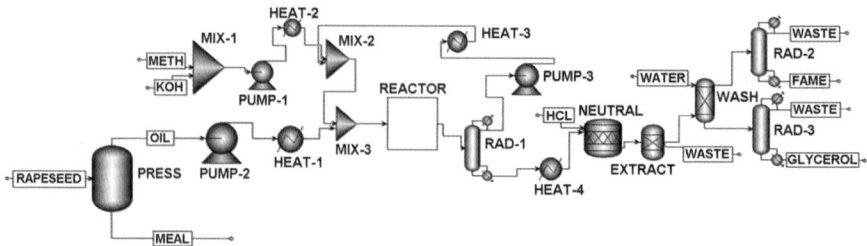

Figure 5.3 Flowsheet for biodiesel production.

triolein is the main component in rapeseed oil accounting for 64.4% of the mass[33] and because considering a single component is simpler.

The other feeds are methanol and potassium hydroxide for the transesterification and hydrochloric acid to neutralise and remove the potassium hydroxide in the form of KCl.

Oil Extraction

Oil extraction is possible through a number of stages, including crushing, pressing, filtering and solvent-based extraction.

A cold mechanical press and filter can extract two-thirds of the oil from rapeseed[34] or up to 80% from Jatropha;[6] however, with the addition of solvent-based extraction 94 or 95% of the oil can be extracted.[26,35]

In the current model we have combined the different processes involved in oil extraction into a single unit with an extraction efficiency of 80% (extracting 80% of the oil). This seems reasonable compared to the above extraction efficiencies for Jatropha oil.

The energy cost of oil extraction is 68 KWh/t based on a cold pressing scheme with a capital cost of €2.56 million (based on 40 000 tonnes per year).[36]

Transesterification

The reaction used here to produce biodiesel is transesterification:

$$\text{oil} + 3 \text{ methanol} \leftrightarrow 3 \text{ FAME} + \text{glycerol}$$

As mentioned above the oil is represented by triolein ($C_{57}H_{104}O_6$) and the corresponding FAME product is methyl oleate ($C_{19}H_{36}O_2$).

The above reaction is broken into three steps when considering the kinetics:

$$\text{triglyceride} + \text{methanol} \leftrightarrow \text{FAME} + \text{diglyceride}$$

$$\text{diglyceride} + \text{methanol} \leftrightarrow \text{FAME} + \text{monoglyceride}$$

$$\text{monoglyceride} + \text{methanol} \leftrightarrow \text{FAME} + \text{glycerol}$$

As these reactions are reversible a high ratio of methanol to oil is used (up to 30:1 methanol to oil) together with approximately 1% catalyst.

The available kinetics for sulfuric acid catalysed transesterification are for soybean oil and they assume pseudo-first-order kinetic expressions, which do not account for varying methanol concentration.[37] However, the kinetics for potassium hydroxide are second-order expressions, which account for methanol concentration and are based on experiments with rapeseed oil.[30]

Using these kinetics in our simulation we find conversions of between 95% and 99%, which fits well with the assumed 95% conversion used by Zhang *et al.*[19]

In order to reduce the amount of methanol added and to remove some methanol from the product stream a distillation column (depicted in Figure 5.4) is used after the transesterification, which recovers and recycles 50–90% of the methanol to be fed back in with the fresh feed.

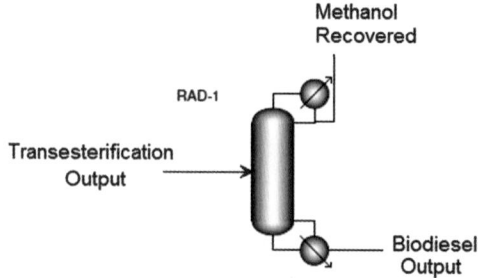

Figure 5.4 Distillation column for the recovery of methanol.

Table 5.2 Capital costs for biodiesel production.

Process unit	Size (diameter × height)	Capital cost (€)
Transesterification reactor	10 m^3	468 914.70
Methanol recovery column	1 m × 10 m (6 trays)	121 734.90
FAME purification column	1 m × 8.5 m (5 trays)	119 346.97
Glycerol purification column	1 m × 11.5 m (7 trays)	144 484.44
Neutralisation reactor	1 m × 1 m	143 472.58
Gravity separator	1 m × 1 m	359 671.19
Water washing column	1 m × 10 m (6 trays)	121 734.85
Fermenter	2 × 120 m^3	3 152 994.86
Heat exchangers	5 × 5 m^2	128 324.36
Pumps	0.466222 kW	19 303.81
Flash	0.5 m × 2 m	5611.42

Capital costs of equipment (shown in Table 5.2) are based on the correlations in Turton et al.,[29] which relate the cost to size, materials, pressure and the price costing index CEPCI.

Catalyst Neutralisation
The potassium hydroxide in the system is neutralised with the addition of hydrochloric acid, which reacts to form potassium chloride. This is then removed using a decanter. However, this process has not been modelled and we have assumed that the units involved perfectly remove the potassium hydroxide from the system.

Biodiesel and Glycerol Purification
As an additional step both the FAME and the glycerol are purified through distillation. This is not strictly necessary for the glycerol and if it can not be sold the glycerol could be treated as a waste stream. The FAME is purified to 99%, although simulations show that purity over 99.9% is achieved.

Materials and Utility Prices
Methanol   €0.202/kg[a,29]
Sulfuric acid   €0.062/kg[a,29]

Potassium hydroxide	€0.2199/kg[38]
Hydrochloric acid	€0.08543/kg[38]
Crude glycerol	€0.0688/kg[b,39] (assumed to be ~80% glycerol)
Purified glycerol	€0.35/kg[25]
FAME	€0.908/kg[b,39]
Rapeseed	€0.279/kg[40]
Rapeseed meal	€0.175/kg[40]
Calcium oxide	€0.04/kg[19]
Process water	€0.000046/kg[a,29]
Potassium hydroxide	€0.220/kg[38]
Hydrochloric acid	€0.085/kg[38]
Electricity	€0.0772/kWh[26]
Cooling water	€0.244/GJ[a,29]
Gas heater	€17.955/MWh[25]

a. converted from $/kg using conversion 0.6879.
b. converted from €/L using MW of 880.2.

5.2.5.1.1 Glycerol Purification Option.

As an alternative to selling crude glycerol we have also considered the purification of glycerol which can be sold for a higher price. This is achieved through distillation as shown in Figure 5.5, where the waste stream contains mostly methanol and water, which are the main impurities in crude glycerol.

It is worth pointing out that even in the crude glycerol case we still purify the glycerol up to 80% purity. However, in this scheme the purity of the glycerol is increased to 95% by changing the specifications of the glycerol purification distillation column.

5.2.5.1.2 Bioconversion of Glycerol to Succinic Acid.

In order to enhance the economics of the process additional steps are considered in order to make better use of the large amounts of crude glycerol produced as a by-product of the transesterification. For computing the economics of this

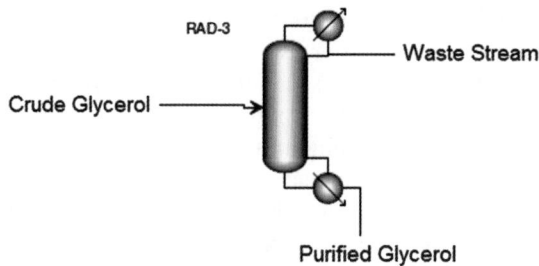

Figure 5.5 Diagram showing the purification of glycerol.

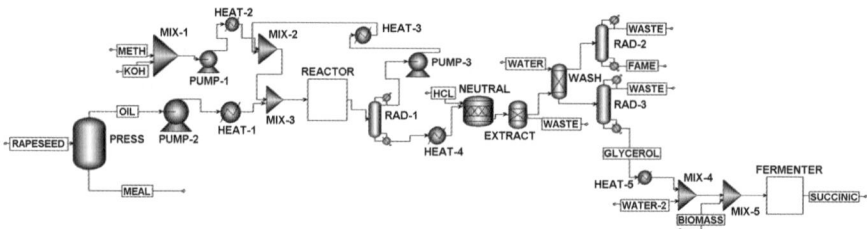

Figure 5.6 Process flowsheet for biodiesel production with co-production of succinic acid through fermentation.

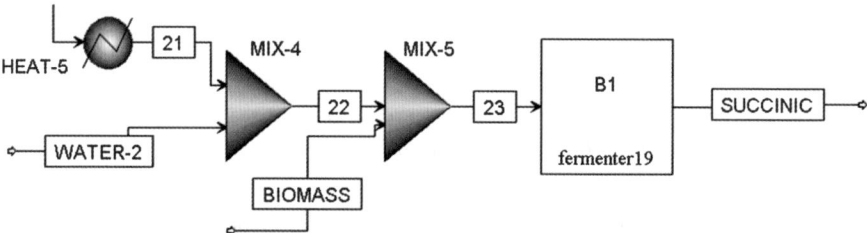

Figure 5.7 Diagram showing the sequence of units added in order to produce succinic acid from glycerol.

process we have considered a succinic acid price of €4.311/kg (which is in the range $5.9 to $9/kg suggested by Song and Lee).[41]

One method considered is to convert the glycerol into succinic acid using fermentation with the bacterium *Actinobacillus Succinogenes*.

The fermentation has been modelled with kinetics derived from experiments performed by Vlysidis et al.[12] These results have been scaled up and applied in simulations of industrial scale bioreactors. However, these kinetics are for batch experiments and in our simulations we have a continuous process.

Unlike the purification option this route requires some extra units, which are added onto the biodiesel flowsheet including a fermenter and an additional heater (see Figures 5.6 and 5.7). The fermenter also requires an additional source of water and biomass in order to operate.

In fact, although the fermenter is represented by a single block in the simulation flowsheet, we have based our design on four smaller batch fermenters, which operate in sequence to give continuous production of succinic acid (at any time two will be filling and the other two will be reacting). This neglects the time required for the reactor to be cleaned, sterilised and emptied, which are assumed to be much faster than the filling and fermentation reaction time. There are two advantages for having multiple smaller fermenters in that this allows them to operate in sequence and, more importantly, because having smaller fermenters requires less capital investment.

5.2.5.2 Biogas Production

In a second major scheme we have also considered the process converting rapeseed straws and rapeseed meal into biogas, which can subsequently be combusted for the production of electricity. This scheme was based on information provided by FORTH and the flowsheet for this process is shown in Figure 5.8.[42]

The Feed

The raw material for this process is in general crop residues. However, in this case we have assumed the feed is either rapeseed meal or straw (or some combination).

The composition of the feed is assumed to be:

10%	moisture
90%	dry matter

This is validated by Diaz *et al.*, who state that air-dried rapeseed straw at room temperature reaches 10% moisture content.[43]

The moisture is assumed to be 100% water. The dry matter is broken down further into:

10%	inert dry matter
90%	volatile dry matter

This assumption might be underestimating the percentage of volatile dry matter since Diaz *et al.* report the composition of rapeseed straw contains only 5.73% ash.[43] However, we believe the assumption that 10% of the matter is inert is reasonable for this work.

So the total feed composition is:

81%	volatile dry matter
9%	inert dry matter
10%	water

These values were chosen based on information and literature provided by FORTH.[42]

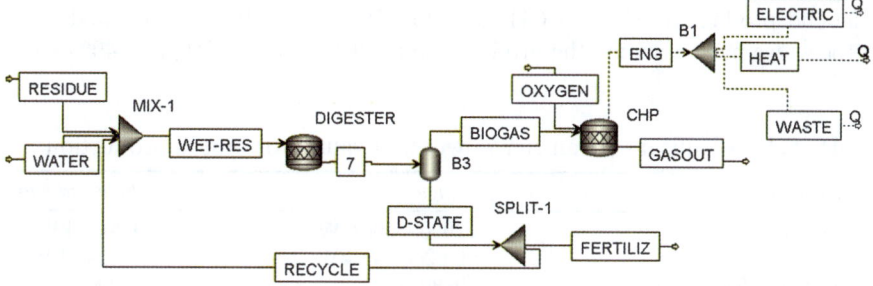

Figure 5.8 The process flowsheet for biogas production and combustion.

In our simulations the volatile dry matter is considered to be the active raw material, which is converted to biogas in the digester.

The properties of the water are well known and can be used easily with Aspen Plus and Aspen custom modeller without any additional input.

However, the properties of the dry matter are not available. For the purpose of this simulation we have assumed that the rapeseed straw and rapeseed meal have the same properties as the seeds. We have also assumed that the inert and the volatile matter have the same properties (although they are treated differently inside the digester).

The bulk density of the seeds is assumed to be $612.1\,\mathrm{kg\,m^{-3}}$ ignoring the small percentage of moisture present in this value.[44]

The Digester and Combined Heat and Power Units

The capital cost and operating costs have been obtained using a correlation to the amount of electricity produced by the combined plant (see Table 5.3). This information was provided by FORTH.[42]

Capital costs: €2000–5000/kW
In this work we have assumed €3000/kW

This is a valid assumption because it is similar to the value used by Walla and Schneeberger who use capital cost = €101,522 + €3500/kW, which is slightly higher than the €3000/kW we have assumed and is within the above range.[45]

Operating costs: €0.02–0.045/kW
In this work we have assumed €0.045/kWh.

These values could be treated as parameters in future work, but so far we have used the above values.

The conversion of the digester is specified as between 70 and 90% and here we have assumed a conversion of 90%.

Also, the feed humidity must be raised from 8–11% up to 70–90%, which is accomplished here with the addition of water.

Digester Reaction

The reaction(s) that occur inside the digester convert the active part of the feed into biogas. In the information provided by FORTH the composition of the end product is: 55–65% CH_4 and 35–45% CO_2.[42] In order to model this process we have assumed the product composition is 60% CH_4 and 40% CO_2.

Table 5.3 Significant parameter ranges and values for biogas production.

Parameter	Range	Selected value
Capital costs	€2000–5000/kW	€3000/kW
Operating costs	€0.02–0.045/kW	€0.045/kW
Conversion (digester)	70–90%	90%
Feed humidity	70–90%	75%

We have also assumed that 100% of the volatile dry matter can be converted into biogas (and 0% of the inert dry matter) subject to the above specified conversion.

In order to account for the inert dry matter and the above product composition we have derived the following reaction:

$$1 \text{ residue} \rightarrow 0.1 \text{ inert residue} + 7.243 \text{ CO}_2 + 29.805 \text{ CH}_4$$

This is the molar reaction is based on the following mass balance:

100% residue → 10% inert matter + 54% CH_4 + 36% CO_2 (taking into account the molecular weight of the different components).

Combining this reaction with the above conversion we can model the production of biogas in the system.

Combined Heat and Power Unit (CHP)

This CHP unit reacts the biogas produced in the digester with air in order to combust the methane (CH_4) in the biogas. The heat produced is then partly converted into electricity with the remainder either sold as waste heat or lost due to inefficiencies. For current processes 30–35% of the heat is converted to electric energy and 40% is recovered for reuse.[34] For our simulations we have assumed that electricity is produced at 35% of the total heat available.

To model this process a relatively simple model is used with the combustion of methane using only one reaction:

$$CH_4 + 2\ O_2 \rightarrow 2\ H_2O + CO_2$$

The conversion of this is reported to be 90–100%[26,46] or close to 100%[34] so we have assumed a conversion of 99%. If we specify the feed (flowrate, composition, temperature, pressure) and the outlet emissions (temperature and pressure) then using the above reaction and conversion we can calculate the amount of heat produced and the composition of the outlet emissions. This is done using Aspen Plus, which has links to the thermodynamic properties of the above common compounds. The biogas is fed in at atmospheric pressure (1 atm) and room temperature (25 °C) and from information provided by WUR the typical outlet temperature is 90 °C.[34]

Splitting and Recycling the Digestate

The digestate is the liquid/solid output from the digester and in this case it consists of water, unused raw material and the inert solid by-product.

This digestate can be used as a source of fertiliser. However, it makes sense to recycle some of the digestate for two reasons:

- To reduce the amount of water required to maintain 75% humidity;
- To improve the productivity and yield by recycling the unused feed.

For these reasons a fraction of the digestate is recycled with the remainder going to become fertiliser.

We have considered recycling between 0 and 99% of the digestate and we have used this parameter in the optimisation of the system. The value of the digestate as fertiliser is dependent on its N, P and K content. As an estimate for the digestate value we have assumed the value is 10% of the value of commercial fertiliser. The price of fertiliser is £127/ha for winter oilseed rape and £60/ha for spring oilseed rape.[26,47]

So we have assumed:

Value of digestate €0.0981 per kg of dry matter;
Value of electricity €0.145/kWh (higher than the value in the above schemes, assuming the plant owners got a good deal).

5.2.5.3 Supercritical Carbon Dioxide Extraction

Supercritical CO_2 extraction involves using CO_2 at high pressures as a clean solvent to extract matter from solid residues. In this biorefinery scheme we are extracting waxes from rapeseed straws. This model (see Figure 5.9) has been developed in Aspen based on communications and associated data provided by UoY.[48]

Units

Pump: increases pressure to 350 bar
Cooler: decreases temperature to 50 °C
Extract: input-output extraction reactor with approximately 1–3% yield
Heater: increases temperature to 50 °C
Valve-1: reduces pressure to 70 bar
Valve-2: reduces pressure to 50 bar
Separate: unphysical separator (removes the wax products)
Split: stream splitting recovering 99.6% of the CO_2 and releasing 0.4%
Mix: combining the recycling CO_2 with the added makeup fed in

This model shows how the solvent interacts with the solid feedstock in order to extract waxes. The important parameters for this system are the size of the reactor and the density of the feedstock, which both affect the productivity of the process. It is possible to increase the density of the feedstock through

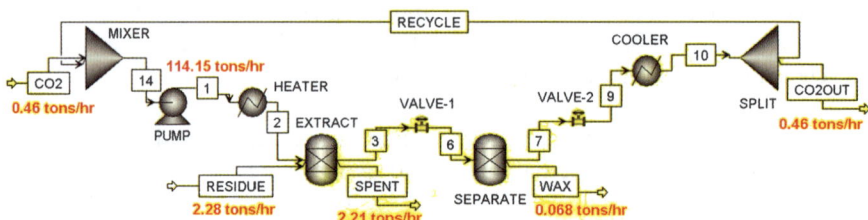

Figure 5.9 Process flowsheet for supercritical CO_2 extraction.

pelletisation at a cost of £38 per metric ton.[49] However, increasing the density to almost $700\,kg\,m^{-3}$ would also reduce the porosity reducing the efficiency of the solvent extraction (with a density of $500–550\,kg\,m^{-3}$ being preferable).

The operating costs are inversely proportional to the size of the plant with a typical 30 000 tons/year plant having operating costs of approximately €350/ton.

This value agrees approximately with the correlation obtained by Brunner for capacity vs. costs.[50]

5.2.5.4 Protein Extraction

Using information from the ENHANCE project supplied by WUR we are able to model the extraction of proteins from rapeseed meal.[34]

This process includes:

- Aqueous protein extraction;
- Precipitation;
- Purification;
- Drying.

The effect of these processes for a year of operating is to convert:

$$7.326\,tDM\ \text{rapeseed meal into}\ (\rightarrow)$$

$$3003.7\,tDM\ \text{fibres} + 1758.2\,tDM\ \text{LMW} + 1098.9\,tDM\ \text{Protein S} + 1465.2\,tDM\ \text{Protein P}$$

This can be modelled as a single reaction:

1 rapeseed meal → 0.41 fibres + 0.24 LMW + 0.15 Protein S + 0.20 Protein P

where S and P are different proteins that can be extracted from the meal.

The values of the outputs are:

Fibres	£133/tDM
LMW	£133/tDM
Protein S	£3300/tDM
Protein P	£800/tDM

Considering the above production per year this gives £5 431 863/year total sales revenue.

The costs for this process include:

7.326 tDM rapeseed meal (£185/tDM) at £1 355 310/year
475 kW electricity (£0.06/kWh) at £218 187/year
345 L/hr of oil to produce steam (£0.5/L) at £1 162 720/year
 plus other costs including capital costs with the total costs reaching £4 137 282/year
giving a net profit of £1 294 580/year.

5.2.5.5 Oil Extraction

5.2.5.5.1 Cold Pressing Oil Extraction. As mentioned above in the biodiesel scheme (Section 5.2.5.1) oil extraction is modelled as a single unit. This allows the extraction to be modelled/simulated using a single parameter: the efficiency of the oil extraction.

For cold pressing (mechanical extraction in the range 20–40 °C) we have used information from CREOL, CETIOM and WUR.[34,36,51]

As we do not have a specific value for cold pressing of rapeseed we have assumed an extraction efficiency of 80%. This is the same as the literature value for cold mechanical pressing of Jatropha.[6] Capital costs of this process are computed based on the cost of a 1400 kg/hr (11 000 t/year) cold pressing unit, which has a capital cost of €703 888.[34,36,51] This is then scaled up using a correlation:

$$\text{increase in cost} = (25\,000/11\,000)^{0.6}$$

to relate cost with size giving €1.15 million for a 25 000 ton/year unit as used in the biodiesel scheme (Section 5.1.5.1).

The energy requirement of a cold pressing unit is 68 kWh/t, which would require 212.5 kW of electricity for a unit processing 3125 kg/hr.[34,36,51]

This formulation gives a very simple input-output model representing a fixed set of conditions. If the process is to be optimised then parameters that can be varied need to be identified (in this case perhaps the amount of water added) so that improved, more economical solutions can be found.

5.2.5.5.2 Hexane-based Oil Extraction. As with the cold pressing scheme hexane-based oil extract is also modelled with a single unit, although here the single unit represents a combination of press and solvent extraction, which make it possible to extract up to 94 or 95% of the oil in agreement with known values.[26,35]

So for modelling purposes we have assumed the unit achieves an oil extraction efficiency of 95%. We have also used information provided by Desmet Ballestra, who have provided us with information about a scheme with solvent extraction in a 1200 tonnes per day unit.[52] This is a considerably larger capacity than the cold pressing scheme (Section 5.2.5.5.1), which has capacity almost 500 times lower. The information provided includes the steam, electricity, compressed gas and water requirements for the steps including: pre-press plant, solvent extraction, acid degumming, silica treatment and thermal stripping.

The requirements of this sequence of units are:

Steam (from a gas boiler)   240 kWh/t (per ton of seeds)
Electricity   50.18 kWh/t

Compressed gas   28.54 kW (or 0.023 kWh/t)
Process water   14.58 kg/t

and additional make-up water for the cooling tower and boiler.

5.2.5.6 Thermomoulding

The information provided by Institut National Polytechnique de Toulouse (INPT) about the uses of sunflower stalks and sunflower meal is used to build a model for thermomoulding of sunflower oil cake into plant pots.[53] This case provided by INPT includes a batch process for converting sunflower oil cake (in pellet form) together with water, glycerol and sodium sulfate into plant pots through the use of a screw press and injection moulding equipment.[53]

This process includes several steps:

1. Grinding of the pellets with maximum capacity of 800 kg/h and 5.5 kWh/t of electricity required, although in this case INPT considered the use of 380 kg of sunflower cake pellets.
2. Twin-screw extrusion combined the ground sunflower cake with the other solid and liquid feeds (glycerol, sodium sulfate and water) with an input rate of 55 kg/h of sunflower cake. This required 167 kWh/kg.
3. Drying (on a continuous belt dryer) at 50 °C to reduce moisture content from 25% down to 4–6%. This requires 40 kW for 20 min in order to dry 460 kg of wet compounds.

These three steps produce the intermediate known as TEGS containing 8% glycerol and 2% sodium sulfite. We assume the remainder is 4% water and 86% sunflower oil cake. For the production of 380 kg of TEGS the costs are:

Raw materials   €116
Utility   €16.5
Labour   €84
Operating costs   €214
Total cost of TEGS €0.56/kg

Constructing plant pots is accomplished with a fourth step:

4. Injection moulding involves using high pressures of 1000 bars to shape the TEGS material into pot shapes. The equipment used here can operate at 1 pot every 30 s or 120 pots/h. Plant pots require 80 g of TEGS so the raw materials cost €0.045/pot (based on the above €0.56/kg).

Energy costs:
 0.1 kWh/pot
 €0.01/pot
 Labour cost €0.021/pot
 Capital costs €320,000
 So total cost is €0.103/pot.

5.2.5.7 Levulinic Acid Production

Levulinic acid can be produced directly from rapeseed straw through the process flowsheet shown in Figure 5.10. The information for this scheme is provided by Biorefinery.de.[54] In this process the straws are reacted with hydrochloric acid to produce formic and levulinic acid. Subsequent units are used to separate and purify the acid produced. This includes evaporation to recycle the hydrochloric acid so the straw:hydrochloric acid ratio can be maintained at 1:10.

Material Prices

Rapeseed straw	€25/ton
Hydrochloric acid	€46/ton
Solvent	€3.20/L
Levulinic acid	€12/kg
Formic acid	€1.15/kg

The above prices and the following conversions were provided by Biorefinery.de and Wagner et al.[54,55] The straw has a cellulose fraction accounting for approximately 45% of the straws fed into the process and the conversion to levulinic acid is specified as:

$$0.45 \text{ kg cellulose} \rightarrow 0.23 \text{ kg levulinic acid}$$

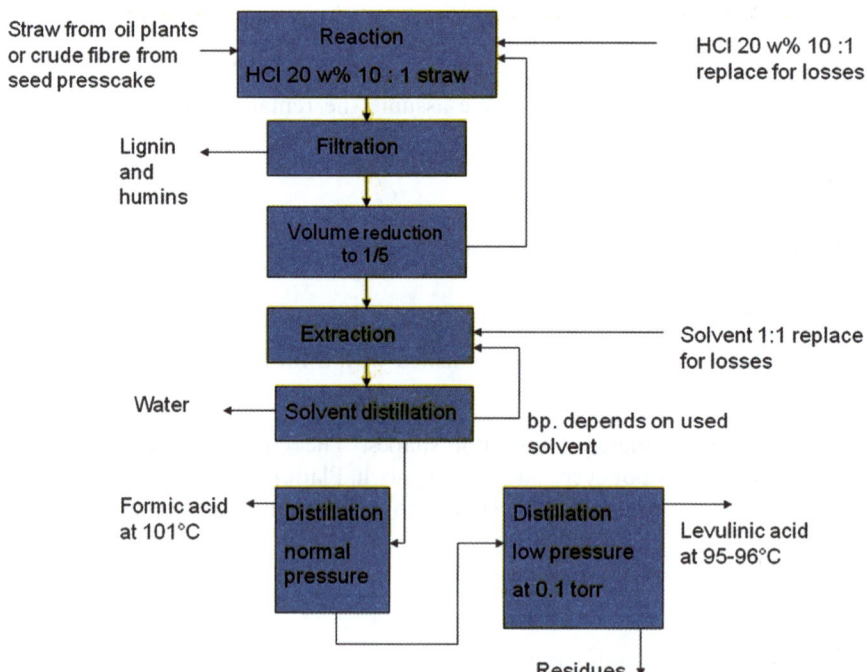

Figure 5.10 Process flowsheet for levulinic acid production.

Assessment of Economic and Environmental Cost-benefits

This correlation is provided by biorefinery.de and Wagner *et al*.[54,55] However, we also know the reaction producing levulinic acid is:

1 glucose → 1 levulinic acid + 1 formic acid + 1 water

So using the molecular weights of levulinic acid (116.115 kg/kmol) and formic acid (46.026 kg/kmol) we find that the overall correlation is:

0.45 kg cellulose → 0.23 kg levulinic acid + 0.091 kg formic acid

5.3 Results and Discussion

5.3.1 Economic Optimisation

For each of the schemes considered we have used the above data to model and optimise the process where we have maximised the profits or minimised the costs of each scheme. Improved operating conditions can be found for cases where there are sufficient data to predict how the system will change in response to modified operating conditions.

5.3.1.1 Biodiesel Production

In such a complicated system (see Figure 5.3) there are many parameters that can be altered or optimised. However, in this case we have limited the optimisation to modifying two parameters that affect the amount of methanol fed into the transesterification reactor. These two parameters are:

- The methanol feed flowrate;
- The fraction of methanol recovered by distillation.

Modifying these two parameters affects the economics and the environmental impacts of the process. Using less methanol would reduce the cost of raw materials but it could also reduce the conversion in the reactor and the quantity of biodiesel produced. Increasing the fraction of methanol recovered by distillation (see the methanol distillation in Figure 5.4) would require more energy and hence higher emissions but would increase the oil:methanol ratio.

The initial values for the significant flowrates are shown in Table 5.4. Most of the input raw material flowrates have been fixed and apart from the potassium chloride none of the product flowrates are fixed. Potassium chloride is the exception because it is produced from neutralising a fixed amount of potassium hydroxide. The other product streams are all affected by the methanol feed either because they contain some of the methanol or because the methanol affects the yield of the products. The values and costs of the different materials are also provided in Table 5.4. This includes the costs of purchasing the raw materials, the values of the final products and the costs of disposing of waste products.

These costs and values were obtained from a variety of sources including literature,[39] books,[29] online resources[25,38,40] and through communications.[26] In addition to the material costs we have also considered the capital costs and

Table 5.4 Overall input and output flowrates.

	Flowrate	Fixed/variable	Cost/value
Raw materials			
Methanol	216.3 kg/h	Variable	€0.202/kg
Rapeseed	3125 kg/h	Fixed	€0.279/kg
Potassium hydroxide	15 kg/h	Fixed	€0.220/kg
Hydrochloric acid	20 kg/h	Fixed	€0.085/kg
Products			
Biodiesel	994.4 kg/h	Variable	€0.908/kg
Crude glycerol	86.68 kg/h	Variable	€0.0688/kg
Rapeseed meal	1.875 kg/h	Variable	€0.175/kg
Wastes			
Potassium chloride	30.19 kg/h	Fixed	€0.0248/kg
Water/methanol	151.6 kg/h	Variable	€0.0248/kg

operating costs. Capital costs were calculated based on correlations[29] that modify the cost based on the size of the equipment and the chemical engineering price cost index (CEPCI), which approximates how costs change with time. The capital costs for this system are fixed at €1 626 988 and we have considered this cost for a biodiesel plant operating for 10 years, 330 days per year and 24 h per day. This reduces the capital cost down to an hourly rate (€20.54/h), which is used in the objective function. The objective function is then computed as:

$$\text{Objective} = -\text{VP} + \text{CR} + \text{EC} + \text{CP} + \text{WC} + \text{EF}$$

where:

VP = Value of products
CR = Cost of raw materials
EC = Energy costs
CP = Capital costs (scaled to €/h based on 10 years' lifespan)
WC = Cost of liquid and solid waste disposal
EF = Error function ($= 1 \times 10^9$ if an error in the simulation occurs otherwise $= 0$)

This form of the objective function is equivalent to 'negative net profit', which has been formulated this way so that optimisation software can make use of minimisation algorithms in order to maximise the profits.

The initial values are (in terms of €/h):

VP = 1230.80
CR = 878.25
EC = 46.23
CP = 35.08
WC = 4.52
EF = 0

giving an initial objective = −266.74, *i.e.* profit = €266.74/h

Assessment of Economic and Environmental Cost-benefits

As mentioned above the methanol feed rate and recovery ratio are the optimisation parameters that were allowed to vary through a sensible range of values with bounds that do not allow the system to be pushed to its limits. These bounds were selected to include a wide range of oil-to-methanol ratios ranging from 1:4.5 to 1:43.9.

Methanol flowrate was constrained to the range: 4.1–8 kmol/h (131–236 kg/h). Methanol recovery was constrained to the range: 50–90%.

Methanol recovery is specified for the distillation column, which is directly connected to the outlet of the transesterification reactor. Increasing the recovery increases the concentration of methanol fed into the reactor. This unit is necessary to maintain a high methanol feed ratio. Modifying the recovery and the feed rate allows us to modify this ratio. The lower bounds above were specified to prevent this ratio from becoming too low. To explore the effects of these two parameters on profit we first performed sensitivity analysis, fixing one of the parameters and varying the other to see how it affected the objective. Here we have fixed the methanol feed rate at 7 kmol/h and we have varied the methanol recovery from 50–90%.

From the results presented in Figure 5.11 we can see that the maximum profits are found with a methanol recovery of 66%. However, there are two parameters available for optimisation so fixing the recovery at 66% and this time modifying the methanol feed rate we obtain the following.

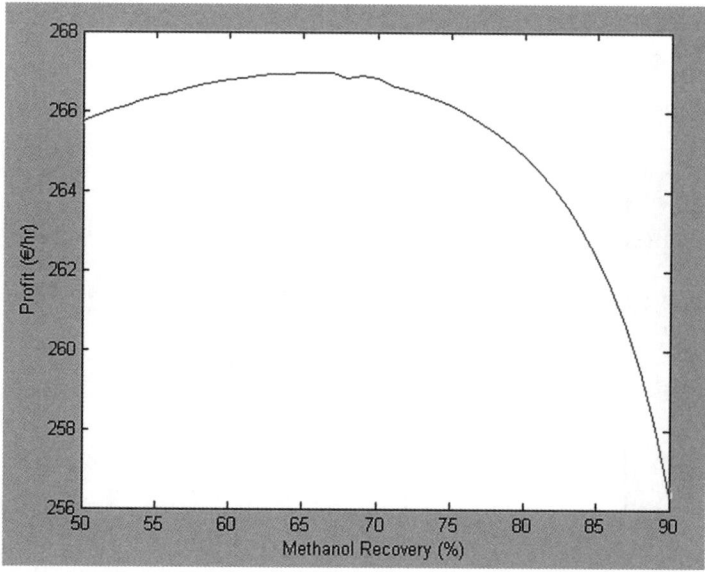

Figure 5.11 Effects of methanol recovery on profits (with methanol feed rate fixed at 7 kmol/h).

The results from Figure 5.12 show that the maximum profits (with methanol recovery fixed at 66%) are found with 6.5 kmol/hr of methanol feed. The results from Figures 5.11 and 5.12 show that the maximum profits are in the middle of the ranges tested. This suggests that very high methanol-feed ratios were not necessary because the maximum conversion is obtained with lower methanol concentrations. However, we also see that if the methanol concentration is too low the conversion will be decreased and less biodiesel will be produced, which accounts for the reduced profit at the lower end of the methanol feed and recovery ranges.

The sensitivity analysis presented above shows how the profits are changed by modifying each of the parameters separately. This gives a general feel for how each parameter will affect the profits and can give a rough optimisation of the parameters by selecting the maximum profit values obtained. However, simultaneous optimisation of both parameters using existing optimisation algorithms can find a better solution in many cases. Here we have performed optimisation using SQP (MatLab function fmincon) and simulated annealing (in-house code) and we have compared these results with the crude optimisation from the sensitivity analysis and the initial values. The optimisation results using these different methods are presented in Table 5.5, where we have also compared the values of the parameters, the profits and the resulting oil:methanol ratio. From these results and from looking at Figures 5.11 and 5.12 we can see that the initial conditions are actually not far away from the optimum values and the maximum profits found by all the methods are similar. Unfortunately in this case the SQP algorithm appears to have failed, possibly

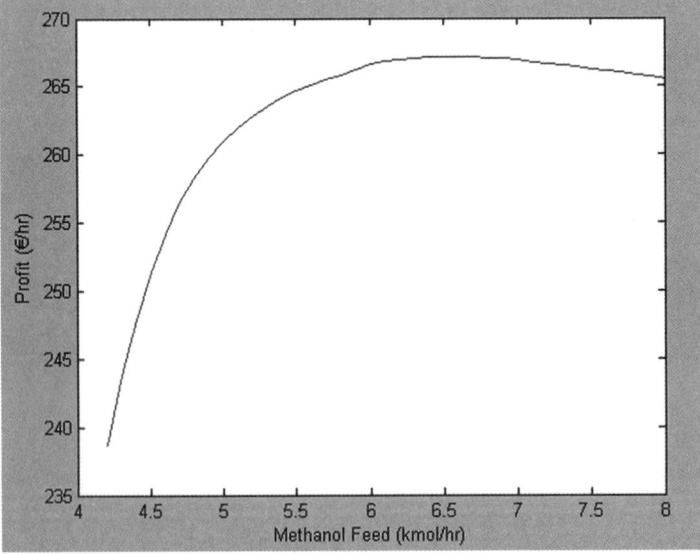

Figure 5.12 Effects of methanol feed rate on profits (methanol recovery fixed at 66%).

Table 5.5 Optimisation results for biodiesel with crude glycerol production scheme.

Parameter	Methanol flowrate (kmol/h)	Methanol recovery (percent recovered)	Oil : methanol ratio	Profit (€/h)
Initial value	6.75	60%	1 : 10.47	266.74
Approximate optimum from sensitivity analysis	6.5	66%	1 : 11.14	267.17
Optimum value (using SQP)	6.747	59.99%	1 : 10.46	266.74
Optimum value (using simulated annealing)	5.178	84.07%	1 : 13.02	268.02

due to the error function in the objective function which applies a penalty to the objective when the connected Aspen simulation fails.

5.3.1.2 Biodiesel Production with Co-production of Purified Glycerol

In addition to the above base case where crude glycerol is sold as an additional product we have also considered the option of purifying the glycerol to a higher grade, which can be sold for a higher value. This option is relatively easy to simulate as it does not require any additional units. Even in the base case we have purified the crude glycerol to some extent through distillation to bring it to a level that could be sold. So to purify the glycerol to a higher grade we simply need to change some parameters in the final distillation column. We have specified the crude glycerol is 80% glycerol and purified glycerol is 95% glycerol with the remainder consisting of mostly water and methanol.

$$\text{Value of crude glycerol} = €0.0688/\text{kg}^{39}$$
$$\text{Value of purified glycerol} = €0.35/\text{kg}^{25}$$

The objective function is also very similar, except that a value for purified glycerol is included instead of the crude values.

As can be seen in Figures 5.13 and 5.14, profit is similar to that using crude glycerol (Figures 5.11 and 5.12) with similar maxima at 66% methanol recovery and at 6.6 kmol/h methanol feed rate (compared to 66% and 6.5 kmol/h, respectively, for the base case). Nevertheless, the optimum profits from the purified glycerol case are higher suggesting purification is a profitable approach. These results are dependent on the values of purified glycerol, which can be lower than the values used here due to the increasing production of glycerol from the biodiesel industry.[39] In addition to the optimum from sensitivity analysis we have also optimised this case using SQP and simulated annealing as above. A comparison of these results and the associated conditions is shown in Table 5.6, which shows that here simulated annealing gives the most profitable conditions.

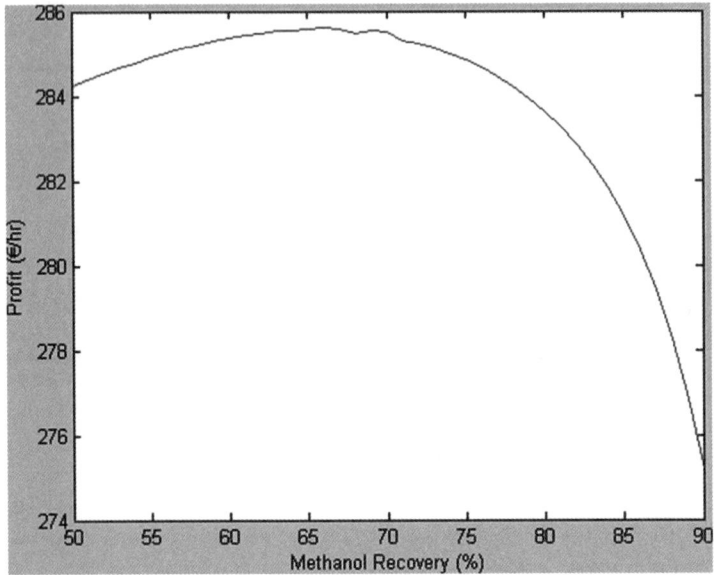

Figure 5.13 Effects of methanol recovery on profits with glycerol purification (with methanol feed rate fixed at 7 kmol/h).

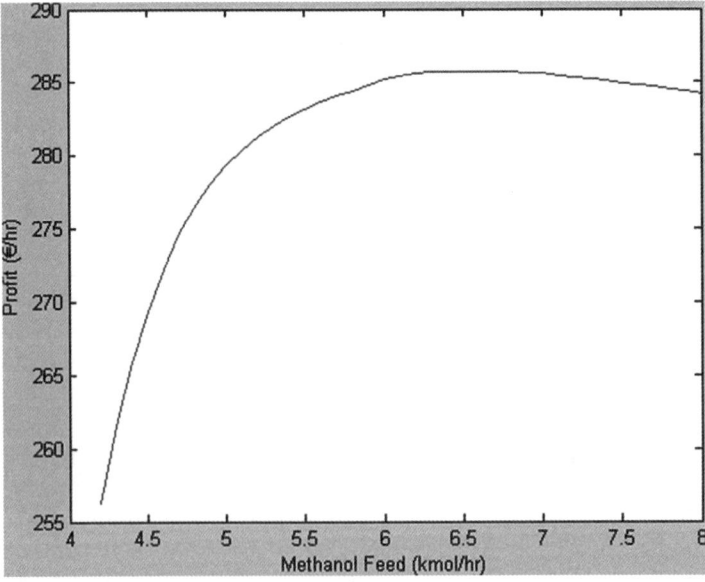

Figure 5.14 Effects of methanol feed rate on profits (with glycerol purification and methanol recovery fixed at 66%).

Assessment of Economic and Environmental Cost-benefits

Table 5.6 Optimisation results for biodiesel with purified glycerol production scheme.

Parameter	Methanol flowrate (kmol/h)	Methanol recovery (percent recovered)	Oil: methanol ratio	Profit (€/h)
Initial value	6.75	60%	1 : 10.47	285.32
Approximate optimum from sensitivity analysis	6.6	66%	1 : 11.40	285.80
Optimum value (using SQP)	6.747	59.99%	1 : 10.46	285.33
Optimum value (using simulated annealing)	5.060	86.94%	1 : 14.40	286.66

5.3.1.3 Biodiesel Production with Co-production of Succinic Acid

A more interesting option is to convert the glycerol produced into a valuable chemical succinic acid. This is possible through a fermentation process, which we are also testing experimentally.[12] In order to optimise this scheme we have kept the same optimum operating conditions as in the base case and we have selected two optimisation parameters, which directly affect fermentation. These are:

- The water feed rate;
- The batch cycle-time of the fermenters.

Water is added to the crude glycerol to dilute it to a level at which the fermentation organism can operate. At higher glycerol concentrations substrate inhibition causes problems in the conversion of glycerol into succinic acid. However, the addition of water increases the size of the required fermenters which has a great effect of the capital costs, as shown in Figure 5.15. The capital cost is calculated using a correlation with volume and chemical engineering plant cost index.[29] The batch cycle time also has a great effect on the fermentation process as it will directly affect the production rate of succinic acid and it will also affect the capital cost because volume is directly related to flowrate and cycle time (this flowrate is the sum of the water and the crude glycerol flowrates).

$$\text{Volume} = \text{flowrate} \times \text{cycle time}$$

However, for continuous operation at least two reactors must be used. The bounds for the cycle time were chosen to be between 10 and 100 h, which are based on the typical ranges of batch times required in experiments.[12]

The bounds for the water flowrate (50 to 120 kmol/h) were based on the range of glycerol concentrations tested in experiments.[12] In general, lower water flowrates and higher glycerol concentrations would be preferable if the organism involved could still process them effectively at those conditions.

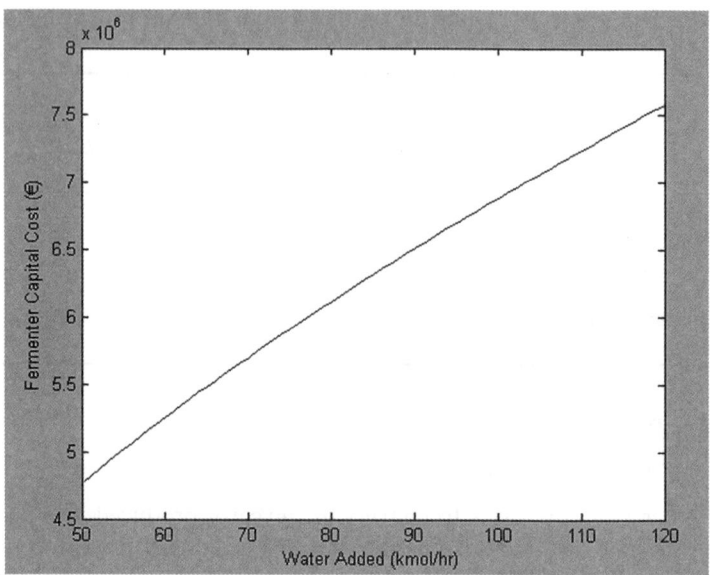

Figure 5.15 Capital costs of four fermenters with fixed cycle time of 100 h, varying the amount of water added.

Initially fixing the cycle time to 100 h and varying the quantity of water added we are able to calculate the profits for a range of different flowrates as shown in Figure 5.16. Increasing the quantities of water added increases the profitability of the process according to the kinetic model used. This appears to be counter-intuitive because adding more water will dilute the glycerol feed; however, this increase could be due to reduced substrate inhibition. The higher flowrate will also mean higher capital costs with larger fermenters; however, the increased succinic acid productivity and the high value of the product counter this increased cost.

Fixing the water flowrate to 120 kmol/h and varying the cycle time of the fermenters we obtain the results shown in Figure 5.17. We see that the highest possible cycle time gives the most profit. This makes sense because the production of succinic acid will also be highest with the longest batch times. However, as with varying the water flowrate, this will also require the largest fermenters.

We have again performed optimisation for this case using SQP and simulated annealing and we have compared the results with the crude optimum found through sensitivity analysis and with the initial conditions. These results can be seen Table 5.7 and in Figures 5.16 and 5.17, where we can see that the maximum profits are found by adding the most water and using the longest batch cycle time. Unlike the previous cases the optimisation algorithms do not find the best solution here, although they do give values close to the optimum and given more iterations they could have reached the same optimum. However, in

Assessment of Economic and Environmental Cost-benefits 231

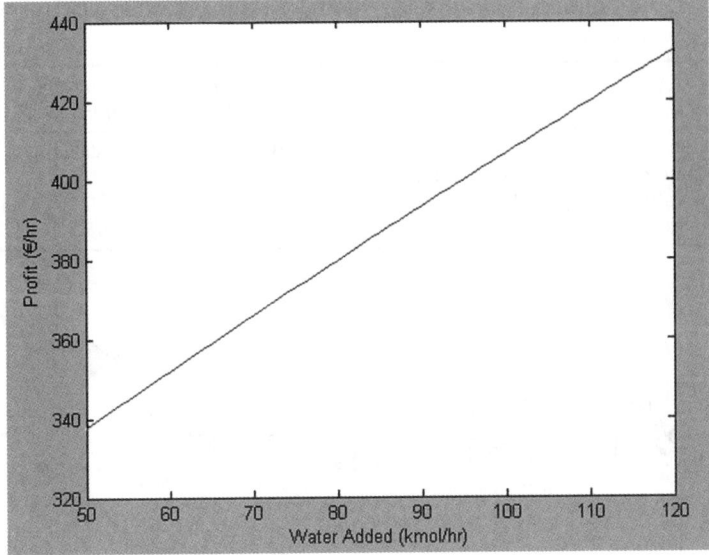

Figure 5.16 Effects of water flowrate on profits.

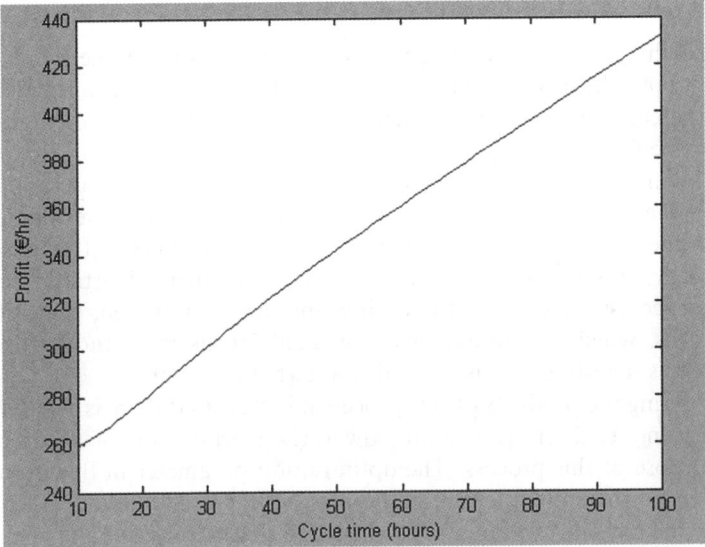

Figure 5.17 Effects of cycle time on profit.

this case the sensitivity analysis of the two parameters reveals the optimal profit conditions. This optimal solution would require the largest and most expensive fermenter; however, the extra production of succinic acid at these conditions makes this the most profitable.

Table 5.7 Optimisation results for biodiesel with succinic acid production scheme.

Parameter	Water flowrate (kmol/h)	Batch cycle time (h)	Profit (€/h)
Initial value	100	80	376.16
Approximate optimum from sensitivity analysis	120	100	432.29
Optimum value (using SQP)	118.64	100	430.57
Optimum value (using simulated annealing)	119.98	95.35	423.93

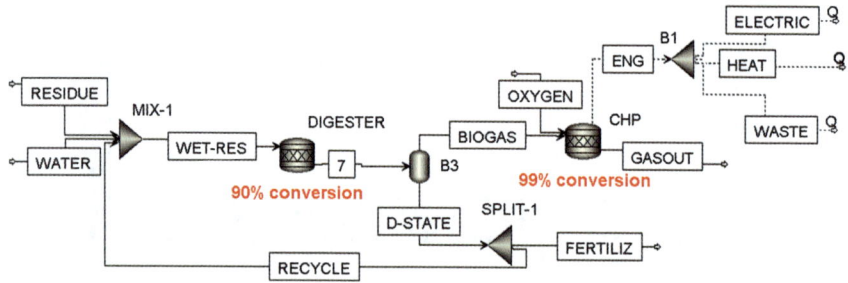

Figure 5.18 Flowsheet for the production and combustion of biogas.

5.3.1.4 Biogas Production

To model this process (in Figure 5.18) we have used stoichiometric reactors to represent both the digester and the combined heat and power unit (CHP). This means the conversions in each reactor are fixed to 90% for the digester and 99% for the CHP unit.

The initial conditions and the costs/values of the different inputs and outputs are shown in Table 5.8. The value of digestate as fertiliser was estimated to be 10% of the value of commercial fertiliser, with commercial fertiliser data provided by FERA.[26] This is not as valuable as commercial fertiliser because it does not necessarily contain the desired mixture of nitrogen, potassium and phosphorus, which are found in commercial fertilisers. Although the exact value of this digestate fertiliser is still somewhat uncertain.

Considering the products of this process it is clear that there is a split between electricity and fertiliser production; however, the production of electricity is the main purpose of this process. The optimisation parameter in this process was chosen to be the split fraction in the unit SPLIT-1 (see Figure 5.19), which controls the amount of liquid/solid digestate that is recycled and the amount that is used for fertiliser.

This digestate contains a combination of water, unspent residue and inert spent residue. The advantage gained by recycling this is that it should increase the overall conversion by recycling the unused raw material and also that it would reduce the amount of water that needs to be added. Theoretically the recycle fraction could be anywhere between 0% and 100%; however, if the recycle was 100% then the inert spent residue would never leave the system and

Assessment of Economic and Environmental Cost-benefits

Table 5.8 Overall input and output flowrates for biogas flowsheet.

	Flowrate	Fixed/variable	Cost/value
Raw materials			
Residue	1350 kg/h	Fixed	None
Water	1928 kg/h	Variable	None
Products			
Fertiliser	198.9 kg/h	Variable	€0.0981/kg
Electricity	3109 kW	Variable	€0.145/kWh
Heat energy	3657 kW	Variable	€0.03/kWh
Wastes			
Carbon dioxide	2336 kg/h	Variable	None
Methane	6.898 kg/h	Variable	None

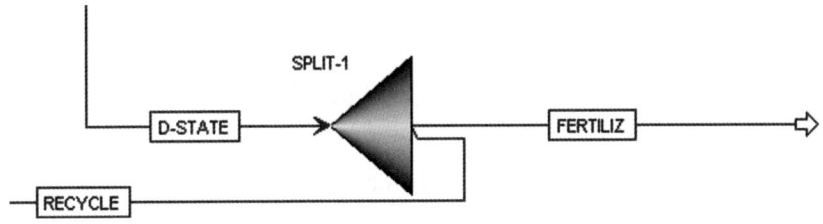

Figure 5.19 Diagram showing the digestate split into fertiliser and recycle streams.

would continuously increase indefinitely. For this reason the recycle fraction is bound to be between 0% and 99%. The humidity of the digester is fixed to be 70% water, which means as the recycle fraction is modified the water feed rate also needs to change in order to maintain this 70% figure.

Looking at the effect of the recycle fraction on profits (Figure 5.20) it can be seen that the maximum profits can be obtained with close to 100% recycle, although the profits do drop off again at higher recycle fractions. To find the optimum value we make use of a nonlinear optimisation algorithm SQP (successive quadratic programming), which, starting from an arbitrary point in the range, can be used to find the optimum value. This was done using MatLab as described earlier with the only constraint being placed on the recycle fraction fixing it between 0% and 99%. We also applied simulated annealing to this problem; however, as there is only one optimisation parameter both these optimisation algorithms should be able to find an optimum with relative ease. In fact using the optimum value obtained from sensitivity analysis also gives a very good solution in this case. These optimum solutions are summarised and compared in Table 5.9.

The most profitable solution is found using the SQP optimisation algorithm and a very similar solution is also found using simulated annealing. The approximate optimum found from the sensitivity analysis also gives a reasonably good optimum value. The benefits of optimisation are clearly seen with the increase in profits compared to the initial conditions with profits increase by €23.74/h (6.7% increase).

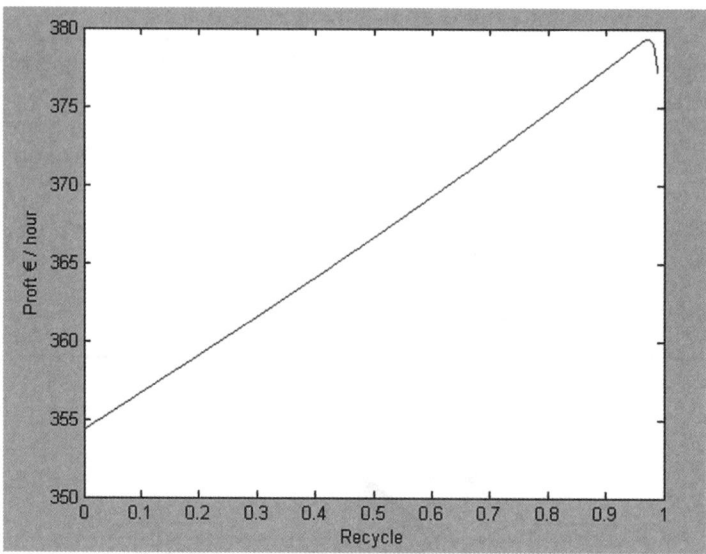

Figure 5.20 Effect of recycle fraction on profits.

Table 5.9 Optimisation results for biogas scheme.

Parameter	Recycle fraction	Profit (€/h)
Initial value	10%	356.04
Approximate optimum from sensitivity analysis	97%	379.72
Optimum value (using SQP)	97.49%	379.78
Optimum value (using simulated annealing)	97.40%	379.77

This optimum value is very high suggesting that to maximise profits most of the digestate should be recycled, which would produce more biogas and more electricity while producing only a very small quantity of digestate fertiliser. If the digestate fertiliser was more valuable then a different solution that involved more fertiliser production and less electricity production could be found. This solution is also highly sensitive to the value at which the electricity can be sold. The current value used (€0.0145/kWh) is a very good value for selling and it implies there is an incentive tariff provided by the local government for the production of renewable energy.

5.3.1.5 Supercritical Carbon Dioxide Extraction

An extraction process is simulated involving the use of supercritical carbon dioxide to recover waxes from wheat straw. This process is modelled using the flowsheet in Figure 5.21 constructed in Aspen Plus. The data for this model and assistance constructing the flowsheet were provided by the University of York.[48]

The important parameters for optimisation are the capacity (flowrate of residue) and the residence time of the extractor. The capacity is the dominant

Assessment of Economic and Environmental Cost-benefits

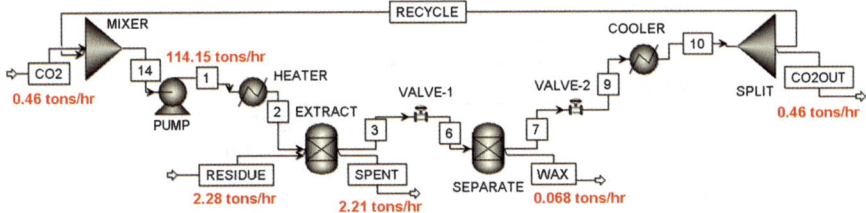

Figure 5.21 Process flowsheet for the extraction of waxes from wheat straws.

Table 5.10 Overall input and output flowrates for straw wax extraction flowsheet.

	Flowrate	Fixed/variable
Raw materials		
Residue	2.28 tons/h	Variable
CO_2 (total)	114.15 tons/h	Variable
CO_2 added	0.46 tons/h	
Products		
Spent straw	2.21 tons/h	Variable
Straw wax	0.068 tons/h	Variable
Recycled CO_2	113.70 tons/h	Variable
Wastes		
Carbon dioxide	0.46 tons/h	Variable

parameter because it controls the size of the plant and the rate at which feedstock (straws) is processed. Initial flowrates, which will change with the capacity of the plant, are given in Table 5.10. The capacity of the plant also controls the operating and capital costs of the plant. Residence time should also affect these costs. However, we have used correlations for operating costs and capital that are based solely on the capacity and we have treated the residence time as a parameter that can affect only the rate of wax production.

Operating costs (including capital) were calculated based on a correlation using the data in Table 5.11 and a linear fit similar to the one used by Brunner.[50]

Using the above linear fitting (Figure 5.22) of data from the literature in MatLab an expression for this relationship can be derived:

$$\text{Log (cost per ton)} = -0.76894 \times \log (\text{capacity}) + 13.807$$

where capacity has units of tons per year. From this relationship the cost per ton can be calculated for the desired range of capacity: 20 000 to 80 000 tons per year as shown in Figure 5.23.

In addition to the costs we have also considered the revenue from waxes produced and how the cycle time from the extraction affects the yield of waxes. The relationship between yield and extraction time is modelled based on information provided by UoY including the values provided in Figure 5.24.[48]

To use this information as part of optimisation we need to be able to calculate the yield for any extraction time inside the known range. To achieve this

Table 5.11 Operating costs for various capacity extractions.[50]

Capacity (tons/year)	Price (€/ton)
200	25 000
500	4500
2000	3500
20 000	500

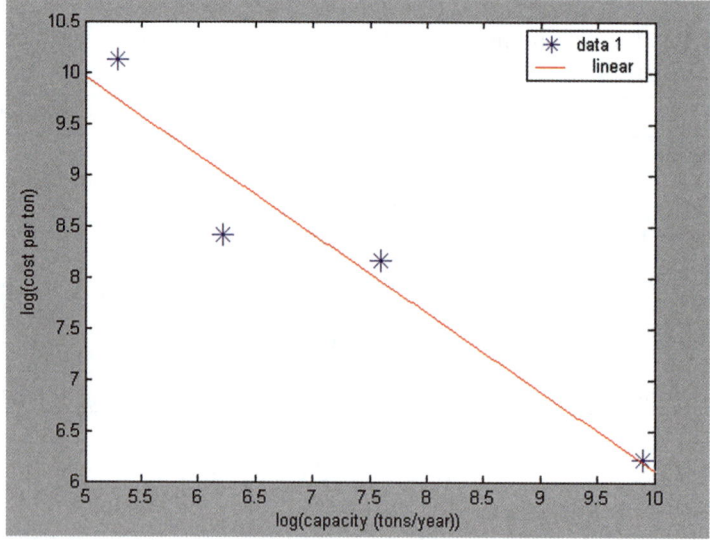

Figure 5.22 Plot showing a linear fit between the log of operating costs and the log of capacity.

we have fitted a seventh-order polynomial to this set of data using MatLab (Figure 5.25).

The polynomial in question is:

$$y = p_1 e^7 + p_2 e^6 + p_3 e^5 + p_4 e^4 + p_5 e^3 + p_6 e^2 + p_7 e + p_8$$

where y = yield and e = extraction time and the coefficients are:

$p_1 = 0.0030775$
$p_2 = -0.06412$
$p_3 = 0.54982$
$p_4 = -2.4944$
$p_5 = 6.3964$
$p_6 = -9.1899$
$p_7 = 7.1223$
$p_8 = 0.0008659$

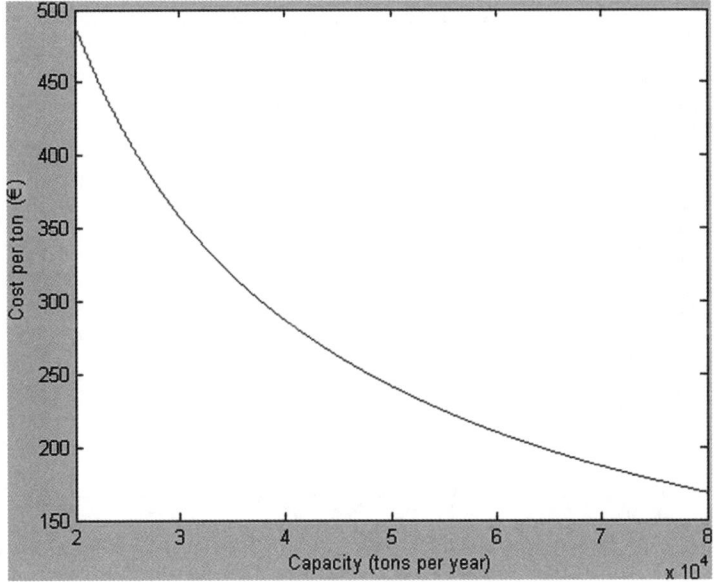

Figure 5.23 Effect of capacity on operating costs calculated using the above correlation.

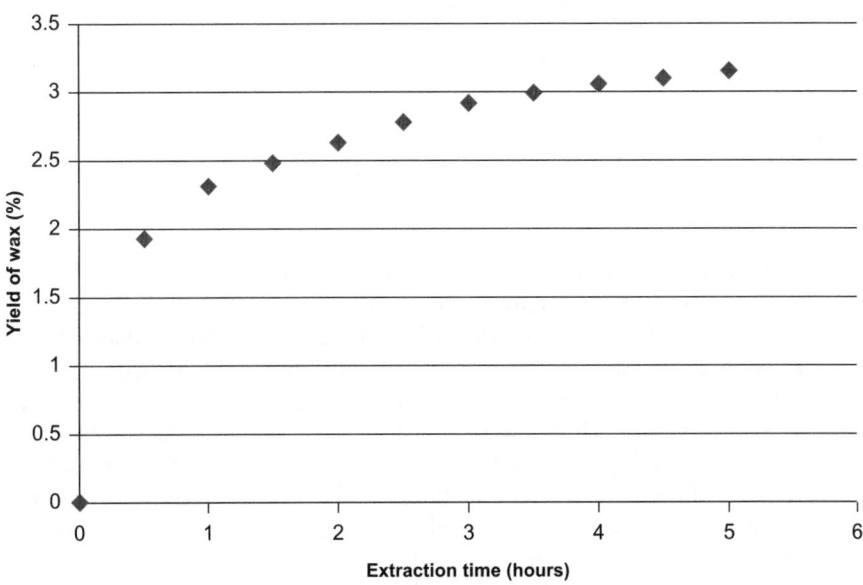

Figure 5.24 Yield of wax achieved with a range of different extraction times.

We have used the above fitting to perform economic analysis of the supercritical CO_2 extraction with respect to fed capacity (tons per year), extraction time (hours) and extractor size (litres). In order to perform the analysis we have

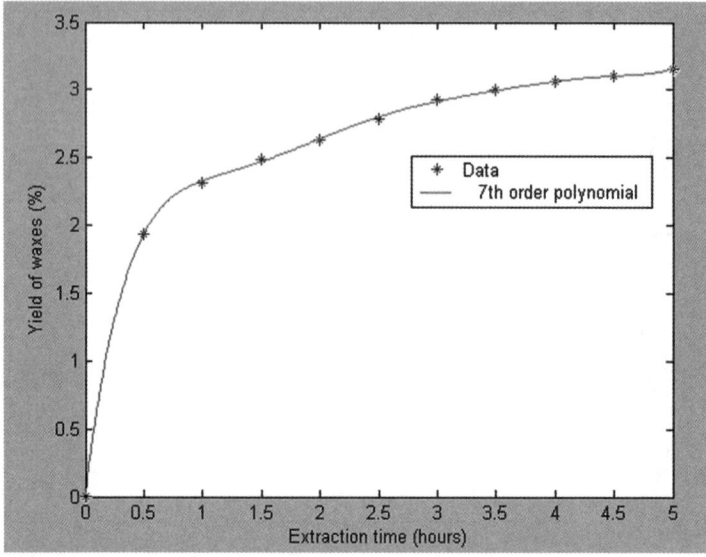

Figure 5.25 Seventh-order polynomial fitted to extraction data.

fixed the extractor size to be 25 000 litres (or 5 × 5000 litre units in parallel), which means we only have one free parameter since extraction time is inversely proportional to fed capacity for a fixed size extractor.

$$Extraction - time = \frac{24 \times (Size/2)}{(Capacity/365)}$$

The total extraction time can be calculated using the above equation where *Size* is the volume of the extractor in m^3 and dividing by 2 gives the capacity in tons assuming the feed has a density of 0.5 tons per m^3 and *Capacity* is the quantity of feed processed per year (tons per year) giving an extraction time in hours. However, if we include 30 min for the change-over between extractions we must also subtract 0.5 h. We have also assumed a value for the waxes of £25 per kg and we can use this together with the above yield correlation to calculate the revenues from sale of waxes.[48] Combining the above correlations for costs and revenues we can obtain the overall profits and we have performed sensitivity analysis on the profit with respect to capacity as shown in Figure 5.26.

As can be seen in Figure 5.26 there are two different peaks with the most efficient plants in terms of profit per ton of feed found at 35 000 tons per year. Increasing the capacity further reduces the yield of waxes extracted due to the reduced extraction time required. However, a second peak is found around 70 000 tons per year due to the costs being reduced faster than the revenues from waxes.

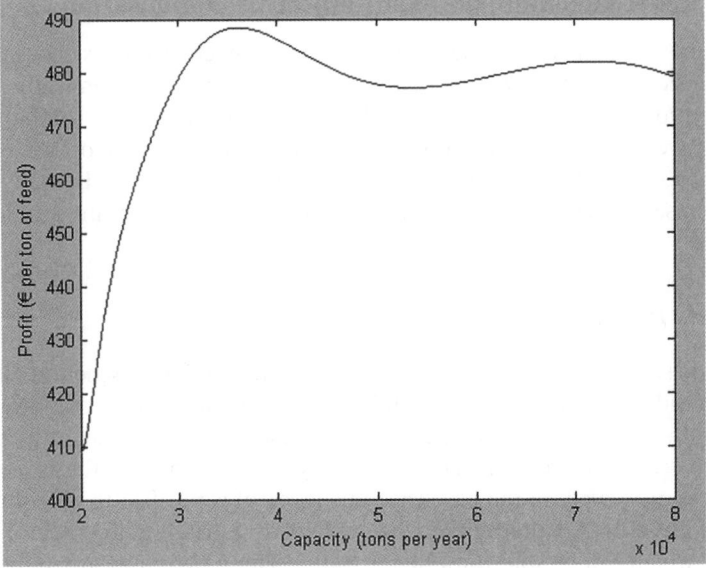

Figure 5.26 Net profits per ton of feed for a range of different capacities.

Table 5.12 Optimisation results for supercritical CO_2 extraction.

Parameter	Capacity (tons/year)	Extraction time (h)	Profit (€/ton)
Initial value	25 000	4.380	455.27
Approximate optimum from sensitivity analysis	36 228	3.023	488.40
Optimum value (using SQP)	71 846	1.524	481.81
Optimum value (using simulated annealing)	36 227	3.022	488.40

Considering these profits in terms of euros per ton of feed we can see that a single plant with a capacity of 70 000 tons per year would make more profit than a single plant operating at 35 000 tons per year. However, these results show that two plants operating at 35 000 tons per year each would be more profitable than the single plant operating at 70 000. To confirm these results we have performed optimisation using SQP and simulated annealing using the above correlations for costs and revenues.

The results summarised in Table 5.12 show that each optimisation method finds a solution which corresponds to one of the two peaks shown in Figure 5.26. SQP being a local method, it identifies the second solution with lower profit. SA and sensitivity analysis both find the better solution at around 36 000 tons per year.

5.3.2 Environmental and Multi-objective Optimisation

In the above optimisation the environmental impact of each scheme has not been considered with the aim being to find the most economically feasible configurations. If we also consider the environmental impact of biorefinery schemes we can either minimise the environmental factors or we can find solutions that satisfy economic and environmental constraints. In many cases a compromise between the economic and environmental optimum solutions can be found.

5.3.2.1 Biodiesel Production

For biodiesel production we have considered three different schemes where we have considered both economic and environmental factors. All three cases involved the production of biodiesel from rapeseed using potassium-hydroxide-catalysed transesterification. However, they vary as to how the glycerol by-product is processed: in the base case this glycerol is purified to 80% then sold as crude glycerol. An alternative scheme involves purifying the glycerol further to 95%, which can be sold for a higher value. These two cases are quite similar as they use the same equipment; however, we have also considered a third case where this glycerol is converted into the high-value product succinic acid.

Liquid and Solid Wastes Produced

In all these three cases there are wastes produced, including the potassium chloride that is created when the potassium hydroxide catalyst is neutralised and removed from the process after the transesterification reaction (see Figure 5.27).

In addition to this potassium chloride waste there are also a number of waste streams associated with the purification of both the biodiesel and the glycerol products.

The purification of biodiesel and glycerol (see Figure 5.28) leaves waste streams consisting mostly of methanol and water with small quantities of the other components involved in biodiesel production. In practice the purified glycerol stream still contains some quantities of methanol and water depending on the level of purification. If the glycerol is purified to a higher grade the quantities of waste methanol and water being produced increase.

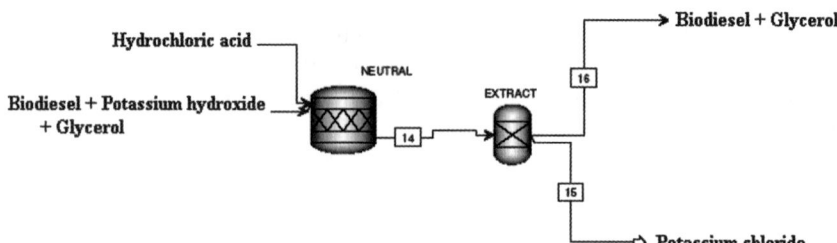

Figure 5.27 Diagram showing how the catalyst is removed as a waste.

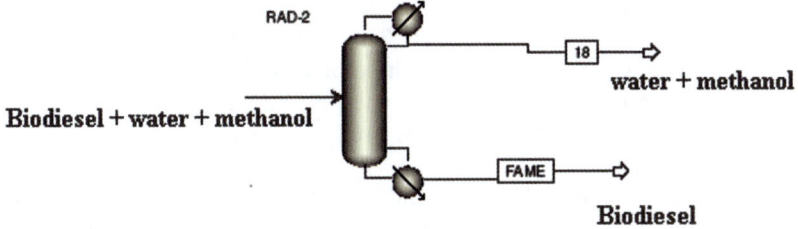

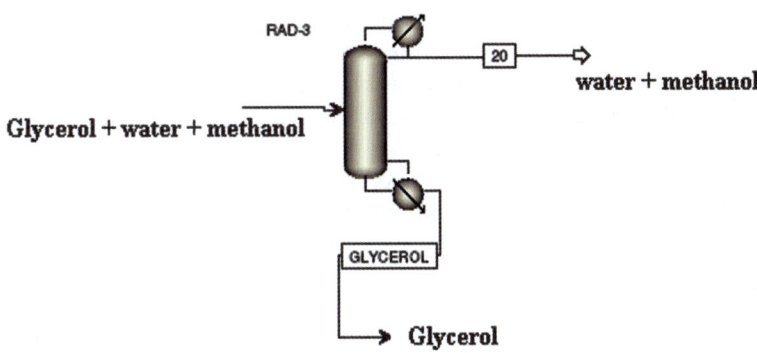

Figure 5.28 Diagram showing the separation units used to purify biodiesel and glycerol.

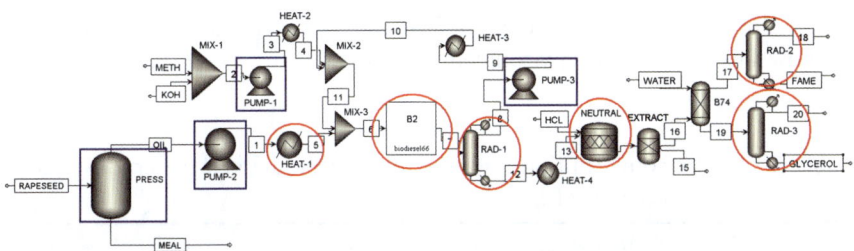

Figure 5.29 Diagram showing the units requiring heat energy circled in red and units requiring electricity highlighted with a blue square.

Emissions from the Biodiesel Production
In the biodiesel production process there are no direct emissions of gases; however, there are indirect emissions and ultimately the end use of biodiesel as a combustion fuel would give some direct emissions. However, these end-use direct emissions of CO_2 are assumed to be balanced by the absorption of CO_2 by the rapeseed crops grown on the farms. The indirect emissions are mainly as a result of the different energy requirements of the process. Heat energy is required by a number of heaters, reboilers inside the distillation columns and to provide heat directly to the reactors. These units requiring heat energy are highlighted in Figure 5.29 by red circles.

In addition to the emissions from the heating system we must also consider emissions due to the electricity usage. The units using electricity include the rapeseed oil extraction press and the three pumps identified by blue boxes in Figure 5.29. Electricity usage is more expensive and gives off more emissions (see Section 5.2.3.1) so it should only be used in cases where the energy can not be provided by heat.

For the purpose of LCA we could also have considered the emissions resulting from transporting the rapeseed from the farm to the biorefinery. However, we have fixed the feed rate of rapeseed into the system, so these emissions will remain fixed and unaffected by the optimisation.

In the previous single-objective optimisation we identified the methanol feed rate and the methanol recovery fraction as the optimisation parameters for this system. These parameters are responsible for controlling the ratio of oil: methanol in the transesterification reactor, which will affect the conversion of oil to biodiesel. In addition to altering the economics of the system, these parameters also affect the energy usage, emissions and the amount of wastes produced. The set of conditions optimising the economics of this scheme is given in Table 5.13.

In addition to this economic optimisation we can also consider the environmental impacts of these initial optimal solutions. As we can see from Table 5.14 the optimum economic solution requires more energy and gives off more emissions than the initial conditions. However, the initial conditions also generate approximately 50 kg/h more wastes than the optimum. To test further how the emissions and the wastes vary, we have also performed sensitivity analysis to see how they are affected by changing the two optimisation parameters. Firstly we fixed the methanol feed and varied the methanol recovery to see how this affected the environmental factors.

Table 5.13 Economic optimisation results for biodiesel with crude glycerol production scheme.

Parameter	Methanol flowrate (kmol/h)	Methanol recovery (percent recovered)	Oil: methanol ratio	Profit (€/h)
Initial value	6.75	60%	1:10.47	266.74
Optimum value	5.178	84.07%	1:13.02	268.02

Table 5.14 Environmental impact of optimisation results for biodiesel with crude glycerol production scheme.

Parameter	Heating required (kW)	Electricity required (kW)	Emissions (kg CO_2/h)	Wastes (kg/h)
Initial value	1642.9	212.938	444.58	182.33
Optimum value	1733.5	212.960	470.83	131.65

Based on these results in Figures 5.30 and 5.31 we can see that the energy requirements increase when we increase the methanol recovery. This occurs because the methanol recovery column uses more energy to recover greater quantities of methanol and because the pump in the recycle stream requires more electricity due to the increased flowrate of methanol, although in this scheme the electricity usage is much lower than the heating energy required. Also the transesterification reactor will require additional heat energy due to the increased flowrates.

The emissions caused by these energy requirements can be calculated using the emission factors from Section 5.2.3.1, giving:

CO_2 emissions (kg/h) = $0.201 \times$ heating energy (kW) + $0.537 \times$ electricity (kW)

This gives the following emissions:

As we can see from Figure 5.32 the emissions increase as the methanol recovery is increased due to the additional heating and electricity requirements (seen in Figures 5.30 and 5.31). The increase in emissions is approximately an additional 180 kg of CO_2 per h comparing the highest and lowest methanol recovery, which is a significant increase giving approximately an extra 40% more emissions. We have also calculated the level of waste production through our Aspen simulations and the effect of methanol recovery on these waste levels is shown in Figure 5.33. Increasing methanol recovery is shown to give a slight reduction in wastes although the level of wastes remains at around 190 kg/h. This slight reduction in wastes is assumed to be due to the increased glycerol

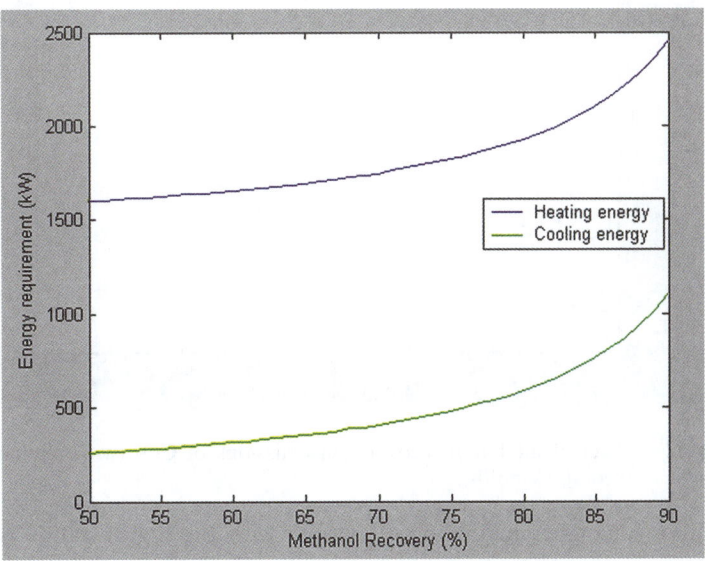

Figure 5.30 Effect of methanol recovery on heating and cooling duties (with methanol feed fixed at 7 kmol/h).

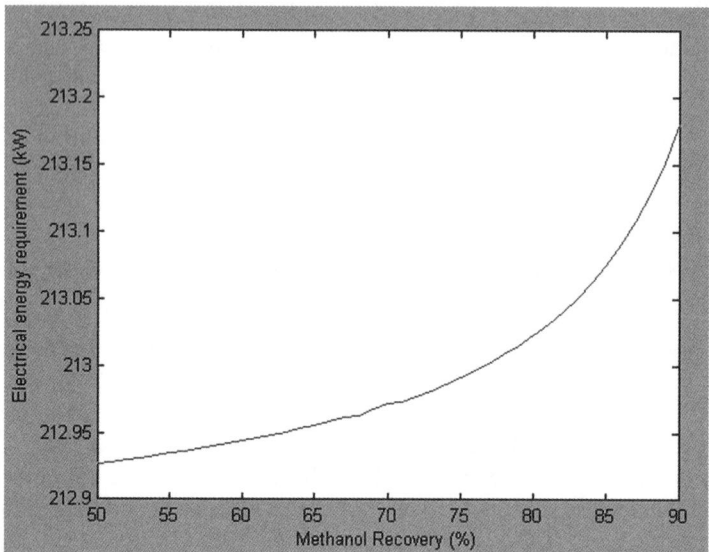

Figure 5.31 Effect of methanol recovery on electricity requirement (with methanol feed fixed at 7 kmol/h).

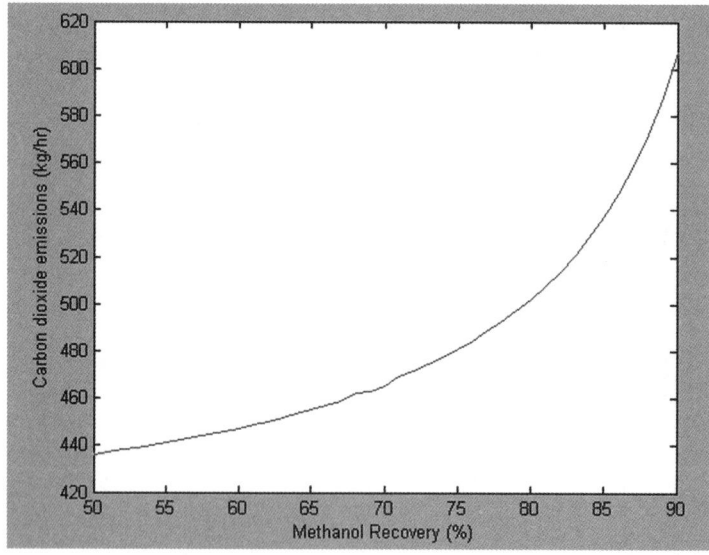

Figure 5.32 Effect of methanol recovery on emissions of CO_2 (with methanol feed fixed at 7 kmol/h).

consumption at higher methanol:oil ratios (with the higher ratios achieved with greater methanol recovery), which will increase the conversion of oil to biodiesel. The improved conversion consumes a little more methanol, which results in less methanol in the waste streams.

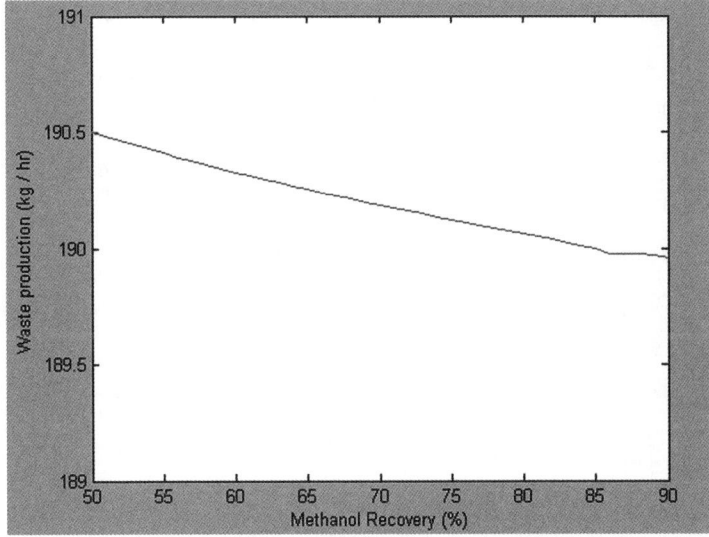

Figure 5.33 Effect of methanol recovery on solid and liquid wastes (with methanol feed fixed at 7 kmol/h).

We have also performed sensitivity analysis with respect to the methanol feed rate where we have fixed the methanol recovery at 66%, which was the optimum economic point obtained from previous sensitivity analysis and we have varied the methanol feed rate from 4.1 kmol/h to 8 kmol/h.

The effect of methanol feed on heating and cooling energy requirements can be seen in Figure 5.34. Compared with the effect of methanol recovery (see Figure 5.30) we can see that increasing methanol feed also increases the energy requirements, although not as dramatically as for the recovery. This increase is likely due to the increased flowrates through most of the process, which increases many of the energy requirements of the units involved.

The electricity requirement seems to increase almost linearly with methanol feed as seen in Figure 5.35. This is directly due to the increased flowrates through some of the pumps that require electricity and is different from the more exponential increase associated with increasing methanol recovery (Figure 5.31), although as before the quantities of electricity are still much lower than the heating requirement.

Combining the heating and electricity values for the range of possible methanol feed rates the emissions can be calculated using the emission factors from Section 5.2.3.1. The effects of varying the feed rate can be seen in Figure 5.36, which shows an almost linear increase in emissions with increased feed rate. There is a small deviation at 6 kmol/h, which could be due to a change in efficiencies at one or more units when they reach certain flowrates or conditions. The effect of methanol feed rate on emissions gave an increase of approximately 45 kg/h (Figure 5.36) comparing the highest and lowest feed rates, which is much lower than the effect of methanol recovery, which gave an increase of approximately 180 kg/h (Figure 5.32).

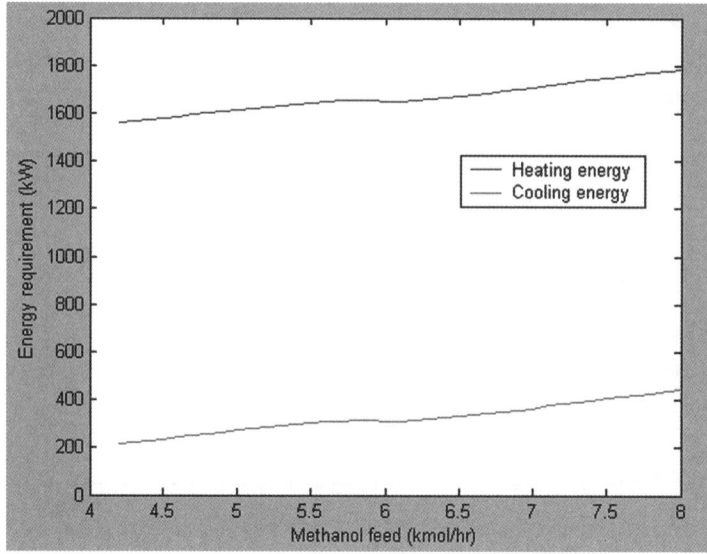

Figure 5.34 Effect of methanol feed rate on heating and cooling duties (methanol recovery fixed at 66%).

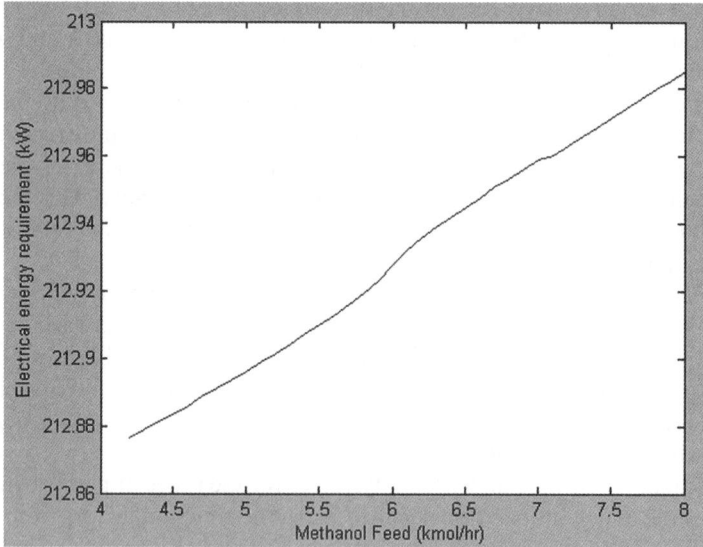

Figure 5.35 Effect of methanol feed on electricity requirement (methanol recovery fixed at 66%).

Methanol feed rate is also shown to have a direct effect on the amount of liquid and solid wastes produced (see Figure 5.37). This is because most of the methanol that is fed into the system ends up in the waste streams and the

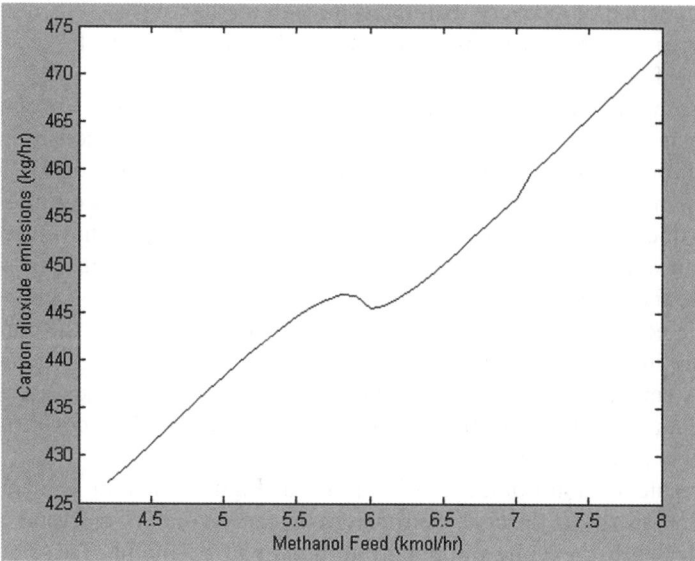

Figure 5.36 Effect of methanol feed on CO_2 emissions (methanol recovery fixed at 66%).

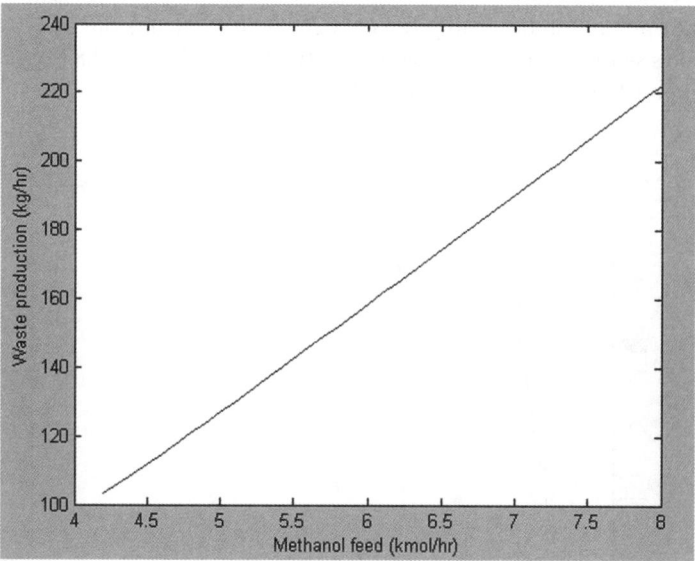

Figure 5.37 Effect of methanol feed on solid and liquid wastes (with methanol recovery fixed at 66%).

consumption only converts a small amount of this methanol. The system is set up in this way to ensure a high methanol : oil ratio to give the most biodiesel possible. The methanol feed has a much greater effect on the level of wastes than the methanol recovery with the change in methanol feed shown in Figure 5.37 increasing the waste by more than 100 kg/h compared to the 0.5 kg/h change in Figure 5.33.

Multi-objective optimisation for this case aims to maximise the profits of this system while considering constraints on the emissions. These constraints were based on the ranges of possible emissions (see Figures 5.32 and 5.36). Since the SQP optimisation algorithm was shown to be unreliable for optimising this system (due to its inability to handle simulation failures) we have used simulated annealing optimisation with multiple different constraints in order to complete the multi-objective optimisation.

Simulated annealing is hence modified so that in addition to the constraints on the optimisation parameters there is also an additional check to make sure the CO_2 emissions do not exceed the imposed constraint. The values of the constraints were chosen to be from 434 kg/h up to 614 kg/h to cover the range of possible emissions given by this system (see Figure 5.32).

The results of multi-objective optimisation (Figure 5.38 and Table 5.15) for this case show that if the constraint placed on carbon dioxide emission is placed above 480 kg/h the same optimal conditions will be found. There are some small variations in these results, which are due to the random nature of simulated annealing, but all these solutions found with constraints above 480 kg/h are approximately the same.

These results fit well with the results from the economic optimisation, which found the optimal solution had profits of €268.02/h and emissions of 470.83 kg/h. So it makes sense that if the emissions constraint was above 470.83 kg/h then we would expect this same solution to be found.

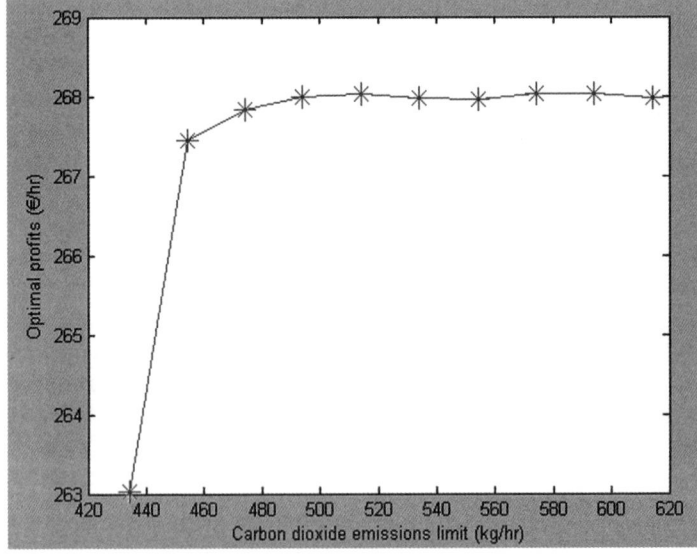

Figure 5.38 Pareto curve showing how emission constraints affect the optimal profits.

Table 5.15 Multi-objective optimisation results for biodiesel production.

CO_2 constraint (kg/h)	Methanol feed (kmol/h)	Methanol recovery (%)	Profit (€/h)
434.11	6.130	50.0	263.02
454.11	6.216	70.1	267.44
474.11	5.718	79.1	267.84
494.11	5.363	82.3	267.99
514.11	5.260	83.7	268.02
534.11	5.427	81.1	267.98
554.11	5.474	80.6	267.96
574.11	5.155	84.7	268.02
594.11	5.232	84.1	268.02
614.11	5.465	81.2	267.97

Table 5.16 Economic optimisation results for biodiesel with purified glycerol production scheme.

Parameter	Methanol flowrate (kmol/h)	Methanol recovery (percent recovered)	Oil : methanol ratio	Profit (€/h)
Initial value	6.75	60%	1 : 10.47	285.32
Optimum value	5.060	86.94%	1 : 14.40	286.66

Table 5.17 Environmental impact of optimisation results for biodiesel with purified glycerol production scheme.

Parameter	Heating required (kW)	Electricity required (kW)	Emissions (kg CO_2/h)	Wastes (kg/h)
Initial value	1 660.6	212.938	448.13	195.82
Optimum value	1 839.9	212.974	484.19	141.39

5.3.2.2 Biodiesel Production with Co-production of Purified Glycerol

To improve the profitability of the above system the option to purify the crude glycerol up to 95% glycerol, which can be sold for a higher price, was also considered through optimisation. The sensitivity analysis showed similar trends compared to the base case (selling crude glycerol) but with higher profit margins.

The set of conditions that optimise the economics of this scheme are given in Table 5.16.

In addition to this economic optimisation we can also consider the environmental impacts of these initial and optimal solutions. As we can see from Table 5.17 the optimum economic solution requires more energy and gives off more emissions than the initial conditions. However, the initial conditions also generate approximately 50 kg/h more wastes than the optimum.

Also comparing with the initial and optimal conditions for the base case (Table 5.14) we can see that the extra purification requires more heat energy and has higher emissions and wastes.

We have also performed sensitivity for this case with purified glycerol to see how the optimisation parameters affect the emissions and wastes in this case. As before we have fixed the methanol feed rate at 7 kmol/h and we have varied the methanol recovery.

In terms of energy this scheme behaves in almost exactly the same way as for the base case (with crude glycerol). As can be seen in Figures 5.39 and 5.40 the energy requirements increase as the methanol recovery is increased. This is mainly due to the extra energy required in the methanol recovery column but also due to increased electricity usage and extra heating required in the reactor.

Since the emissions calculated here are linked to the energy usage we can see in Figure 5.41 that the emissions increase from 440 kg/h up to almost 620 kg/h of CO_2 emitted. Although it is difficult to see here the emissions are slightly higher due to the increased energy required by the higher purification of the glycerol. This can be seen from comparisons of Tables 5.14 and 5.17 with the additional purification leading to approximately 4 kg/h of additional CO_2 emitted.

Another consequence of this purification can be seen through comparisons of Figures 5.33 and 5.42, which show how the amount of wastes produced varies with the recovery for the base case and for the purified cases respectively.

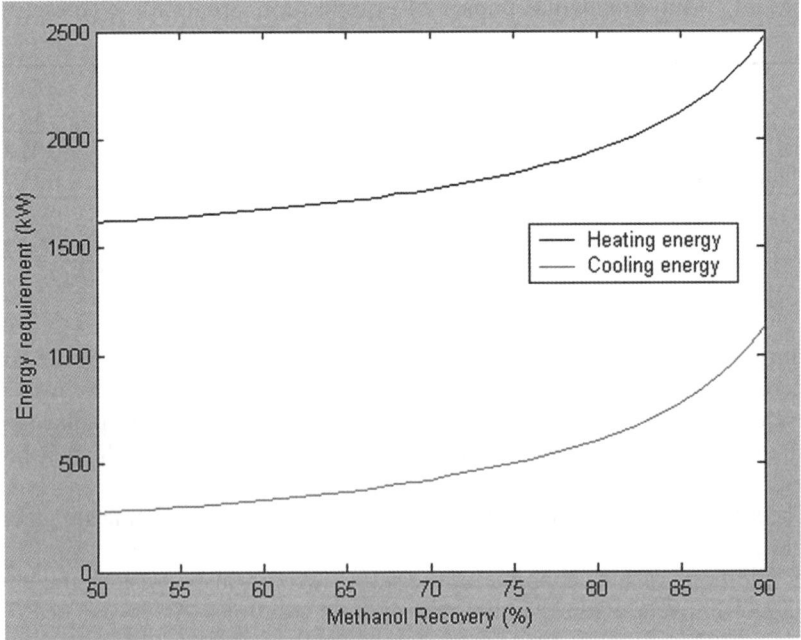

Figure 5.39 Effect of methanol recovery on heating and cooling duties (with methanol feed fixed at 7 kmol/h).

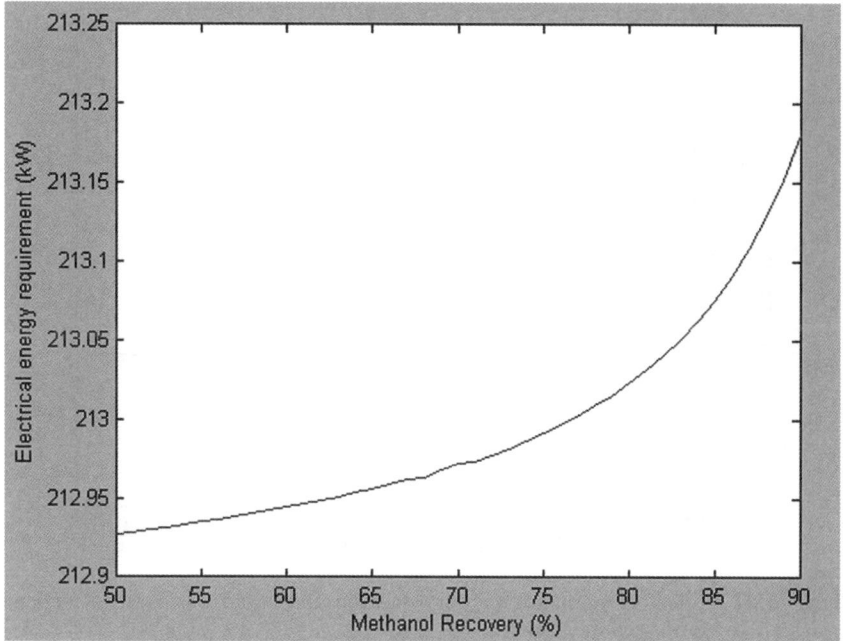

Figure 5.40 Effect of methanol recovery on electricity requirement (with methanol feed fixed at 7 kmol/h).

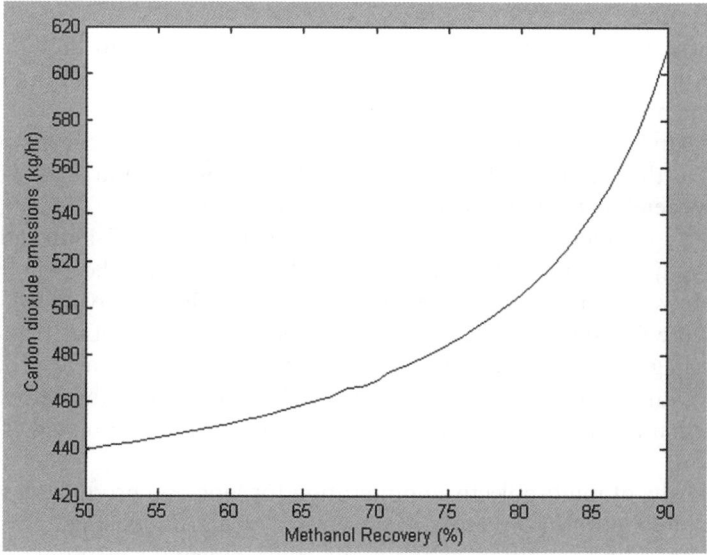

Figure 5.41 Effect of methanol recovery on emissions of CO_2 (with methanol feed fixed at 7 kmol/h).

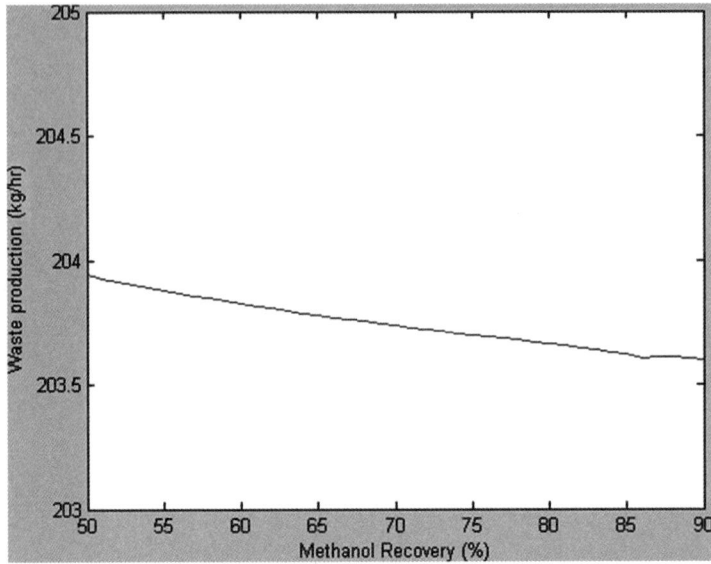

Figure 5.42 Effect of methanol recovery on solid and liquid wastes (with methanol feed fixed at 7 kmol/h).

In both cases the wastes are reduced slightly by increasing the methanol recovery; however, in the purified case there are approximately 14 kg/h extra wastes produced. This is directly due to the extra purification, which removes methanol and water from the crude glycerol stream and redirects it into the waste stream.

We have also performed sensitivity analysis with respect to the other optimisation parameter (methanol feed rate) to see how this will affect emissions and waste production.

The heating and cooling duties in addition to the electricity consumption all increase as the feed of methanol is increased (Figures 5.43 and 5.44) following the same trend as for the base case with crude glycerol.

Similarly the emissions also increase as shown in Figure 5.45 including the same deviation at 6 kmol/h as the base case (Figure 5.36) and keeping the same trend. However the emissions here are still slightly higher giving off around 4 kg/h more carbon dioxide every hour across all the values of methanol feed.

Increasing the methanol feed has a direct effect on the waste production (Figure 5.46) with most of the methanol added going into the waste stream. The extra methanol will be helping to improve the conversion into biodiesel up to a point.

The results of multi-objective optimisation for biodiesel production with co-production of purified glycerol are shown in Table 5.18 and Figure 5.47. The optimal operating conditions without any environmental constraints are similar to those for the scheme with crude glycerol with the optimal profits of €286.66/h and emissions of approximately 484 kg/h of CO_2. So we can see that

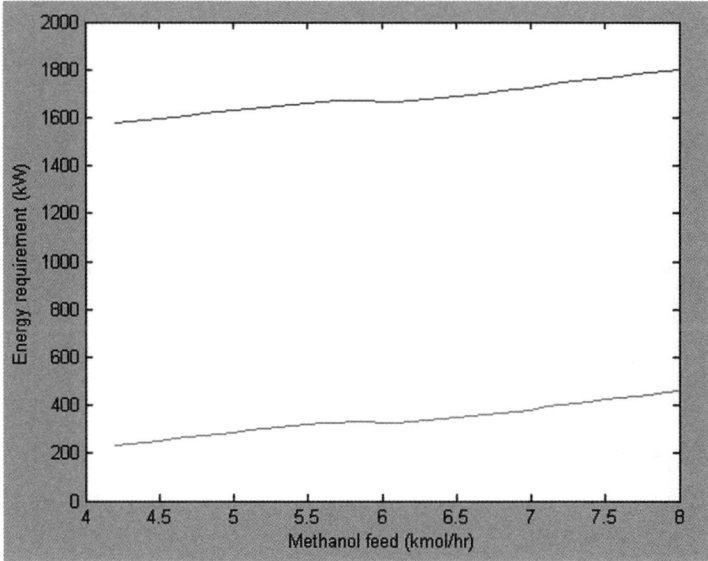

Figure 5.43 Effect of methanol feed rate on heating and cooling duties (with methanol recovery fixed at 66%).

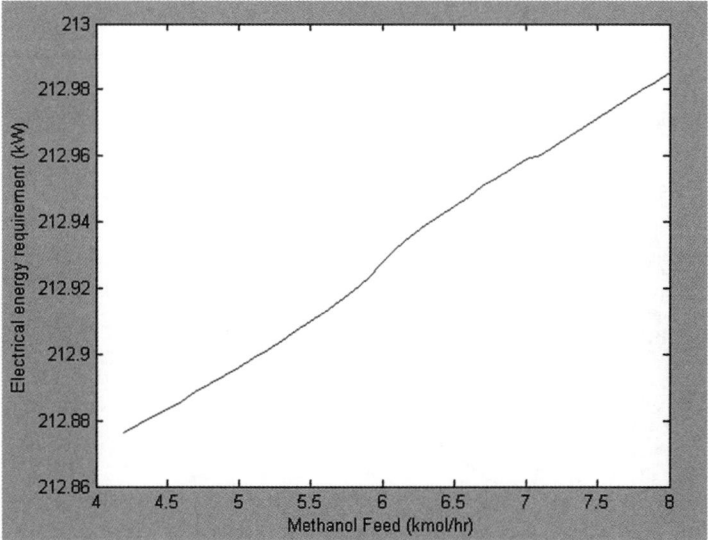

Figure 5.44 Effect of methanol feed on electricity requirement (with methanol recovery fixed at 66%).

using emissions constraints lower than this value a solution with lower profits is obtained. However, with constraints higher than 484 kg/h the same optimal solution is found because the higher constraint has no effect on the optimum.

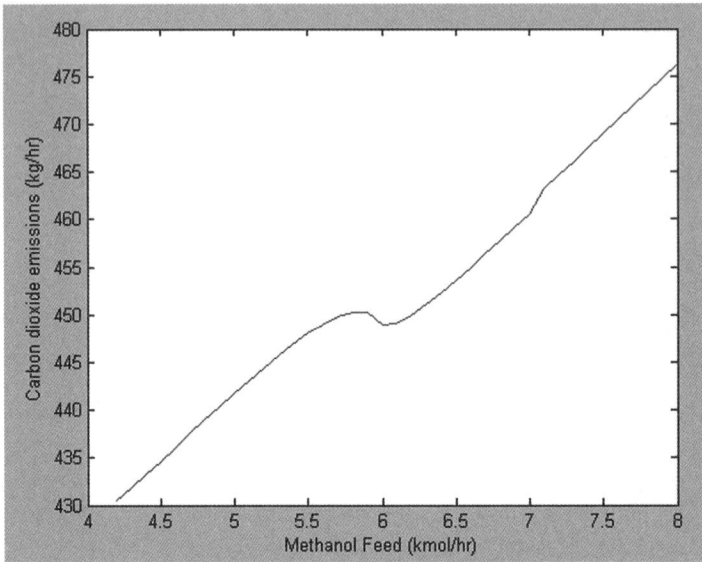

Figure 5.45 Effect of methanol feed on emissions of CO_2 (with methanol recovery fixed at 66%).

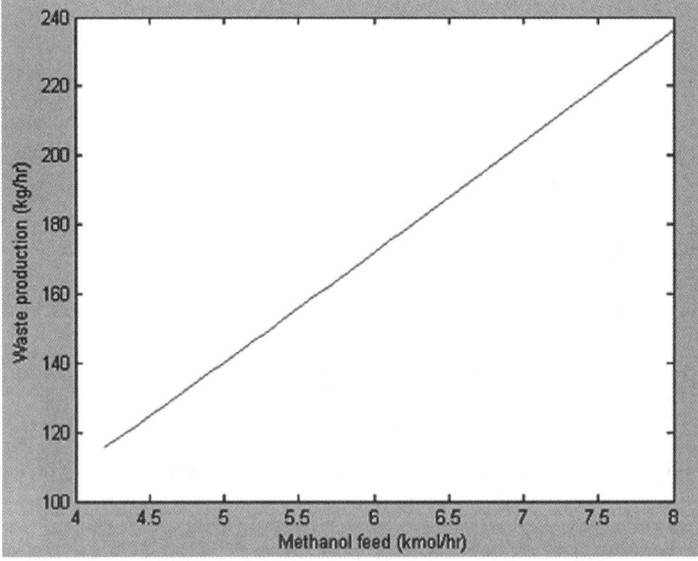

Figure 5.46 Effect of methanol feed on solid and liquid wastes (with methanol recovery fixed at 66%).

Table 5.18 Multi-objective optimisation results for biodiesel production with co-production of purified glycerol.

CO_2 constraint (kg/h)	Methanol feed (kmol/h)	Methanol recovery (%)	Profit (€/h)
434.11	5.535	50.5	277.11
454.11	6.194	68.1	285.88
474.11	5.707	78.8	286.54
494.11	5.220	85.5	286.66
514.11	5.062	86.7	286.69
534.11	5.200	84.2	286.74
554.11	5.179	84.5	286.73
574.11	5.515	80.2	286.63
594.11	5.226	84.5	286.73
614.11	5.431	83.4	286.61

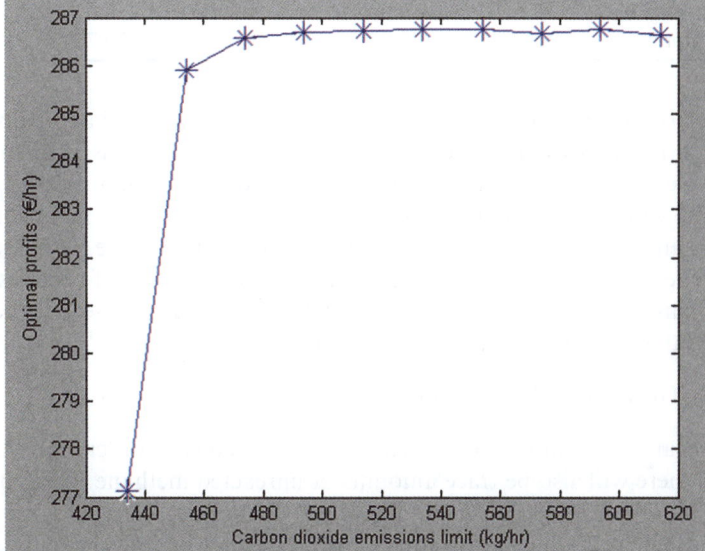

Figure 5.47 Pareto curve showing how emission constraints affect the optimal profits.

5.3.2.3 Biogas Production

In this scheme (Figure 5.48) the aim is to produce biogas from residues, which is then used in combustion to produce electricity and heat. We have assumed the residues in question are rapeseed straws, although other options are available including rapeseed oil press cake. Another product of this process is the digestate, which is the liquid/solid mix that comes out of the digestion reactor along with the biogas.

For economic optimisation of this scheme we considered the fraction of digestate recycled as the only optimisation parameter. This parameter controls

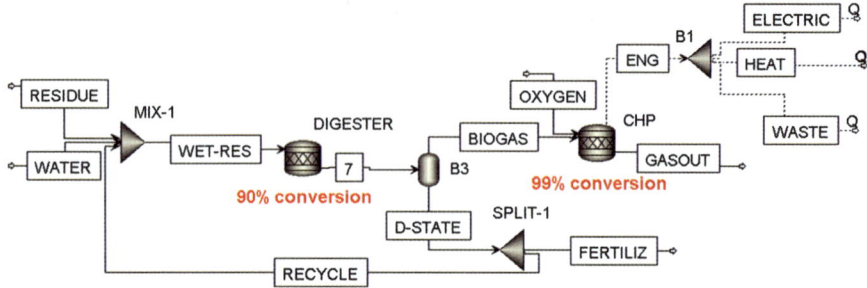

Figure 5.48 Flowsheet for the production and combustion of biogas.

Table 5.19 Economic optimisation results for biogas scheme.

Parameter	Recycle fraction	Profit (€/h)
Initial value	10%	356.04
Optimum value	97.49%	379.78

how much of the digestate is removed from the system and used as fertiliser and how much is fed back into the reactor. The digestion process is assumed to require no energy; however, the combined heat and power unit produces energy from combustion of the biogas derived. This combustion is the main source of emissions and it is the only source of emissions considered here. Other sources of indirect emissions could include those due to the transport of residues to the biogas plant. However, this has been neglected because these will be fixed quantities that are not affected by the optimisation.

The combustion reaction is: $CH_4 + 2\,O_2 \rightarrow 2\,H_2O + CO_2$

This means that most of the emissions will consist of carbon dioxide, although there will also be trace amounts of unreacted methane (CH_4) present. The stream labelled 'waste' in the flowsheet (Figure 5.48) actually represents wasted heat energy, which is lost due to the inefficiency of converting the energy of combustion into electricity and useful heat energy. There are actually no liquid or solid wastes in this system because the only by-product is the part of the digestate that is sold or used as fertiliser. The results of economic optimisation are shown in Table 5.19 for this scheme. This shows that the highest profits are found when most of the digestate is recycled and only a small fraction is released as fertiliser. These conditions give greater quantities of biogas produced and hence greater quantities of electricity produced. This can be seen from the emissions values given in Table 5.20 where we see that the optimum conditions emit approximately 226 kg/h extra carbon dioxide (due to increased biogas combustion).

To see how the recycle fraction affects the emissions in more detail we have performed sensitivity analysis to see how emissions vary across the range of possible recycle fractions (see Figure 5.49). This shows much greater emissions

Table 5.20 Environmental impact of optimisation results for biogas scheme.

Parameter	Carbon dioxide emissions (kg/h)
Initial value	2236.3
Optimum value	2459.0

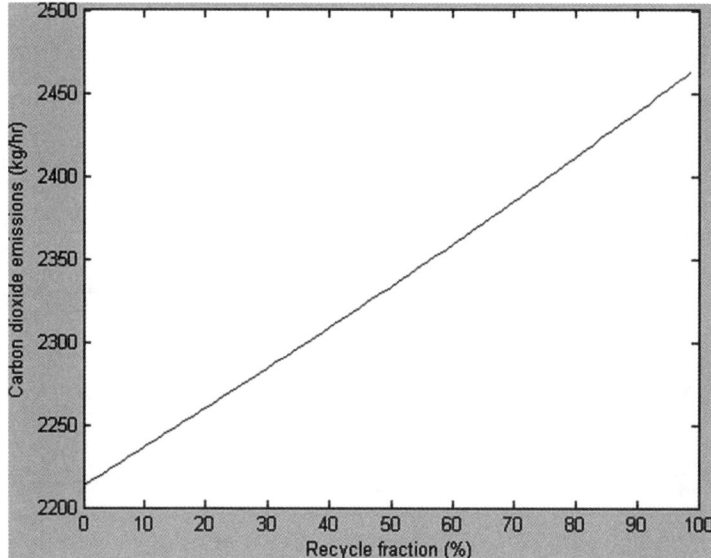

Figure 5.49 Effect of the recycle fraction on emissions of CO_2.

than the biodiesel production cases; however, the biodiesel figures do not include the emissions derived from using biodiesel as a fuel in trucks, buses *etc*. Also since this is a renewable energy source we can say that the emissions from biogas combustion are balanced by the carbon dioxide absorbed by the crops grown to give the residues.

The results of multi-objective optimisation for biogas production are shown in Table 5.21 and Figure 5.50. As we can see the optimum profits increase as the limits on the emissions are lifted higher until this limit passes 2450 kg/h of CO_2. After this point the optimum profits remain constant because the economic optimum with no constraints had emissions of 2459 kg/h of CO_2. So emissions limits higher than this should give the same optimum with profits around €379/h.

5.3.2.4 Supercritical Carbon Dioxide Extraction

The objective of this scheme (Figure 5.51) is the extraction of waxes from wheat straws by using high-pressure (supercritical) CO_2 as a solvent. To optimise it we

Table 5.21 Multi-objective optimisation results for biogas production.

CO_2 constraint (kg/h)	Recycle fraction (%)	Profit (€/h)
2250	15.85	357.51
2275	26.35	360.19
2300	36.65	362.88
2325	46.75	365.56
2350	56.63	368.25
2375	66.32	370.92
2400	75.81	373.60
2425	85.12	376.28
2450	94.25	378.96
2475	97.47	379.78
2500	97.47	379.78
2525	97.47	379.78
2550	97.47	379.78
2575	97.47	379.78
2600	97.47	379.78

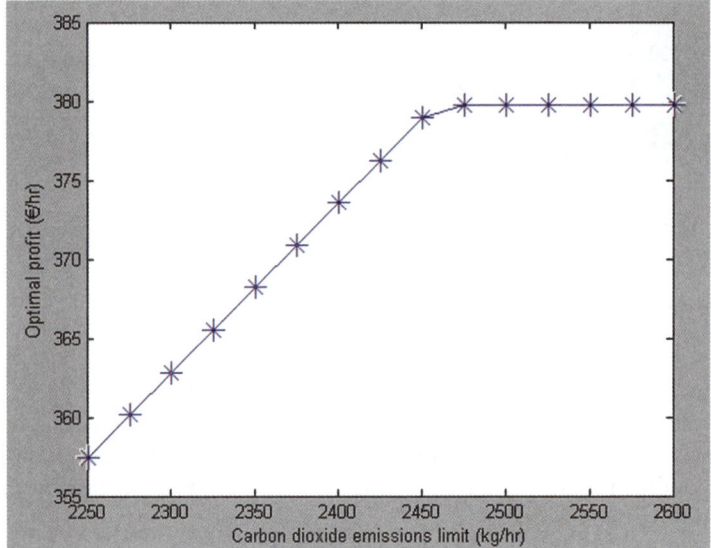

Figure 5.50 Pareto curve showing how emission constraints affect the optimal profits.

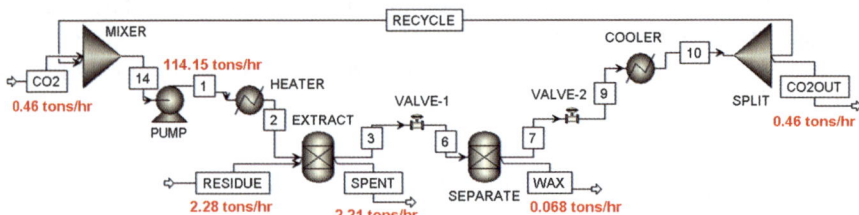

Figure 5.51 Process flowsheet for extraction of waxes from wheat straws.

have fixed the size of the extractor to 25 000 litres (or 5 × 5000 litres vessels) and we have used the capacity as the only optimisation parameter. The capacity is defined as the total mass of feed into (and out of) the extractor per year. Supercritical CO_2 extraction is another case where there are no liquid or solid wastes since the only solid output from the system is the spent straws (after extraction of waxes). These are considered just as valuable as the pre-extraction straws and hence can be sold or used in other processes. This extraction process has both direct and indirect emissions with the direct emissions resulting in CO_2 being lost during the depressurisation of the extractor. The amount of CO_2 lost is assumed to be directly related to the quantity of straw feed in the extractor with mass of CO_2 lost = 20% of the mass of straws in the extractor.

In addition to the direct emissions there are also indirect emissions due to the heating and electrical energy required by the unit 'HEAT-1' and by the pump. The results of economic optimisation are summarised in Table 5.22, where we can see that the optimum profits are €488.40/h operating with a capacity of 36 227 tons/year. Profits drop off at higher capacities because the required extraction time (which is inversely proportional to capacity) drops lower, which results in lower yields of waxes and hence less profit. The extraction costs also reduce as capacity increases; however, the optimum in Table 5.22 is shown to give the highest profit in terms of euros per ton of feed.

In Table 5.23 the environmental impacts of the initial and economic optimum conditions are calculated. The heating energy, electrical energy and the indirect emissions are calculated using Aspen Plus with the flowsheet in Figure 5.51. The direct emissions are calculated as 20% of the capacity. Sensitivity analysis studies reveal the effect of capacity on heating and electrical energy requirements as shown in Figures 5.52 and 5.53.

The energies required are proportional to the capacity. It is worth noting that the quantities of electricity required here are much greater than for the other cases due to the high demand for electricity from the pump. Indirect emissions calculated using the energy requirements and direct emissions calculated from

Table 5.22 Economic optimisation results for supercritical CO_2 extraction.

Parameter	Capacity (tons/year)	Extraction time (h)	Profit (€/ton)
Initial value	25 000	4.380	455.27
Optimum value	36 227	3.022	488.40

Table 5.23 Environmental impact of optimised cases for supercritical CO_2 extraction.

Parameter	Heating required (kW)	Electricity required (kW)	Indirect emissions (kg CO_2/h)	Direct emissions (kg CO_2/h)
Initial value	6842.5	1642.6	2257.4	631.31
Optimum value	10 008.7	2286.9	3239.8	914.82

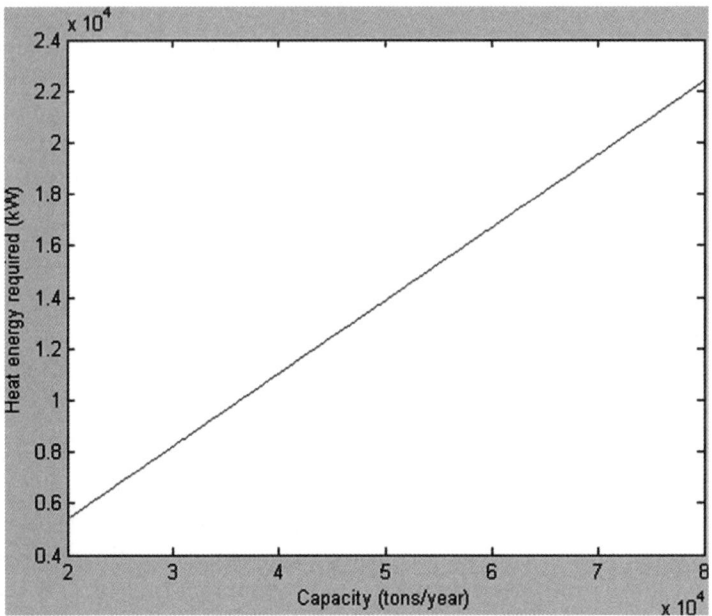

Figure 5.52 Effect of capacity on the heat energy required.

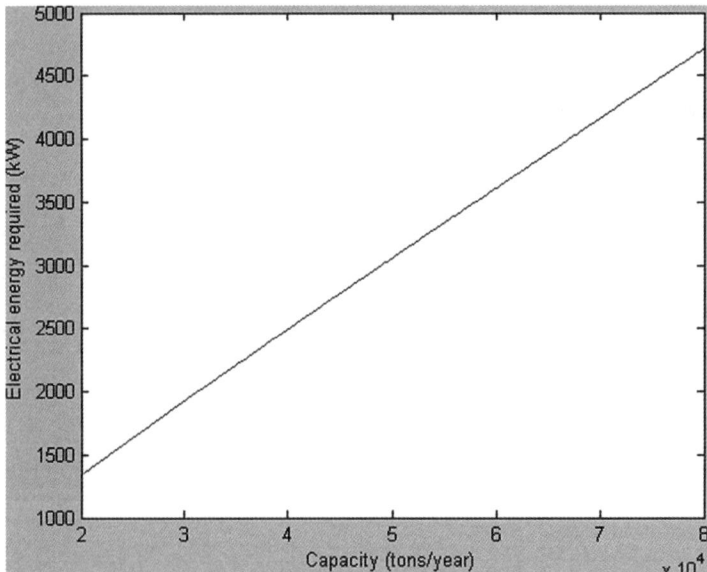

Figure 5.53 Effect of capacity on the electrical energy required.

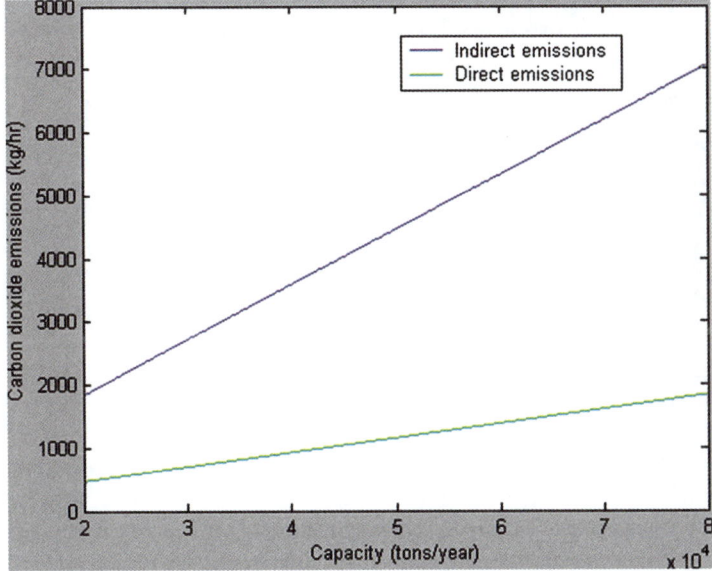

Figure 5.54 Effect of capacity on indirect and direct emissions.

the capacity (due to CO_2 loss from the system) are shown in Figure 5.54. To simplify calculations of emissions we have fitted a linear correlation between indirect emissions and capacity.

Indirect emissions (tons of CO_2 per year) = (capacity (tons per > year)
$\times$ 0.76269) + 736.47

Here the emissions are given in terms of tons/year instead of kg/h for simpler comparisons with capacity. Using this equation allows indirect emissions to be calculated quickly without the need to call Aspen Plus to simulate the entire process.

The total emissions (including the direct emissions, which are 20% of the capacity) can then be calculated as:

Total emissions (tons of CO_2 per year) = (capacity (tons per year)
$\times$ 0.96269) + 736.47

Using this equation the total emissions for the range of capacities are shown in Figure 5.55.

Combining the previous economic optimisation with constraints on the emissions using the above linear correlations allows multi-objective optimisation to proceed relatively quickly since we do not need to call the simulator. Multi-objective optimisation results for supercritical CO_2 extraction are shown in Table 5.24 and Figure 5.56.

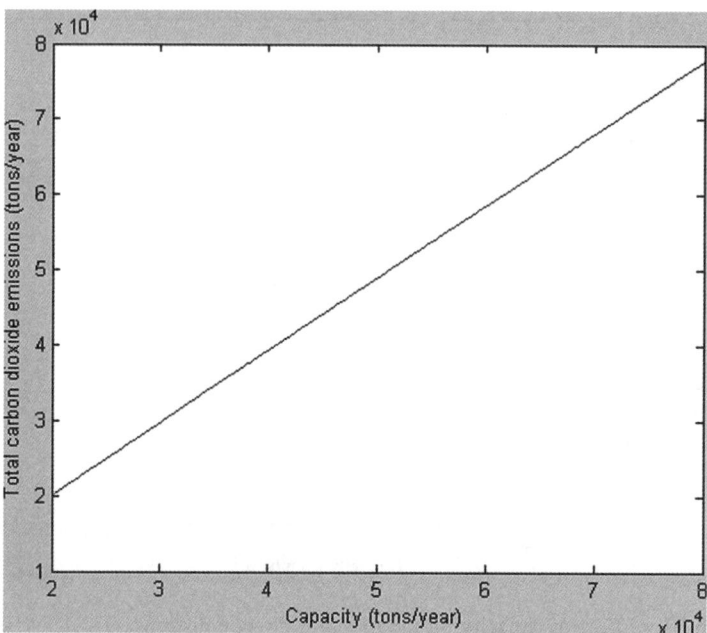

Figure 5.55 Effect of capacity on the total CO_2 emissions.

Table 5.24 Multi-objective optimisation results for supercritical CO_2 extraction.

CO_2 constraint (tons/year)	Optimal capacity (tons/year)	Profit (€/hr)
20 000	20 000	409.07
22 500	22 607	435.81
25 000	25 204	456.51
27 500	27 801	469.89
30 000	30 398	479.86
32 500	32 995	486.00
35 000	35 591	488.32
37 500	36 228	488.40
40 000	36 228	488.40
42 500	36 228	488.40
45 000	36 228	488.40
47 500	36 228	488.40
50 000	36 228	488.40
52 500	36 228	488.40
55 000	36 228	488.40
57 500	36 228	488.40
60 000	36 228	488.40
62 500	36 228	488.40
65 000	36 228	488.40
67 500	36 228	488.40
70 000	36 228	488.40
72 500	36 227	488.40
75 000	36 228	488.40

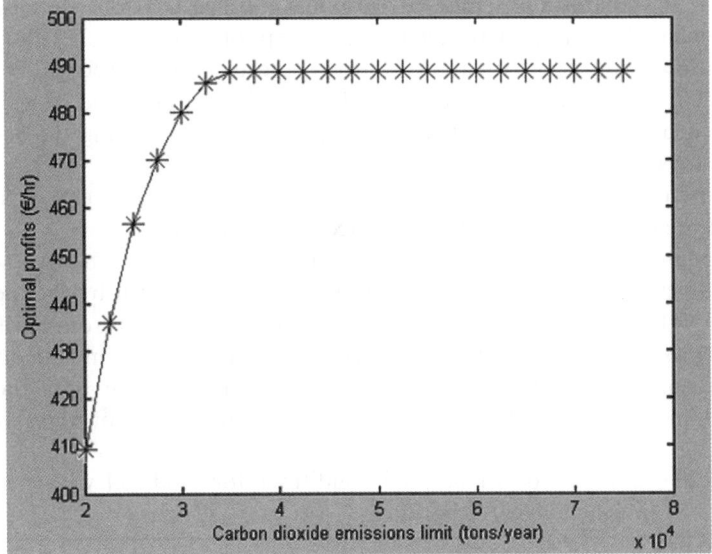

Figure 5.56 Pareto curve showing how emission constraints affect the optimal profits.

As we can see the optimum profits increase as the limits on the emissions are lifted higher until this limit passes 32 500 tons/year of carbon dioxide. After this point the optimum profits remain constant because the economic optimum with no constraints had emissions of approximately 32 900 tons/year of carbon dioxide. So emissions limits higher than this should give the same optimum with profits around €488/h.

5.3.3 Holistic Comparisons of Process Options

A holistic assessment attempts to consider all aspects of a scheme including economics, environmental impact and various other effects the scheme would have. For this reason we have considered each scheme and weighed up the various advantages and disadvantages of applying each process to their respective feedstocks. In order to assess each process we have calculated the economic and environmental impacts for given quantities of their respective feedstocks. This means if we have a ton of biomass, how much profit would we make from applying each of the different schemes? Also how much impact would this have on the environment?

5.3.3.1 Comparison of Biodiesel Production Options

For the schemes involving biodiesel production the feedstock is rapeseed. For optimisation purposes the feed rate of rapeseed was considered fixed at 3125 kg/h. To better analyse the different biodiesel schemes considered we have recalculated all the significant values in terms of the rapeseed feed.

5.3.3.1.1 Comparison of Crude Glycerol and Purified Glycerol Options.

The comparison of crude and purified glycerol options (Table 5.25) shows that purification adds approximately €6/ton of feed to the process. However, this is at the expense of an additional 4 kg of CO_2 emitted and 3 kg of waste produced per ton of feed. This comparison is only considering the economic optima for each case.

A more detailed comparison is presented in Figure 5.57 where we see how varying the optimisation parameters affect the differences.

Although we can see from Figure 5.57 that the profits of the two schemes follow a very similar trend there is still some change caused by the methanol recovery. This is shown in Figure 5.58, where the added profit from purification increases as the methanol recovery is increased. The increase in emissions due to increasing methanol recovery is shown in Figure 5.59 for both the crude glycerol and purified glycerol cases. This shows an increase of approximately

Table 5.25 Comparison of optimal conditions for biodiesel with crude and purified glycerol options.

Parameter	With co-production of crude glycerol	With co-production of purified glycerol
Methanol recovery (%)	84.07	86.94
Methanol feed (ton/ton of feed)	0.0531	0.0519
Profit (€/ton of feed)	85.77	91.73
Emissions (tons CO_2/ton of feed)	0.151	0.155
Waste (ton/ton of feed)	0.0421	0.0452

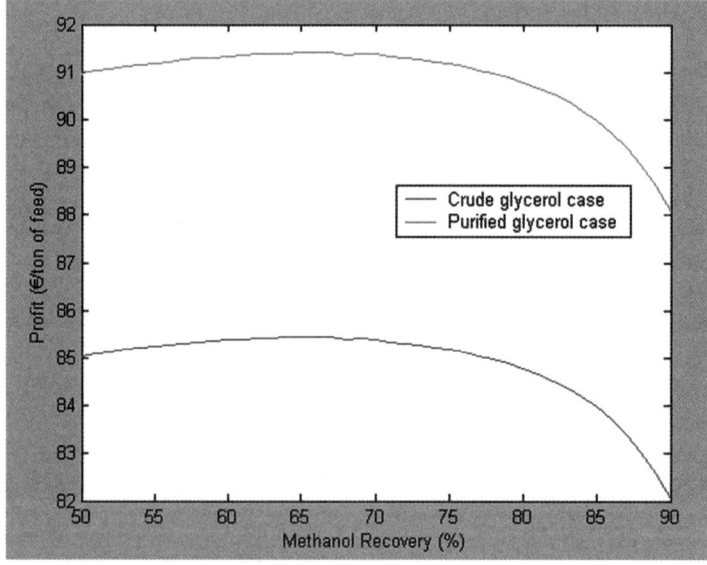

Figure 5.57 Effect of methanol recovery on profits from crude glycerol and purified glycerol cases.

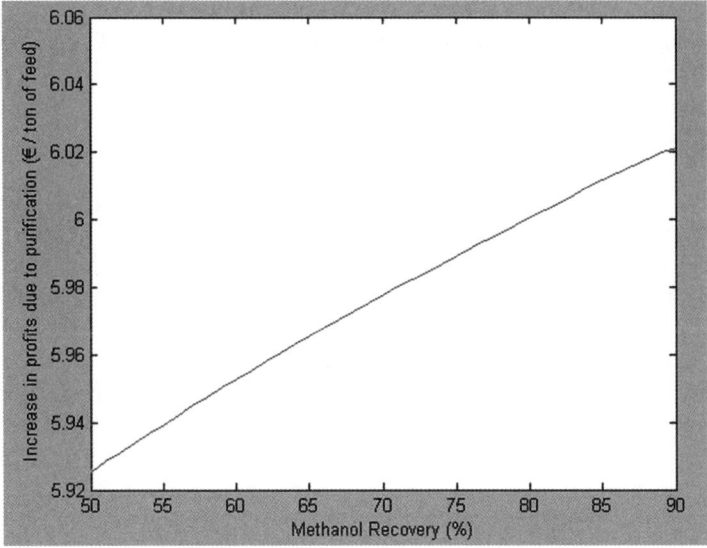

Figure 5.58 Effect of methanol recovery on the increase of profits from purification.

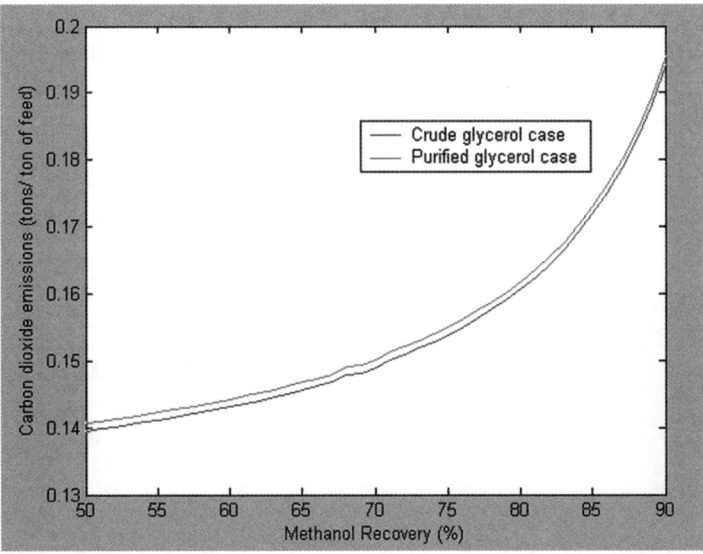

Figure 5.59 Effect of methanol recovery on CO_2 emissions (crude/purified glycerol options).

50 kg/ton of extra CO_2 emitted comparing the lowest and highest methanol recovery. However, the difference between purified and crude glycerol remains fairly constant, staying at around 1 kg/ton of feed extra CO_2 emitted in the purified case. We can also see how the methanol feed affects these differences.

From Table 5.25 and Figures 5.57–5.64 we can see that the effect of added purification seems to add an almost fixed increase of profits, emissions and wastes. However, adding the extra purification does give slightly increased added profits (see Figures 5.58 and 5.62) at higher methanol concentrations (high methanol feed and high methanol recovery).

5.3.3.1.2 Comparison of Succinic Acid with Glycerol Producing Options. In addition to the two options above we have also considered converting the glycerol produced by the biodiesel process into succinic acid. This option required the addition of a fermenter with water and biomass (including the organism) added to enable the fermentation process. For the production of succinic acid the optimal conditions for the crude glycerol producing case are used. This is because the succinic acid production uses this crude glycerol as its feed and further purification beforehand is assumed to be unnecessary. The methanol present might inhibit the fermentation process if the concentrations were too high. The optimal operating conditions for succinic acid production also include a water feed rate of 120 kmol/h and a batch cycle time for the fermenter of 100 h.

The most evident result from the comparison in Table 5.26 is that the optimal profits for biodiesel with succinic acid are more than 50% higher than those for biodiesel with crude glycerol. This does give slightly increased emissions (an extra 1 kg of CO_2 released for every ton of feed). Nevertheless the emissions are not as high as those for the purified glycerol case. Also worth noting is that separation processes after fermentation have not been considered, which would add to the cost, wastes and emissions.

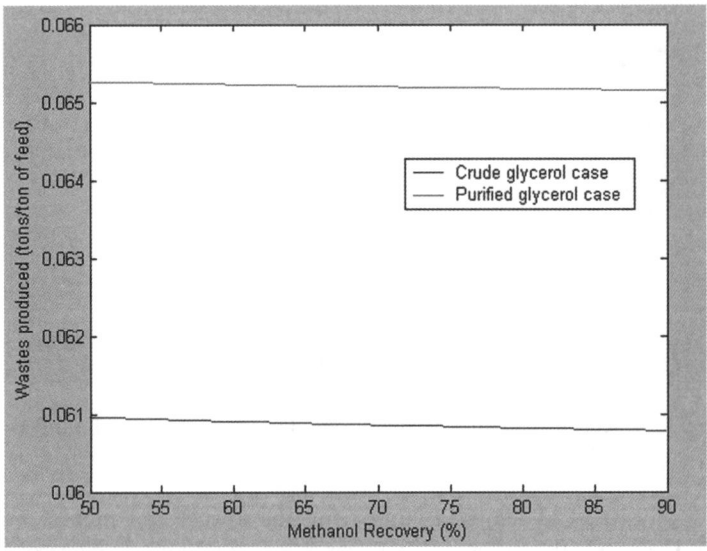

Figure 5.60 Effect of methanol recovery on waste production (crude/purified glycerol options).

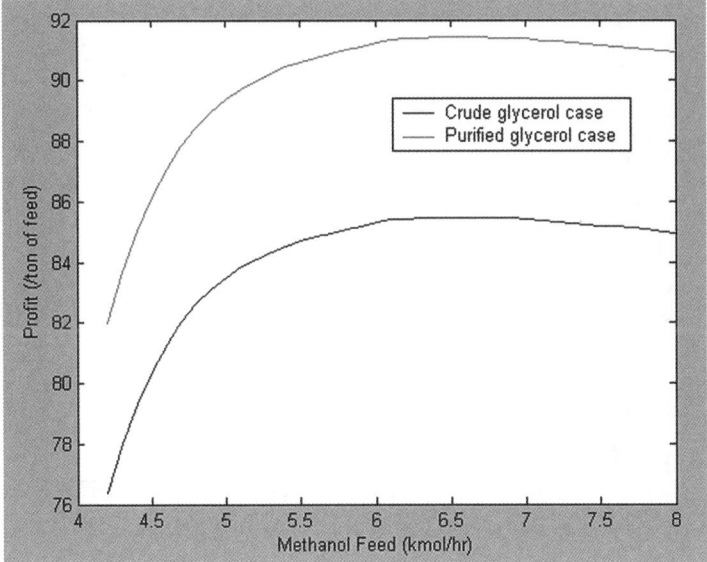

Figure 5.61 Effect of methanol feed on profits (crude/purified glycerol options).

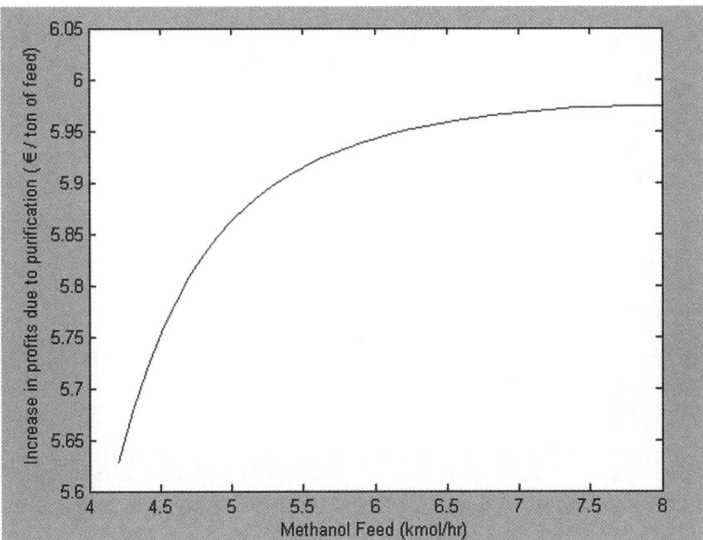

Figure 5.62 Effect of methanol feed on the increase of profits from purification.

5.3.3.2 Comparison of Oil Extraction Options

The extraction of oil from seeds is an important step in the biodiesel production process discussed above. However, oil extraction is a significant process on its

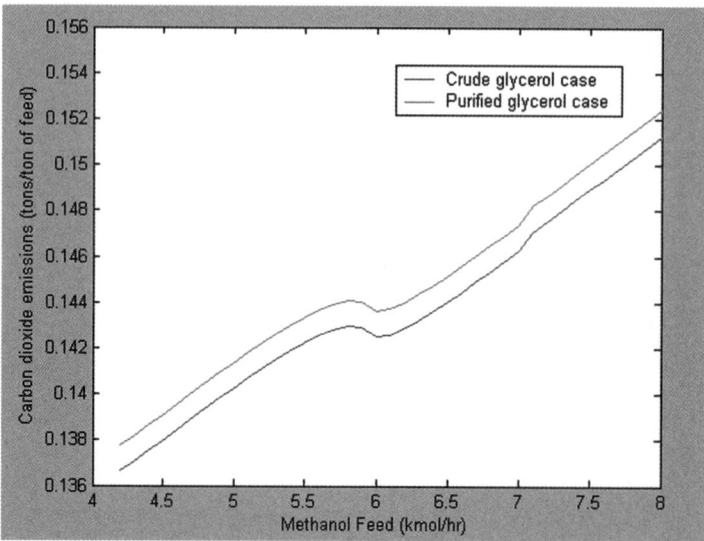

Figure 5.63 Effect of methanol feed on CO_2 emissions (crude/purified glycerol case).

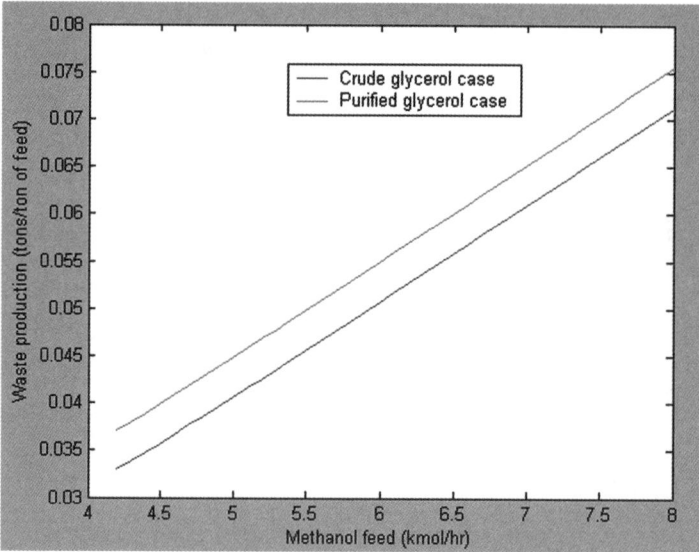

Figure 5.64 Effect of methanol feed on waste production (crude/purified glycerol case).

own and it is applied to various different oilseeds for different end uses. In the biodiesel process above we have used a relatively simple model for cold pressing in order to extract the oil to convert to biodiesel. However, in addition to cold mechanical pressing there is also another option involving the use of hexane as a solvent.

Table 5.26 Comparison of optimal conditions for biodiesel with succinic acid, crude and purified glycerol options.

Parameter	With co-production of succinic acid	With co-production of crude glycerol	With co-production of purified glycerol
Methanol recovery (%)	84.07	84.07	86.94
Methanol feed (ton/ton of feed)	0.0531	0.0531	0.0519
Profit (€/ton of feed)	138.33	85.77	91.73
Emissions (tons CO_2/ton of feed)	0.152	0.151	0.155
Waste (ton/ton of feed)	0.0421	0.0421	0.0452

Table 5.27 Comparison of costs in cold pressing and hexane-based oil extraction.

Parameter	Cold-pressing option	Hexane-based option
Capacity (tons/year)	11 000	1 000 000
Operators	3	70
Extraction efficiency (%)	83.3	95
Rapeseed cost (€/ton)	279	279
Capital costs (€/ton)	9.14	10
Labour costs (€/ton)	8.18	1.91
Energy costs (€/ton)	3.68	8.46
Maintenance (€/ton)	1.92	3.5
Finance (€/ton)	8.01	3.05
Other costs (€/ton)	2	2
Total cost (€/ton)	311.93	307.92

5.3.3.2.1 Economic Comparison. In a comparison of the different oil extraction options we have used data provided by CETIOM, CREOL and Desmet Ballestra (2009).[36,51,52]

The economic data summarised in Table 5.27 shows a comparison of two different oil extraction schemes. The scale of the plants is very different with the hexane-based method having capacity almost 100 times higher. Hence, the comparisons are made in terms of costs per ton of feed assuming the processes could be scaled up or down appropriately. Comparison of the costs involved show that hexane-based oil extraction involves lower costs. The breakdown of these costs shows that the two schemes are quite different. Although capital costs are similar, the labour costs are much higher for the cold pressing, possibly because the larger scale hexane-based plant operates more efficiently (14 000 tons/operator/year *vs.* 3600 tons/operator/year). Nevertheless, even if costs of cold pressing could be reduced by scaling up the capacity the most important disadvantage is the lower oil extraction efficiency. The added value from the higher oil extraction can also be calculated approximately based on the feed rate of 3125 kg/h with oil content of 40%.

With 83.3% efficiency:

$$\text{Oil extracted} = 3125 \times 0.4 \times 0.833 = 1041.25 \, \text{kg/h}$$

With 95% efficiency:

$$\text{Oil extracted} = 3125 \times 0.4 \times 0.95 = 1187.5 \, \text{kg/h}$$

Switching from cold pressing to hexane-based extraction nets 146 kilograms of extra oil per hour. Assuming an oil value of €630/ton this adds an extra €91.98/h. At the optimum conditions the yield of biodiesel is approximately 99% meaning the extra biodiesel produced is 144.8 kg/h, which adds an extra €131.47/h additional revenue from biodiesel. Nevertheless, the energy requirements would also increase; however, the lower cost of hexane-based extraction and the added value from the extra biodiesel should outweigh this extra cost.

5.3.3.2.2 Environmental Comparison. In addition to considering the economics of the situation we have also considered the energy usage and the emissions from each process (see Table 5.28). There should be negligible wastes produced in both cases (the hexane is recycled).

As we can see the cold pressing uses a large quantity of electricity in order to operate the mechanical press and extract the oil. However, the hexane-based option also uses a large quantity of heat energy (used as steam). As a result of these energy requirements the emissions can be calculated for both cases and we can see that despite using higher quantities of electricity the overall emissions for cold pressing are half those found for the hexane-based option. So while using hexane is shown to be the most economical approach it is also has the highest environmental impact.

5.3.3.3 Comparison of Straw Consuming Options

In addition to processes that start from oilseed we have also considered processes that use straws as their primary feedstock. We have considered three processes that deal with crop straws:

- Production of biogas (digestion);
- Supercritical CO_2 extraction;
- Production of levulinic acid.

Table 5.28 Comparison of energy usage of emissions for cold pressing and hexane-based oil extraction options.

Parameter	Cold-pressing option	Hexane-based option
Capacity (tons/year)	11 000	1 000 000
Operators	3	70
Extraction efficiency (%)	83.3	95
Electrical energy required (kWh/ton)	68	38.2
Heat energy required (kWh/ton)	0	240
CO_2 emitted (tons/ton of feed)	0.0365	0.0688

A comparison of the two schemes optimised is shown in Table 5.29 (levulinic acid is considered in Section 5.3.3.4). This reveals that supercritical CO_2 extraction gives approximately 70% more profit per ton of feed. However, it is worth mentioning that the two different cases use different straws (rapeseed for the biogas and wheat for the supercritical extraction). Also, these two options are not necessarily in competition since the spent straws from the extraction could be used as feed for the biogas process. Comparison of the emissions of the two processes shows that the biogas option emits approximately twice as much CO_2 compared to supercritical CO_2 extraction. Supercritical CO_2 extraction is an interesting case because it is the only scheme where we have considered feed capacity as an optimisation parameter. Considering how the profits change with capacity we obtain two very different graphs depending on whether we calculate total profits or profits per ton of feed (see Figures 5.65 and 5.66). Increasing the capacity increases the total profits of the plant; however, if

Table 5.29 Comparison of biogas production and supercritical CO_2 extraction options.

Parameter	Biogas production	Supercritical CO_2 extraction
Capacity (tons/year)	10 854	36 227
Profit (€/ton)	281.32	488.40
Emissions (tons/ton)	1.821	0.983
Total costs (€/ton)	166.96	309.51
Total revenues (€/ton)	448.28	797.41

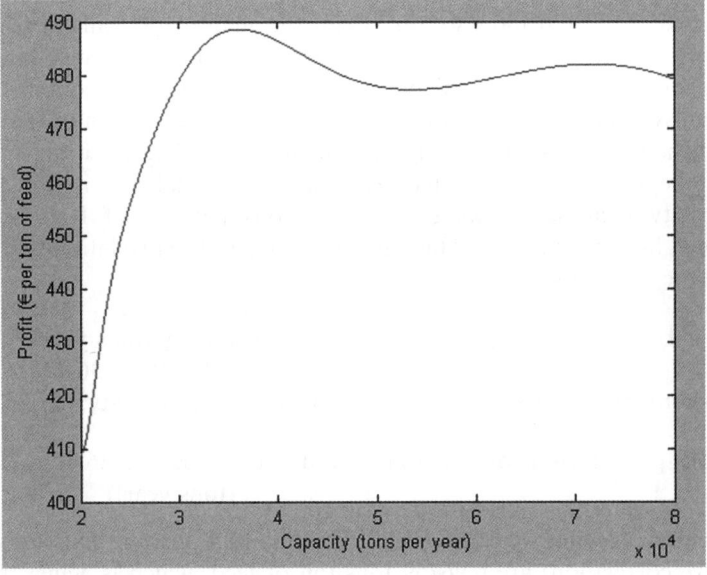

Figure 5.65 Profit per ton of feed for supercritical CO_2 extraction considering a range of different capacities.

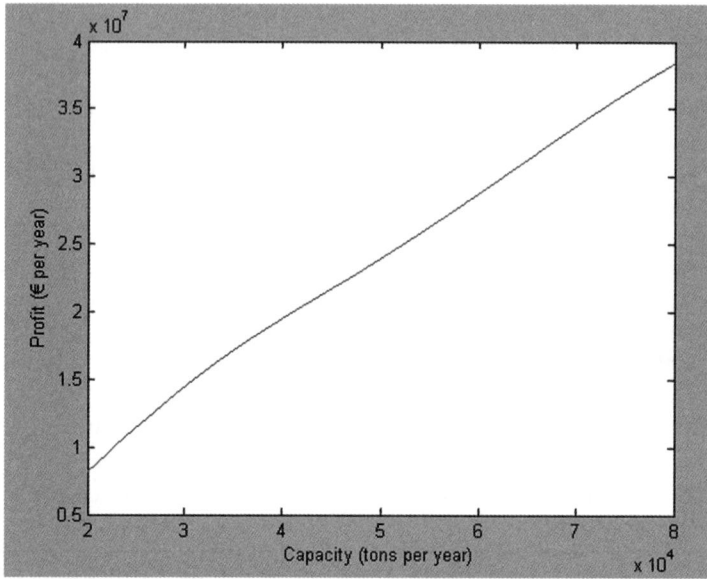

Figure 5.66 Profit per year for supercritical CO_2 extraction considering a range of different capacities.

profits per ton are considered these profits reduce after around 36 000 with a second lower profit peak at around 70 000 tons/year.

The decisions on plant capacity should be taken considering both graphs (Figures 5.65 and 5.66) and also considering the quantity of available feedstock. Depending on the availability and on other local factors a single large-scale plant or a number of smaller plants could give the optimum economics.

The emissions from the process can also be calculated per ton of feed or as a total for the plant as shown in Figures 5.67 and 5.68. There is a big difference between the total and the per ton of feed values. The total emissions increase as the capacity increases; however, the emissions per ton of feed processed decrease at higher capacities. This can be seen from the correlation that is used to calculate the emissions:

Total emissions (tons of CO_2/year) = (capacity (tons/year) × 0.96269) + 736.47

So to convert to emissions per ton we divide by the capacity:

Emissions per ton (tons of CO_2/ton of feed) = (0.96269) + (736.47 ÷ capacity (tons/year))

From this equation we can see that the minimum possible emissions (using the above correlation) are 0.96269 tons/ton of feed, which is achievable at a very high capacity. However, this efficiency saving in terms of emissions is balanced by an efficiency loss of profit per ton of feed at higher capacities. Also

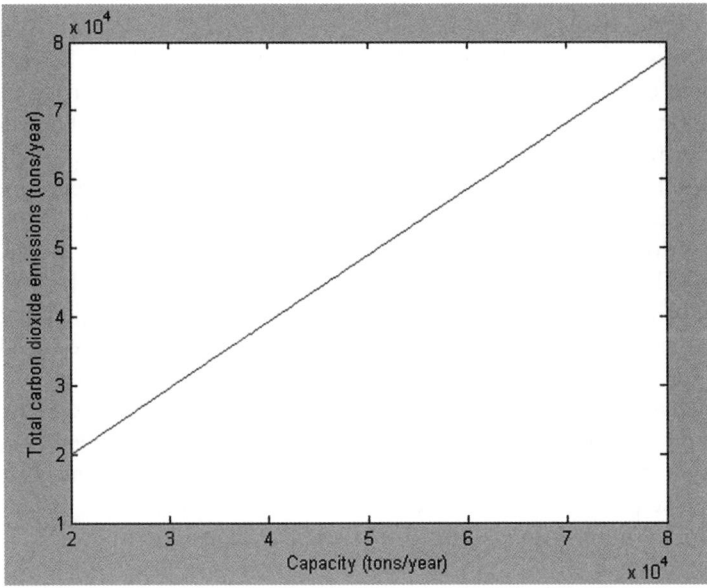

Figure 5.67 Effect of capacity on the total carbon dioxide emissions for supercritical CO_2 extraction.

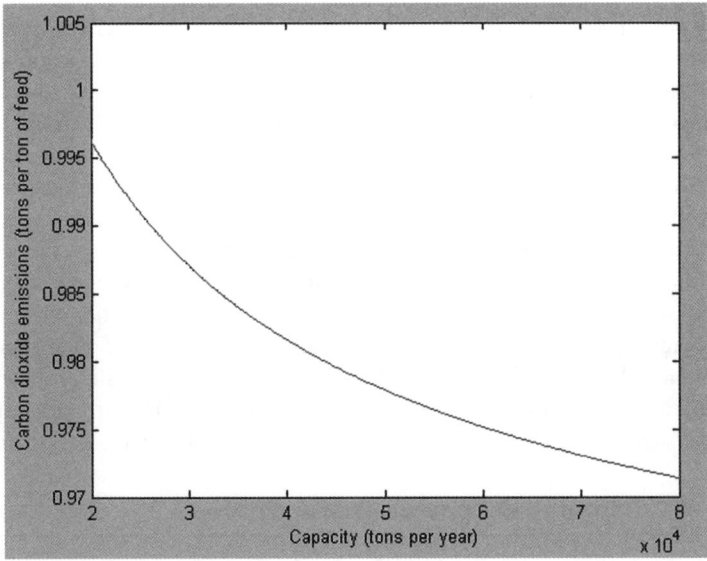

Figure 5.68 Emissions per ton of feed for supercritical CO_2 extraction considering a range of different capacities.

at the optimum economic conditions (profit/ton) the emissions are only at 0.983 tons of CO_2 per ton of feed, which is only approximately 20 kg/ton of feed higher than the theoretical minimum.

5.3.3.4 Comparison of Small-scale Processes

The schemes considered here involve lower capacities of feed; however, they can still be compared in the same way. Calculated values for profit emissions and electricity usage per ton of feed are shown in Table 5.30. The values for protein extraction are similar to those for supercritical CO_2 extraction and biogas production in Table 5.29. Profits are much higher for thermomoulding and levulinic acid production. For levulinic acid this value is actually too high because it does not include capital or operating costs or the costs of the solvent involved. Thermomoulding is an interesting case as it involves the lowest feed capacity of all the schemes considered. The process produces plant pots at a cost of approximately €0.10/pot and the overall profits above are based on the assumption that the pots can be sold for €0.12 each.

The actual selling prices could be higher or lower, but at larger scales the production costs might also be reduced. The current value assumed gives a very good profit per ton of material consumed. This is a very promising process; however, the energy usage and the emissions are also much higher than for any other process (indirect emissions from electricity consumption).

5.3.3.5 Economic Comparison of Process Options

In addition to the above comparisons we have also compared all cases together in terms of economics. These comparisons are based on profits per ton of feed, which is perhaps not a fair comparison because different feedstocks will have different values. However, this does allow us to compare the profitability of schemes at different scales.

The profit margins per ton of feed are shown in Table 5.31 along with the capacity of the schemes considered. The levulinic acid profits are much higher than the other schemes because the capital and operating costs have not been considered. Profits from thermomoulding are also very high, partly because we have assumed a value of €0.12 per plant pot produced by this scheme. However, this scheme could potentially be the most profitable if the capacity can be increased and if there is sufficient market capacity. For the other schemes with profits ranging from €33/ton up to €488/ton the capacities of the schemes are much higher than for thermomoulding. From these cases supercritical CO_2 extraction has the greatest immediate potential (high profits and high capacity). The other methods have lower profit per ton of feed; however, the profitability

Table 5.30 Comparison of protein extraction options.

Parameter	Protein extraction	Thermomoulding	Levulinic acid production
Capacity (tons/year)	7326	69.70	
Profit	201.45	1413.20	2758.95*
Emissions (tons/ton of feed)	0.945	4.762	
Electricity (kWh/ton)	475	8867	

*For levulinic acid the profits are calculated without considering capital or operating costs.

Assessment of Economic and Environmental Cost-benefits

Table 5.31 Comparison of the optimal profits for different biorefinery schemes.

Process	Optimal profits (€/ton of feed)	Scale of plant (tons/year)
Cold pressing oil extraction	33.03	11 000
Hexane-based oil extraction	52.28	1 000 000
Biodiesel with crude glycerol	85.77	25 000
Biodiesel with purified glycerol	91.73	25 000
Biodiesel with succinic acid	138.33	25 000
Protein extraction	201.45	7326
Biogas production	281.32	10 854
Supercritical CO_2 extraction	488.40	36 227
Thermomoulding	1413.20	69.70
Levulinic acid production	2758.95*	

*For levulinic acid the profits are calculated without considering capital or operating costs.

Table 5.32 Comparison of the emissions for different biorefinery schemes.

Process	Optimal profits (€/ton of feed)	Emissions (kg CO_2/ton of feed)
Cold-pressing oil extraction	33.03	0.0365
Hexane-based oil extraction	52.28	0.0688
Biodiesel with crude glycerol	85.77	0.151
Biodiesel with purified glycerol	91.73	0.155
Biodiesel with succinic acid	138.33	0.152
Protein extraction	201.45	0.945
Biogas production	281.32	1.821
Supercritical CO_2 extraction	488.40	0.983
Thermomoulding	1,413.20	4.762
Levulinic acid production	2,758.95*	

*For levulinic acid the profits are calculated without considering capital or operating costs.

of the processes also depends on the availability of the different feedstocks and the demand for the end products. Also, some methods could become more efficient at higher scales of production.

5.3.3.6 Environmental Impact of Process Options

Although there are more than one environmental factors that can be considered, the emission values of CO_2 have been used here to compare the different schemes. Keeping the same order as in the previous table (increasing profit) we have shown a comparison (Table 5.32) of the emissions alongside the profits for each scheme.

These results clearly show that the most profitable processes are also the worst in terms of emissions. All the cases with profits over €200/ton of feed also emit at least 940 kilograms of CO_2 for every ton of feed. The emissions are high in these cases either because of high energy demands (*e.g.* in supercritical CO_2 extraction) or because the end product is combusted to provide energy (*e.g.* in biogas production). The emissions of the oil extraction and biodiesel production might be higher if we had included the end-use emissions from biodiesel combustion. However, the high emissions cases could also be improved in terms of economics and emissions if their energy usage could be reduced or made more efficient.

5.4 Conclusions

This report includes life cycle analysis and optimisation results for various different biorefinery schemes.

Process schemes were modelled using simulation software Aspen Plus combined with Aspen custom modeller. Optimisation was made possible by linking Aspen software with MatLab, allowing simulated annealing and sequential quadratic programming optimisation algorithms to be employed.

For cases where the system is not already optimised the biorefinery schemes have been optimised with respect to economics to find the most profitable operating conditions. This was possible for the biodiesel, biogas and supercritical CO_2 extraction schemes where improved profits were obtained. For biodiesel production an optimal oil:methanol ratio of around 1:13 to 1:14 was found to be optimal for the transesterification reaction (producing the biodiesel). The optimum conditions for the biogas case involved recycling most of the digestate, which led to a high productivity of biogas. Supercritical CO_2 extraction was found to be most profitable at around 36 000 tons/year capacity.

To implement multi-objective optimisation we considered constraints placed only on the emissions of each process because the solid and liquid wastes already appear in economic objectives (a cost is included for waste disposal).

The results of multi-objective optimisation show that the economic optimum (with no environmental constraints) and the emissions of the associated conditions gives useful information because they tells us the point at which any constraint would start having an effect. If the emissions are constrained to a value lower than the emissions of the economic optimum then the resulting multi-objective optimum will have a lower profit. This can be seen in the pareto curves for each scheme where the profits drop off as the environmental constraints are tightened.

For the biodiesel production schemes the penalty for imposing environmental constraints varies for the three different cases, reducing profits by €5/h for the base case with crude glycerol. However these differences are higher for the purified glycerol case where the drop is €10/h.

Emissions from the biogas and supercritical CO_2 extraction schemes were much higher than in the biodiesel cases considered. This is partly due to the combustion of the biogas and the high heating and electrical demands of the supercritical extraction. However, the biodiesel emissions would probably have been comparable if we had included the end-use emissions from using the biodiesel as fuel in trucks and buses *etc.*

For the biogas case lowering the carbon dioxide emissions constraint resulted in optimal solutions with lower recycle fractions with only 15% of the digestate being recycled back into the reactor using the tightest emission constraints. This shows that using more of the digestate results in fewer emissions; however, this also reduces the profits as the fertiliser produced is less valuable than the electricity.

The economic optimum for the supercritical CO_2 extraction involved using a capacity of 36 227 tons/year. However, placing constraints on the emissions, multi-objective optimisation gave optimal solutions with lower capacity and lower profit. This was found to reduce the optimal capacity to 21 735 tons/year, which reduces the optimal profit by approximately €63/h.

These results show that it is still possible to make a profit with these emission constraints; however, the potential profits can be reduced significantly.

Comparison of the different biorefinery schemes (Table 5.31) shows that the most profitable schemes in terms of profit per ton of feed were those cases with the lowest capacity. These lower capacity schemes could be very profitable if they can be scaled up and if there is sufficient feedstock available. Comparison of the emissions for these cases (Table 5.32) also shows that the most profitable schemes also have the highest emissions. This is perhaps because these cases either involve high-energy usage (usually with larger quantities of electricity) or they involve combustion.

References

1. F. Cherubini, *Energ. Convers. Manag.*, 2010, **51**, 1412.
2. H. Uellendahl and B. K. Ahring, *Biotechnol. Bioeng.*, 2010, **107**(1), 59.
3. R. H. Wijffels, M. J. Barbosa and M. H. M. Eppink, *Biofuels, Bioprod. Biorefin.*, 2010, **4**, 287.
4. E. Vlad, C. S. Bildea, V. Plesu, G. Marton and G. Bozga, *Rev. Chim. (Bucharest, Rom.)*, 2010, **61**(6), 595.
5. A. B. M. S. Hossain, A. Salleh, A. N. Boyce, P. Chowdhury and M. Naqiuddin, *Am. J. Biochem. Biotechnol.*, 2008, **4**(3), 250.
6. W. M. V. Achten, L. Verchot, Y. J. Franken, E. Mathijs, V. P. Singh, R. Aerts and B. Muys, *Biomass Bioenergy*, 2008, **32**, 1063.
7. A. Bouaid, M. Martinez and J. Aracil, *Bioresour. Technol.*, 2010, **101**, 4006.
8. A. Demirbas, *Energ. Sourc., Part A*, 2010, **32**(16), 1547.
9. J. H. Clark, F. E. I. Deswarte and T. J. Farmer, *Biofuels, Bioprod. Biorefin.*, 2009, **3**, 72.
10. B. Kamm, C. Hille and P. Schonicke, *Biofuels, Bioprod. Biorefin.*, 2010 **4**, 253.
11. Y. Nakagawa, Y. Shinmi, S. Koso and K. Tomishige, *J. Catal.*, 2010 **272**, 191.

12. A. Vlysidis, M. Binns, C. Webb and C. Theodoropoulos, *Energy*, 2011, DOI: 10.1016/j.energy.2011.04.046.
13. X. Zhang, K. T. Shanmugam and L. O. Ingram, *Appl. Environ. Microbiol.*, 2010, **76**(8), 2397.
14. J. A. Siles, M. A. Martin, A. F. Chica and A. Martin, *Bioresour. Technol.*, 2010, **101**, 6315.
15. J. J. Bozell and G. R. Petersen, *Green Chem.*, 2010, **12**, 539.
16. S. Kim and B. E. Dale, *Biomass Bioenergy*, 2005, **29**, 426.
17. G. J. Harmsen, *Chem. Eng. Process.*, 2004, **43**, 677.
18. M. Emmerich, M. Grotzner and M. Schutz, *Evol. Comput.*, 2001, **9**(3), 329.
19. Y. Zhang, M. A. Dube, D. D. McLean and M. Kates, *Bioresour. Technol.*, 2003, **89**, 1.
20. S. P. Han, *J. Optimiz. Theory Applic.*, 1977, **22**(3), 297–309.
21. S. Kirkpatrick, C. D. Gelatt and M. P. Vecchi, *Science*, 1983, **220**(4598), 671.
22. J. J. Grefenstette, *IEEE Trans. Syst. Man Cyber.*, 1986, **16**(1), 122.
23. X. Pelet, D. Favrat and G. Leyland, *Int. J. Therm. Sci.*, 2005, **44**, 1180.
24. A. Azapagic and R. Clift, *Comput. Chem. Eng.*, 1999, **23**, 1509.
25. GEMIS database, http://www.oeko.de/service/gemis/en/. Accessed 30 June 2009.
26. FERA, 2009, Personal communications.
27. EC, 2009, EU Energy and Transport in Figures: Statistical Pocket book. http://ec.europa.eu/energy/publications/statistics/doc/2009_energy_transport_figures.pdf. Accessed 13 October 2009.
28. DEFRA, 2009, Market transformation programme, BNXS01: Carbon dioxide emission factors for UK energy.
29. R. Turton, R. C. Bailie, W. B. Whiting and J. A. Shaeiwitz, *Analysis, Synthesis, and Design of Chemical Processes*, 3rd edn, Pearson Education, Inc., 2009.
30. K. Komers,, F. Skopal, R. Stloukal and J. Machek, *Eur. J. Lipid Sci. Technol.*, 2002, **104**, 728.
31. SAC, Oilseed Rape, http://www.sac.ac.uk/mainrep/pdfs/osr2008northuk.pdf. Accessed 14 July 2009.
32. A. L. Stephenson, J. S. Dennis and S. A. Scott, *Process Saf. Environ. Protect.*, 2008, **86**, 427.
33. F. Ma and M. A. Hanna, *Bioresour. Technol.*, 1999, **70**, 1.
34. WUR, Wageningen University, 2009, Personal communications.
35. A. Rosenthal, D. L. Pyle and K. Niranjan, *Enzym. Microb. Technol.*, 1996, **19**, 402.
36. CREOL, 2009, Personal communications.
37. B. Freedman, R. O. Butterfield and E. H. Pryde, *J. Am. Oil Chem. Soc.*, 1986, **63**(10), 1375.
38. www.icis.com/staticpages/a-e.htm. Accessed 20 April 2010.
39. A. A. Apostolakou, I. K. Kookos, C. Marazioti and K. C. Angelopoulos, *Fuel Process. Technol.*, 2009, **90**, 1023.
40. www.Agricommodityprices.com. Accessed 14 December 2009.

41. H. Song and S. Y. Lee, *Enzym. Microb. Technol.*, 2006, **39**, 352.
42. FORTH, Foundation for Research and Technology – Hellas, 2009, Personal communications.
43. M. J Diaz, C. Cara, E. Ruiz, I. Romero, M. Moya and E. Castro, *Bioresour. Technol.*, 2010, **101**, 2428.
44. S. Calisir, T. Marakoglu, H. Ogut and O. Ozturk, *J. Food Eng.*, 2005 **69**, 61.
45. C. Walla and W. Schneeberger, *Biomass Bioenergy*, 2008, **32**, 551.
46. Chicago climate exchange, http://cpoffsets.com/pdf/agricultural_methane.pdf. Accessed 13 October 2009.
47. J. Nix, *Farm Management Pocket Book*, 38th edn, Imperial College London, London, UK, 2008.
48. UoY, University of York, 2009, Personal Communications.
49. Charles Jackson, 2009, Personal Communications.
50. G. Brunner, *J. Food Eng.*, 2005, **67**, 21.
51. CETIOM, 2009, Personal Communications.
52. Desmet Ballestra, 2009, Personal Communications.
53. INPT, Institut national polytechnique de Toulouse, 2009, Personal Communications.
54. Biorefinery.de, 2009, Personal Communications.
55. S. Wagner, N. Graf, H. Bochzelt and H. Schnitzer, 2005, Nachwachsende rohstoffe fur die chemische industrie, berichte aus energie- und umweltforschung, Graz, www.nachhaltigwirtschaften.at.

CHAPTER 6
Modelling Stakeholders' Interplay and Policy Scenarios for Biorefinery Implementation

PIERGIUSEPPE MORONE,[a] CATERINA DE LUCIA,[b] ANTONIO LOPOLITO[c] AND MAURIZIO PROSPERI[d]

[a] Associate Professor of Economics, Department of Economics, Mathematics and Statistics, University of Foggia, Largo Papa Giovanni Paolo II – 71121 – Foggia, Italy; [b] Lecturer in Economics and Research Fellow in Environmental Economics, University of York, Environment Department, Heslington, York, YO105DD, UK and Department of Economics, Mathematics and Statistics, University of Foggia, Largo Papa Giovanni Paolo II – 71121 – Foggia, Italy; [c] Research associate, Department of Production and Innovation in Mediterranean Agriculture and Food Systems (PrIME), University of Foggia, Via Napoli 25 – 71122 Foggia, Italy; [d] Researcher, Department of Production and Innovation in Mediterranean Agriculture and Food Systems (PrIME), University of Foggia, Via Napoli 25 – 71122 Foggia, Italy

6.1 Introduction

This chapter is structured in two independent and self-contained sections. However, both sections deal with the very same problem—that is, assessing and modelling policy scenarios for the development of biorefineries. The difference lies in the fact that the problem is targeted from two rather distinct perspectives: micro and macro. The first section (6.2) will adopt a micro-economic perspective and develop a methodological approach for policy modelling and policy evaluation. On the other hand, in the second section (6.3) we shall adopt a macro-economic perspective, presenting, through a computable general equilibrium (CGE) model, well-established behavioural economic interactions of a biorefinery economy.

Both the micro- and the macro-economic approaches respond to two key policy objectives envisaged in the development phase of the SUSTOIL project in June 2008: (1) to develop various policy scenarios for biodiesel and biorefinery production through the analysis of social network interactions across stakeholders and consumer welfare at micro-level; (2) to understand the evolution of various environmental, economic and policy constraints within a sustainable development framework for biofuel and biorefinery energy sectors.

In the summing-up section (6.4) we shall reconcile these two approaches, underlining their mutual relevance for a holistic approach to policy actions, which should, as we believe, consider both the micro and the macro nature of a pervasive socio-technical change such as switching from fossil to green fuel production.

6.2 The Micro-economic Approach to Policy Modelling for Biorefineries

This is a methodological section aiming at defining an appropriate protocol for modelling policy scenarios with respect to the development of biorefineries. Biorefineries can be intended as 'integrated bio-based firms, using a variety of different technologies to produce chemicals, biofuels, food and feed ingredients, biomaterials (including fibers) and power from biomass raw materials'.[1] We shall propose an empirical methodology, which should allow researchers and policy makers to assess the development status of a biorefinery and identify the most suitable policy actions to promote its development. We will first sketch out the theoretical framework within which our empirical approach is nested. Subsequently we shall present a three-steps methodology, which will serve the purpose of defining a common approach for modelling and evaluating policy action aiming at promoting the development of biorefineries. Finally, we will conclude this section briefly discussing possible applications of the proposed methodology to specific case studies.

6.2.1 The Theoretical Framework

The management of technological progress is a long-standing question. Historically, there are two main policy concerns in this field: (1) to control the possible deleterious effects of new technologies and (2) to encourage technologies that bear wider social benefits.[2] In various periods and in various places such concerns have assumed several forms, such as social equity,[3] gender equality and[4] reduced unemployment.[5] Along with these traditional issues, the concept of 'sustainable development' has recently gained momentum among scholars and policy makers, affecting also the most recent debate on technological change. As a result, a wider vision of innovation has been promoted, where the aim has become to reshape entire technological configurations so that they are more responsive to environmental signals and ecological principles.[6]

In order to better understand and make more effective this vision, technological transition theorists have developed a multi-level approach (MLA)—i.e. a model that can be seen as a nested framework formed by three linked levels: socio-technical regime (henceforth ST-regime), socio-technical landscape and innovation niches.[6] The ST-regime represents the meso-level unit of analysis. It can be defined as a relatively stable configuration of institutions, techniques and artefacts (such as hardware and infrastructures), as well as rules, practices and networks that determine the 'normal' development and use of technologies.[7] The ST-regime is identified by the socially valued function it fulfils. Electricity and water supply, health care, education, food supply and building trades are all examples of socially valued functions that identify specific ST-regimes. This concept accounts for both technical and sociological elements of the community involved in these functions. In other words, technical trajectories are not only influenced by engineers and their practices, but also by users, policy makers, societal groups, suppliers, scientists, banks *etc*. Within such a regime, the activities of these different groups are aligned to each other and coordinated.[8]

The ST-regime concept sums up the main features of this complex set of community and activities and represents it as 'a configuration that works'.[9] The term 'configuration' refers to the alignment of the heterogeneous elements involved in the technological trajectories and the expression 'that works' means that this set of elements fulfils a specific function.[7]

The activities carried out by any ST-regime need to be situated in a wider context which influences its performance. Such a context consists of a set of deep structural trends and variables that go beyond the direct influence of regime actors. Oil prices, economic growth, wars, emigration, broad political coalitions, cultural and normative values or environmental problems are some examples of such exogenous variables.

In the MLA these elements form the macro-level and represent a whole socio-technical landscape, which is the second level of the analysis. A landscape is an external structure or context for interactions of actors.[8] Kemp and Rotmans define landscape as a set of '... background variables such as the material infrastructure, political culture and coalitions, social values, worldviews and paradigms, the macro economy, demography and the natural environment which channel transition processes and change themselves slowly in an autonomous way' (p. 7).[10] Geels notes that in this stream of literature the term 'landscape' is used 'because of the literal connotation of relative "hardness" and material context of society, *e.g.* the material and spatial arrangements of cities, factories, highways, ...' (p. 1260).[8]

Finally, the innovation niche level represents the micro-level. Niches are 'protected spaces for the development and use of promising technologies by means of experimentation, with the aim of (a) learning about the desirability of the new technology, and (b) enhancing the further development and the rate of application of the new technology' (p. 186).[11] We can see niches as 'incubation rooms' for novelties because they are protected or insulated from 'normal' market selection in the regime.[12] Their operation relies on three main mechanisms: (1) the expectations mechanism; (2) the power mechanism; and (3) the knowledge mechanism.

The first mechanism depends upon the convergences of actors' *expectations* towards a common view. This is crucial for the emerging of an innovation niche. One of the main barriers to the adoption of a new technology is that, often, its advantages are not clearly understood by all possible adopters.[11] Indeed, actors participate in projects on the basis of their expectations;[13] diverging expectations can affect the way goals are defined and prioritised.[14] This initial obstacle can be overcome only through the development of a robust and shared vision among the actors potentially involved. Such convergence legitimates actors to invest time and resources into a new technology that does not yet have any market value.[15]

Along with converging expectations, involved actors need an adequate level of *power* to act. This issue concerns the second niche formation mechanism, which involves a networking process. Considering the niche as a 'small network of dedicated actors' (p. 400)[9] it is fundamental for its formation that powerful actors join the network. Their support is crucial to gather and mobilise the resources required to guide the technical change in a desirable way.[14] Indeed, no single actor has sufficient resources on her/his own to coordinate the innovation process.[14] Following Smith *et al.* (p. 1506), 'one major group of resources concerns control over financial revenues or capital stocks. Others include the ability to control material artefacts, such as hardware and infrastructures, or the production of salient knowledge, through research and marketing. [...] A further type of resource is embodied in the command of legitimacy, credibility or other recognised sources of authority in making demands upon the behaviour of others. Examples here might include a competence in developing or passing legislation, or in implementing regulations'.[14]

As a result, niche members are dependent upon each other for vital resources. Access to (or denial of) such resources depends crucially on the existing power relations in transition pathways.[14] Consequently, niche members can be seen as actors of a relational network possessing those complementary resources needed to control the innovation process.

Once all needed resources have been acquired by niche members, then *knowledge* for development and implementation of the new technology is required. This points out the importance of the last mechanism for niche creation, which concerns the formation of patterns of learning interactions occurring among agents in the niche. A large part of this process is deeply informal as tacit and uncodified knowledge can only be acquired and shared by means of intensive and direct interactions. Moreover, some noted that knowledge flows more intensively within a core group of firms characterised by advanced absorptive capacities.[16] Hence the niche structure and actors' individual characteristics are both key factors in assessing the intensity of knowledge flows.

These three mechanisms represent the pre-conditions that need to co-exist for the formation of an innovation niche. The multi-level approach sees transitions as outcomes of the co-evolution of these three levels. Geels and Schot sum up the basic processes of this co-evolutionary dynamics as follows: '(a) niche innovations build up internal momentum, [...] (b) changes at the landscape level create pressure on the regime and (c) destabilisation of the regime creates windows of opportunity for niche innovations' (p. 405).[9] Indeed, transition is enabled from the alignment of these processes that facilitates the breakthrough of novelties into mainstream markets and can be seen as changes from the incumbent regime to another.[9] In Figure 6.1 we summarise this alignment process, paying due attention to the niche mechanisms required for an adequate niche development.

In line with the MLA, the case of biofuels technologies can be seen as a specific element of the nested model. Such technologies are developed within the field of energy supply, while a major component of the energy supply field is represented by the fuel regime. Such a regime, in turn, spans a variety of specific nested and subordinated configurations (niches), such as those based on renewable energy. In this perspective the community and the activities involved in the production of biofuel can be seen as one innovation niche that competes with other renewable energy innovation niches within the fuel energy regime. Conversely, the fuel regime itself may be seen as nested within a global energy landscape, organised primarily around the extraction of, trade in and combustion of fossil fuels.

Hence, a policy for the promotion of the development of biorefineries could be understood as an activity aiming at stimulating the innovation niche formation process itself. This is particularly true if we aim at defining a set of

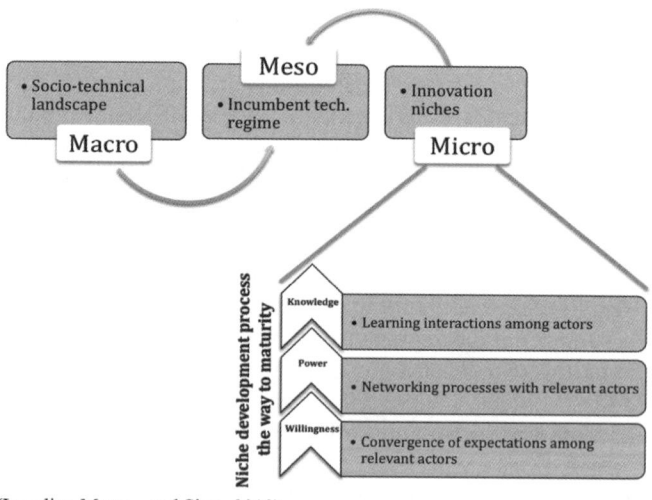

(Lopolito, Morone and Sisto, 2010)

Figure 6.1 Multi-level approach and the niche formation mechanisms.

Modelling Stakeholders' Interplay and Policy

policy actions which could be implemented locally: although unable to affect the landscape (macro) level of this nested model, a local policy could indeed have an impact at the niche (micro) level, with a set of measures specifically crafted to influence the three niche mechanisms above described.

Along this line of reasoning, in what follows we shall concentrate on the formation process of a stable innovation niche, as this is, in our view, the core of the problem for local policy actions. In fact, the proposed three-steps methodology aims at defining a protocol to assess the niche formation status and to define a set of suitable policy actions.

6.2.2 A Three-steps Methodology

In this section an innovative methodological approach to policy design for the development of the biorefinery is proposed. As schematically represented in Figure 6.2, the methodology is based on three basic steps, in which stakeholders play an active role in the definition of the policy actions.

The first step aims at assessing the development status of a biorefinery—that is, an innovation niche as discussed in Section 6.2.1. Subsequently, the second step proposes a modelling framework to define and simulate the impact of various policy actions aiming at promoting the full development of such an innovation niche. Finally, the third step is concerned with the evaluation of the proposed policy strategies.

Such a methodology represents, in the authors' view, a novelty and a useful approach in understanding and modelling stakeholders' interplay and policy scenarios for biorefinery and biodiesel production. In fact, it is assumed that without the direct involvement of local stakeholders, and without considering their expectations, perceptions and system values, it is hardly possible to define a long-term stable and sustainable development process for the biorefinery project.

The relevance of the proposed multi-steps investigation stems directly from the complexity of the phenomenon under investigation. This complexity is due to three main characteristics of the biorefinery industry. First, it has a strong local nature since its developing conditions are highly context-dependent. Second, there are relatively few developed and documented biorefinery systems in operation. Third, biomasses are often by-products generated in other sectors (agriculture, forestry, wood processing and so on).[17] As a result, the biorefinery system is characterised by plenty of interacting elements, evolving towards uncertain status through unknown paths. Uncertainty affects the selection of

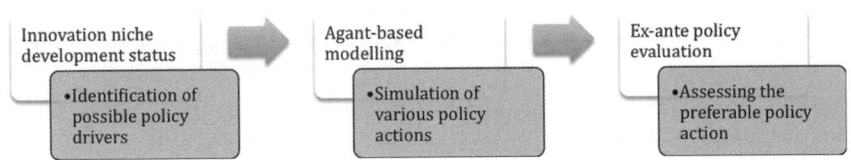

Figure 6.2 A three-steps methodology.

new raw materials, the adoption of the technology necessary for the industrial processing and manufacturing and the development of new intermediate products directed to traditional industry.

Such complexity makes it difficult to understand fully and rationally model the multi-dimensional features affecting the object of the analysis. In this sense, the assumption underlying our multi-steps approach is that relevant knowledge and information on complex systems can be best obtained by analysing the perception of people directly involved in the phenomenon.[18-20] More generally, a case-study-based investigation can help in 'building up generalization from data, and to test competing models, to evaluate policies and to forecast the effects of new policies or modifications of existing policies' (p. 3)—which is exactly what we shall try to do.[21]

6.2.2.1 *First Step—Assessing the Niche Development Status*

As we discussed above, the development of a biorefinery can be understood as an innovation niche formation process. In order to stimulate such a process through policy actions, it is particularly important to have an analytical tool to assess its development status and to collect, evaluate and systematise stakeholders' perspectives upon such a development process. In what follows we shall present the first step of our proposed methodology, which aims at collecting such information through a thorough investigation of the case study under investigation. It is composed of two key phases: the first phase is designed to collect all information needed to assess the development status of the niche; the second phase is meant to integrate such data with a deep understanding of the perception of various stakeholders on the niche development process.

(1) First phase
Various elements make part of the formation of a well-established niche. Some of these are the availability of venture capital, technology cost and the extent to which the existing regime is locked in. However, as pointed out above, social interconnections and network interactions among involved actors are key factors in socio-technical transitions. Such social elements characterising the niche creation process have been categorised into three fundamental mechanisms introduced in Section 6.2.1.

Since these fundamental mechanisms have a wide network nature, deeper insights on the emergence of an innovation niche can be obtained by the application of social network analysis (SNA), which recently came to the front of organisational studies as a key technique for empirical investigations. We believe SNA is highly suitable for these tasks. Specifically, in this first step of our proposed methodology, we shall suggest using SNA to define a methodological conceptualisation of each of the three niche mechanisms discussed above.

Expectation mechanism. The first mechanism requires a strong convergence of expectations of agents on the future development of the niche. This represents the initial condition that could lead to the niche formation. The SNA concept that best highlights the link between the development of a common view and the formation of a cohesive group is the so-called *social circle*.[22] A social circle is not yet a network, since it does not necessarily imply social interaction. It is, rather, a social entity characterised by the sharing of similar interests among its members. Thus, a social circle could be seen as a group of people linked by these common interests. It acts as a network incubator since the existence of a common set of interests provides chances for interactions and, therefore, facilitates the emergence of a network structure.[23]

In the process of formation and development of a biorefinery niche, social circles play a central role. A developing biorefinery could be seen as a social circle in which the central interest is the development of the new technology. Thus, when assessing the convergence of expectations, the presence and structure of such a social circle should be one of the primary concerns addressed.

Power mechanism. The second mechanism of niche formation requires an actual networking process with powerful actors. Power is by any means an important topic within SNA as it is a fundamental property of any social structure. This approach emphasises that power is a relational concept: an individual does not have power in a void; he/she can exert power only in relation with other actors. Thus power arises from occupying advantageous positions in networks of relations. Furthermore, since power is a consequence of relational patterns, the amount of power in social structures varies according to the social architecture within which interactions occur. *Centrality* is an SNA conceptual and analytical tool, especially useful in analysing power in social networks.[24] At the individual level, this concept refers to the properties of actor location in a social network that make an actor prominent or important. These properties concern the whole myriad of ties owned by the individual actor. When a group of actors reaches a sufficient amount of power, this could be seen as a group property. Such a result is reached only when the group is involving those actors individually detaining the needed resources. Under such circumstances, powerful actors act as a resource for the whole group. Hence, the niche has enough power to act.

Knowledge mechanism. The third niche mechanism refers to learning patterns. These require that the network representing the niche is characterised by substantial flows of knowledge. A way to capture these flows is to account for the presence of all possible *communication relations* among actors that could allow knowledge to flow from one agent to another. The SNA allows classifying some communication interactions such as formal collaboration agreements and face-to-face knowledge exchange. In this context, a niche could be understood as a network in which such communication interactions take place. However, measuring knowledge flows is a major challenge for applied researchers.[25] Recently, several empirical studies have focused their attention on clusters and

Table 6.1 Questionnaire questions for niche mechanisms investigation.

Niche mechanism	Questions	Response
Expectation mechanism	A project for biofuel development and use is going to be planned for your region. The plan envisages experiments in the production of raw material for biodiesel, the establishment of facilities for production of biofuel and tax exemptions for some categories of users. Would you like to join the project?	Please indicate your level of expectation of the project: 5 = very high, 4 = high, 3 = medium, 2 = low, 1 = very low
Power mechanism	In your opinion, in order to make this project successful, the participation of whom is important?	Please indicate the reason: 1 = because of its capacity to mobilise people, 2 = because of its technological and knowledge resource, 3 = because of its capacity to coordinate the local development strategy, 4 = because of its capacity to control financial resources, 5 = because of its capacity to control material infrastructure, 6 = because of its raw materials production capacity – please indicate if you already interact with this actor
Knowledge mechanism	Could you mark, among the following actors, those with whom you have/had knowledge exchange on biorefinery?	Please indicate if you: 1 = receive information, 2 = exchange information, 3 = transmit information

networks, considering them as the locus of knowledge diffusion. However, some have stigmatised 'the role of fuzzy social relationships and ill-defined spillover mechanisms as the basis of knowledge flows and learning processes within territory-bounded communities' (p. 48),[16] and have consequently proposed more direct and reliable way of defining and measuring knowledge flows which is largely based on SNA.[26–28]

In order to collect the required data to perform such social network analysis with respect to the development of a biorefinery, we suggest to use the questions reported in Table 6.1.

The findings based on the use of these indicators could provide researchers with valuable information on the overall readiness of a niche, highlighting its potential to gain internal momentum and, eventually, replace the existing regime. Specifically, we suggest a four-fold classification:

1. The analysis can reveal the absence or a negligible presence of the fundamental mechanisms of niche formation, where, in other words, the

niche itself is completely inexistent. This requires what Kemp *et al.* call a socio-technical alignment policy focusing on specific interventions for aggregation and enabling actors.[11] In this case, first of all, the lacking of vision and expectation through the new technology should be addressed. This means stimulating interest and attention of potential developers, investors, regulators, producers and users, through the development and the spreading of a clear vision about the new technology. This should include a suitable communication plan encouraging other actors to develop their visions and expectations.

2. Only a minor level of convergence of expectation is detected and there is a basis for the formation of a group sharing the same vision about a specific innovation technology, which could act as an incubator for an actual niche. Under these circumstances there is a need for networking. The attention should thus be focused on the strong local actor, previously identified by means of a resource generator survey. The measures taken could envisage regular meetings and discussions orientated to users, scientists and societal organisations as well as producers. Dedicated networking experts could be employed in order to build and maintain such a network.

3. The convergence is attained and a networking process begins. In this case we just have what could be called a proto-niche; that is a network of actors with a clear vision where a certain power to act is already mobilised but there is still the need to build a stable pattern of communication. Possibilities to share new knowledge can be created through interaction platforms, regular meetings and conferences. A monitoring programme with public access could be put in place using information technology solutions.

4. All the three mechanisms have taken place and the niche is at a good level of development. Specific efforts should be dedicated to reinforce and encourage the niche's mechanisms by maintaining the activities and measures mentioned above and with economic protection such as subsidies or tax exemption, or by ensuring R&D funding commitments by private actors participating in the experiment.[29]

(2) Second phase
In order to complete the first step analysis of the biorefinery niche, we shall now focus on the possible development paths that, being accepted by the whole community, will benefit from synergetic interactions between the stakeholders and, therefore, have the highest chance of success. In this second phase we shall define an empirical methodology suitable to elicit stakeholders' view regarding the niche development process as well as the policy actions suitable to stimulate such process.

As stated before in this chapter, the biorefinery industry is a very complex system with a number of interacting elements, evolving conjunctly towards uncertain status. Many approaches have been used to deal with such complex systems.

Among the variety of methods available to challenge this task (*e.g.* neural networks, genetic algorithms, nonlinear equation systems and agent-based modelling), fuzzy cognitive maps (FCMs) have been chosen to investigate the specific domain of biorefinery development in the context of rural areas.

The advantages in using this method are several. First of all, FCMs are easy to build from the knowledge of local people; they are very powerful in dealing with dynamic systems (since their structure is based on variables and mutual relationships existing among them) and allow drawing a comprehensive picture of the elements of the problem under analysis.[19] Secondly, FCMs deal with qualitative information, which can be obtained from local stakeholders, allowing the researcher to overcome the lack of reliable quantitative data. Thirdly, although the method does not provide any prediction in quantitative terms, it is suitable for providing support in the long run, through the investigation of all the effects exerted on the whole system by the changes that may occur under certain conditions.[30]

In technical terms, FCMs are signed fuzzy digraphs made of elements (*e.g.* concepts, events, project resources) with causal relations among them. The idea behind this logical structure is the fuzzy logic, which is a general name for logical theories that attempt to apply the principles of logical thinking on fields where the classical logic seems inappropriate.

Cognitive maps have been introduced for the first time by Tolman,[31] as an application for psychology research. Later, FCMs have been applied in several fields, such as anthropology, to represent different social communities in human society,[32] to study the relationships among benthic organisms[33] or to model policy scenarios.[18] More recently, Ozesmi and Ozesmi built a model to study the effects of institutional changes on a lake ecosystem, based on the perception of local stakeholders.[30] Coban and Secme modelled the effects of privatisation policies on the distilled alcohol sector of Turkey, based on the perceptions of the employees of alcohol factories, civil servants and other social groups.[19]

In order to apply FCMs to identify and select the relevant variables affecting the development of the biorefinery in rural areas, we proceed as follows:

(a) Introduction (warming-up): self-presentation of each participant, in order to facilitate the interaction within the discussion group;
(b) Description of the purpose of the analysis: participants are informed of the domain of the analysis and provided with only a few essential elements, helping them to understand the objective of the analysis, while avoiding any biased information;
(c) Main research question: presenting a clear and explicit direct question to the participants. This is a crucial aspect, as if the question is not clear, participants will easily disguise the objective of the analysis, and the whole process will be disrupted;
(d) Brainstorming (in turn, for each person): needed in order to identify the system variables and their relationships. This step requires asking a group of stakeholders about their perceptions and expectations from the object of the analysis. Participants are asked to write three brief

statements in three separate tags. Once everyone has finished, the group moderator collects all tags, and provokes an interactive discussion on each statement. This phase is necessary in order to share information in the group. According to existing related literature, different categories of stakeholders can be considered.[17,34–36] The most recurrent groups reported in the literature are farmers, private entrepreneurs, researchers, technological transfer agents, consumers, local citizens, policy makers and institutions;

(e) Conceptual coding: the moderator asks the group to simplify the statements, in order to perform an aggregation of similar concepts. The handling of a limited number of concepts (*e.g.* a range of 10 to 30 items) will facilitate the next step of the analysis;

(f) Building-up of the cognitive map (co-operative group): can be done with a participatory approach or by means of individual questionnaires. With the participatory approach the stakeholders are asked to specify the sign (positive or negative) and the intensity of the causal relationship among the variables, according to three increasing degrees: weak, moderate and strong. When stakeholders reach a sufficient consensus on the relationship between two variables, the link is drawn in the map. Alternatively, each stakeholder can perform this task individually. Then, individual maps are condensed in a social cognitive map on the basis of specific criteria. For instance, the relationship between two variables is established if at least three participants agree upon it; its intensity is equal to the mode of the individual values;

(g) Network analysis: the toolkit of the social network analysis allows analysing the structure of the social cognitive map, in its matrix form. The variables are listed both on the vertical axis and on the horizontal one forming a matrix, technically called the *adjacency matrix*.[37] This matrix shows the existing connections between each couple of variables. To analyse the features of a cognitive map, *punctual* and *network* indices can be calculated. Three punctual indices (in-degree, out-degree, centrality) and five network indices (number of variables, number of connections, density, index of complexity) are commonly used in order to describe whole network features. For further details regarding the SNA indicators applied to FCMs, refer to Lopolito *et al.*[38]

In particular, for the purpose of our methodology, punctual indices are very useful to classify each variable of the system into (i) sender (not affected by other variables, but affecting other variables), (ii) transmitter (receiving and sending stimulus to other variables) and (iii) receiver (affected by other variables, but do not send stimulus to other variables).

The identification of sender variables provides important information, since from a policy point of view, they may play a key role as possible 'drivers' of development of the whole system and, therefore, may be used as policy 'instruments'.

6.2.2.2 Second-Step—an Agent-based Model Design

In what follows we shall propose an agent-based model (ABM) design, which could be of use to investigate and simulate the development process of a biorefinery niche. The model conceptualisation represents the second step of our investigation: its architecture stems directly from the theoretical framework discussed above and the empirical investigation conducted in the first step. Specifically, the system depicted is a productive system formed by artificial agents representing firms, whose behaviours and 'initial conditions' are based upon the fieldwork investigation.

As a general starting point we consider two alternative options between which each firm can choose when producing:

$O1_{regime}$: the regime option, which employs a dominant and consolidated technology;

$O2_{niche}$: the niche option, which produces goods by means of an innovative technology.

The purpose of the model is to investigate the converging process towards $O2_{niche}$, departing from an initial status in which all the agents use a regime technology (*i.e.* $O1_{regime}$). Such a converging process should eventually lead to the full niche development and can occur completely endogenously – *i.e.* firms' interactions automatically produce an evolution trajectory that, eventually, leads to the niche development – or can be stimulated and facilitated by policy actions. In both cases, convergence towards the niche option requires the occurrence of the three network mechanisms described in the literature review (see Section 6.2.1) and empirically assessed in the first step (see Section 6.2.2.1). In order to understand and model the operational aspects of such mechanisms, we assume that firms make choices as network members rather than as autonomous entities. That is, the performance of the network to which firms belong strongly influences their production decisions.

In what follows we shall present an accurate description of the proposed AB model, first describing the constituent elements of the model itself, then concentrating our attention on the main behavioural rules governing the system. Finally, we shall briefly assess the key channels through which policy actions could be designed and implemented.

The Constituent Elements

Agents: the unit of analysis of this model is the *firm*. In this model we assume a population of N firms, where N is defined according to the first step analysis. Other kinds of agents could be introduced: for instance, *institutions* can be modelled in order to take into account finding of the fuzzy analysis.

Social space: the environment within which agents act (and interact) is defined as a social space represented as a 2-dimentional, finite, regular grid of cells. These cells represent a social space rather than a physical space: this spatially explicit feature of the model serves to structure the interactions of firms, which is based on social proximity. Each cell carries also information on the risk associated to the production choice of the firms, as it will be clarified later on.

Time frame: the unit of time we define in our model is called the *time-step*. As we will discuss later on, in this model the social network evolves over time as firms interact with each other. Each time-step every agent will have the opportunity to interact with the other agents within a social proximity range.

Web of connections: whenever two firms 'meet', they establish a *tie*, which can be conceived as a channel through which they share resources. The strength of the tie depends on the shared view upon the niche technology. As more ties are established, a network of firms emerges as an endogenous feature of the system.

Behavioural Rules and the Niche Development Process
As already mentioned, all firms initially produce a generic good, using $O1_{regime}$. Under the assumption of perfect competition, every firm producing with the regime technology will have extra profits equal to zero. Profits might rise as firms switch to $O2_{niche}$. Indeed, as time goes by, firms have to decide whether to keep producing using the incumbent technology (*i.e.* the regime option) or switch to the innovation niche option. Such a decision would depend on the performance of the niche technology (in terms of comparative profitability), which, in its turn, depends upon the three niche mechanisms discussed in the theoretical background and analysed empirically in the first step.

We shall now try to define a general framework for modelling the three niche mechanisms.

Expectations mechanism: each firm is characterised by a level of expectation. It describes the preference of a firm towards $O2_{niche}$. The higher it is, the more likely it is that the firm will switch to the new technology. At initialisation phase, all firms have an expectation set to a specified starting value which should be fed in from the first step analysis.

Once a firm reaches a high level of expectation of the new technology, it becomes a supporter of the niche option. Whenever two supporting firms interact they establish a link. The tie will initially endure until the following time-step. Stability of ties is related to the power of the network (see later). Over time a network of relations among supporting firms can emerge. Such a network, which is the emerging innovation niche, can be seen as the socio-economic space within which firms develop the new technology by means of sharing their knowledge.

Note that being part of the niche network does not imply that such firms have necessarily switched to the niche technology. For this to happen supporting firms must find it convenient to produce with the niche technology; this occurs any time their expected profit is greater than zero (and therefore higher than the profit obtained producing with the incumbent technology).

The expected profit is calculated as the difference between expected revenue and the expected cost in $O2_{niche}$. Both expected revenue and cost are influenced by a firm's expectation towards $O2_{niche}$. This introduces a bias in the calculation of expected profit, as firms will tend to underestimate the potential revenue attached to the niche technology.

Expectations of active agents (*i.e.* those who have switched to the niche technology) can increase or decrease overtime. Specifically, they will increase any time the actual profit in $O2_{niche}$ exceeds the expected profit and *vice versa*. The actual profit depends on the probability that the firm will obtain the highest profit, that is a function of the risk associated to the new technology, and depends on the attribution of the cell on which the firms are located. This probability captures the risk associated with production under $O2_{niche}$, which stems from the lack of knowledge available on the new technology.

Power mechanism: individual power (I^{power}) describes the firms' endowment of strategic resources. Intuitively, a strategic resource is a resource that can be used in order to develop and promote a new technology. For instance, an R&D laboratory is a resource that could serve the purpose of developing a new technology. A wide-ranging proxy of such resources could be firms' turnover as, in general, larger firms are also the most powerful. As for the expectation mechanism, also in this case data on power's level is fed into the model from the first step analysis. This attribute can increase or decrease according to the profit obtained producing under $O2_{niche}$, since this extra profit is added to the individual resource of the firm.

It is assumed that each time two supporting firms establish a link, the total amount of their respective resources flows through this link. Thus, each link has a feature called energy, which is the sum of the resources of the agents on either end of the link. The total sum of links' energy represents, in turn, the overall network power (N^{power}).

Both individual and network power have an impact on the cost structure faced by those firms operating under $O2_{niche}$. On the one hand, we assume that increasing individual power will allow active firms to engage in cost reduction activities (*e.g.* by investing extra profits in R&D, firms could introduce process innovations). On the other hand, as the network power increases, active firms will have access to a growing amount of external resources. In other words, we maintain that resources accumulated by other firms can be exploited by means of spillovers within the emerging social network.

Knowledge mechanism: Each firm is initially assigned a level of knowledge with respect to the new technology. The amount of knowledge possessed by individual firms is supplied by the first step analysis. Each time the firm produces using the new technology its knowledge increases. This captures learning-by-doing activities.

Moreover, any supporting firm can learn from those firms with whom it has established a link. Every time-step a (randomly determined) proportion of knowledge flows among every pair of firms connected by an active link. Such a knowledge mechanism represents the idea of developing expertise; a link will provide an opportunity to perfect the technology by means of learning-by-interacting.

It is quite important to observe that, as the overall level of firms' knowledge on the niche technology increases, the probability of obtaining the highest profit increases. This is because, overall, agents become more knowledgeable on the niche technology and therefore the risk associated with the production involving such new technology decreases. This is a system feature that affects also firms currently not involved in the niche option; in fact, if they will switch to the niche option they will get the highest profit with a higher probability. Specifically, we assume that the probability of obtaining the highest profit increases in a linear fashion.

Modelling Policy Actions

As mentioned earlier, the niche emerging process could occur endogenously or can be activated/stimulated by policy actions. We shall now consider two key directions along which a policy action could develop: (1) the information action – *i.e.* policy makers could increase the amount of information available on the niche technology (both in terms of expectations and in terms of technical knowledge); (2) the subsidy action – *i.e.* policy makers could stimulate the initial switching-over phase by introducing subsidies that should be withdrawn once the network has emerged and the initial constraints associated with the lack of technical knowledge have been overcome.

We believe this model provides an effective framework for modelling such policy scenarios. Moreover, it also provides a useful tool to assess the impact of such policy actions upon the desired target. It should be noted that combined policy actions might be the best way to achieve a desirable target, such as a fast and stable convergence towards the innovation niche at the minimum cost (*i.e.* the social cost of the policy actions). Hence, in order to define an efficient policy-mix and identify the *most* efficient one, we shall move on to the third step of our proposed methodology.

6.2.2.3 Third Step – Evaluating Policy Actions

The last step of the proposed methodology serves the purpose of evaluating the impact of various policy actions and identifying the best policy-mix to achieve a fast and stable convergence towards the innovation niche, bearing in mind the desired target of policy costs minimisation (*i.e.* the social cost of the policy actions).

The problem can be specified as follows: the public cost can be expressed in terms of financial budget to provide subsidies to private firms and/or the personnel cost to provide extension services through institutions aiming at diffusing information related to the niche technology. The number of firms switching to the new technology over time represents the benefits of the policy.

The cost-benefit analysis is among the most popular methodologies adopted to select the most efficient policy among a set of alternatives. However, since it is a monetary method, every variable (public cost and switching firms) should be evaluated, and the assumption of perfect information (*i.e.* distribution of costs and benefits along time horizon; social preference rate for risk in the long run) represents the main obstacle for its application. In fact, when evaluating the development of the niche, because of the complexity of this process (novelty, mutual interdependency of site-specifics characteristics, indeterminacy), economic agents involved will necessarily operate under incomplete information conditions.

Alternatively, multi-criteria methods may also be adopted. But in order to facilitate the interaction with the decision maker, the reduction of the number of policy-mix into a bundle of possible solutions is recommended.

For these reasons, we propose two methods of analysis. First, we suggest the Data Envelopment Analysis (DEA),[39] in order to rank the policy options in terms of technical economic efficiency. Secondly, a statistical analysis on the simulated data should be performed in order to analyse the converging dynamics and the stability of the policy achievements.

Application of the DEA
The Data Envelopment Analysis (DEA) is a technique based on the application of a linear programming algorithm, aimed to find the most suitable weights for each variable such that the ratio of outputs on inputs of several data sets is made as close as possible to one.[39] Once the most suitable weights are found, from each data pattern (performance of a decision making unit, or outcomes of each policy) the relative efficiency index is calculated (the most efficient would be equal to 1, while others would have a lower index).

In order to find the most efficient policy, we consider efficiency conceived in terms of inputs (*i.e.* subsidy and information-spreading institutional agencies) necessary to reach the same output (*i.e.* number of switching firms at various points in time). In this sense, the most efficient policy-mix will be able to consume less input, for the same number of switching firms.

As a result this analysis would provide a clear rank of various policy-mix, ordered in terms of technical economic efficiency. Subsequently, the decision maker will be able to reduce the number of alternatives, and will be facilitated in the final decision of choosing the most suitable policy.

Statistical Analysis on the Simulated Data
The DEA allows identifying a set of efficient policy-mix. We shall now attempt to suggest a rule for selecting the most preferable one. As mentioned earlier, we suggest doing so by investigating the converging dynamic along with the system stability.

The simulation exercise developed in the second step produces a lot of information about the dynamics of firms' decision making over the time reference. In order to obtain a clear picture of the switching process, a statistical analysis of time series referred to efficient policy-mix should be performed. We suggest the following basic model: $N = f(t)$, where N is the number of firms deciding to switch to the new technology and t is the time-step. From this model, it should be possible to draw the ideal performance of each policy-mix, over time. For instance, the policy maker may select a policy-mix that is more costly, but performs better at an earlier stage or, a less costly one that performs better in the longer run. The model allows estimation of parameters of performance of the policy – that is, the time response in terms of number of switching firms.

6.2.3 Wrapping-Up – an Application of the Proposed Protocol of Analysis

In order to show an application of the methodology to a real problem, the main results referred to the development of a biorefinery in the province of Foggia (Apulia region, Italy) are presented in Box 6.1. This area has been chosen because it is one of the largest agricultural areas in the south of Italy, with high potential for producing agricultural raw materials (co-products and by-products) suitable for biorefinery. In addition, this case study shows an

Box 6.1 Foggia case study.

First step—assessing the niche development status

➢ *1st Phase: Social Network Analysis:*

Convergence of expectations : Present
Networking with powerful actors : Absent
Learning process : Absent

In light of these findings, the area of Foggia can be considered a proto-niche.

➢ *2nd Phase: Fuzzy Cognitive Maps: five possible 'drivers' identified:*
(1) Subsidies for biorefinery
(2) Public information and stakeholders training
(3) Competition between food/non-food crops
(4) Geographic dispersion of biomass sources
(5) Availability of biomass from spontaneous species

'Drivers' (1) and (2) have been identified as possible policy instruments and have been used in the simulation in the second step.

Second step – modelling policy actions

➢ *Agent Based Modelling:*

Nine different policy-mixes based on two inputs have been simulated: (a) public subsidies (driver 1), (b) information-spreading institutional agencies (driver 2). Follows the specification of the nine policy-mixes: p1 (+3% subsidies; n.3 information-spreading institutional agencies), p2 (+5%; n.3), p3 (+7%; n.3), p4 (+3%; n.5), p5 (+5%; n.5), p6 (+7%; n.5), p7 (+3%; n.7), p8 (+5%; n.7), p9 (+7%; n.7).

The output of the policy simulation is the number of firms switching to the niche option. Their efficiency has been assessed in the third step.

Third step – evaluating policy actions

➢ *Data Envelopment Analysis:*
 The nine policy mixes have been ranked based on their efficiency: four are efficient (p1, p2, p3, p4) while five are not (p5, p6, p7, p8, p9).

➢ *Statistical Analysis:*
 The number of switching firms is expressed as a function of the time-steps. This analysis enriched our information on each policy-mix.

$$N = \alpha + \beta_1 t + \beta_2 t^2 + \beta_3 t^3 + \varepsilon$$

The estimated trends of firms switching process are described by the parameters:

	α	β_1	β_2	β_3	ε
p1	0.448	−0.0054	0.0000145	0.0000000650	0.257
p2	0.625	−0.0090	0.0000264	0.0000000759	0.286
p3	0.846	−0.0151	0.0000797	0.0000000114	0.247
p4	1.471	−0.0402	0.0002471	−0.000000202	0.412

These parameters show that the effectiveness is the result of the opposite effects of the linear (β_1) and the quadratic (β_2) components, while the effect of the cubic (β_3) component is negligible.

From the analysis of these figures, it is clear that there is no dominant policy-mix and, therefore, the final choice should be made according to the decision maker(s) constraints and/or perceptions.

example of investigation of the aspects related to the development of biorefinery in rural areas.

Provided that the development of the innovation niche implies the lack of full information, the final decision requires a multi-criteria decision making, through which the whole society plays a determinant role for the success of the biorefinery. Therefore, from this example it is clear that the main objective of the methodology is not the definition of the *best policy*, but the production of additional information, providing decision maker(s) and stakeholders with the awareness and science-based facts in a multi-faceted approach. This will facilitate public debate and the negotiation process necessary to reach the consensus of the rural community, which is the basic condition to the successful development of the biorefinery.

6.3 The Macro-economic Approach: a CGE Model with the Inclusion of Biorefineries in the Production Process

To study the effects of renewable policies on a variety of economic and environmental aspects, current literature focuses on the use of Computable General Equilibrium (CGE) models. These are based on Arrow and Debreu theories where agents interact in competitive markets by setting up optimum quantities and prices, satisfying all markets and agents equilibrium conditions.[40] These theories, through a converging process, find an equilibrium around a fixed point satisfying Walras' law.[41]

The rationale of the second part of this chapter is to illustrate a theoretical CGE model with the inclusion of biorefineries in the production process. The remainder of this section is structured as follows: Section 6.3.1 illustrates an overview of CGE models applied to biofuels; Section 6.3.2 describes a theoretical model of a biorefinery economy; and Section 6.3.3 concludes.

6.3.1 Application of CGE Models to Biofuels

Literature on modelling biofuels economics is recent. Most studies focus on projections and policy analysis in the short and medium term given the lack of time series data on biofuels. Many existing models generally focus on extending food processes to include biofuels commodities.[42,43] Banse *et al.* consider the impact of biofuels production on land use in OECD countries. Main results suggest that assuming no changes in current production technology or in international trade in biofuels, the USA, Canada and the EU would require 30–70% of their crop lands to replace 10% of biofuels in transport activities.[42] An OECD study focuses on the effects of the EU Biofuels Directive 2003/30/EC on agricultural markets. Main findings show a break (or even a reversed trend) in the decline of real world prices for agricultural commodities when crops are used for biofuels activities. This would lead to an increase in land prices (with a consequent decrease in international competitiveness) in the EU with a limited capacity for farmers and other economic agents to comply with the Directive. Only the effect of a mandatory biofuels subsidy would help in reaching the objective of the Directive by the end of 2010 (although this could suggest distortionary economic effect in the long run). Improving competitiveness by increasing R&D activities (in particular enhancing second- and third-generation biofuels) would necessarily result in the matching optimal development path of biorefineries for the economy and the environment.[43]

Further studies on the implementation of the Biofuels Directive in EU-15 consider the agricultural sector and its full behavioural design.[44,45] Main results reveal that an increase in demand for biofuels would have positive effects on arable land in EU Member States. Furthermore, domestic production would meet the requirements of ethanol and biodiesel only when high import tariffs are applied to world prices. This would help EU-15 farmers to increase their income and create new employment opportunities. Rajagopal and Zilberman provide an interesting review of the interactions between economic, environmental and biofuels policies.[45] The main aim of this study is to highlight the need of biofuels use and production in developing countries and to help rural communities to reduce GHG emissions, creating new jobs and improving welfare conditions. Birur *et al.* focus on liquid biofuels produced from food or feed crops in various OECD and non-OECD countries and their effects on agro-ecological zones to account for land use impacts.[46] This study extends current consumption and production structures to incorporate biofuels use and technology. Main results suggest optimal substitutability between biofuels and petroleum products when crude oil prices increase. Also, demand for biofuels feedstock would mainly affect the EU, Brazil and the USA, with land use impacts higher in Brazil.

Early studies on the use of third-generation biofuels analyse the effects of replacing biomass (switchgrass) with standard crude oil technologies in the USA.[47] To include switchgrass in the Global Trade Analysis Project (GTAP) database, simulations on optimal input-output coefficients are conducted such that intermediate input coefficients correspond to 70% of those of cereals commodities. Replacing switchgrass with crude oil use in the oil industry would increase the world price of cereals and decrease that of other types of crops and feedstock.

A recent generation of CGE models for bioenergy purposes is hybrid in nature. This means that these models integrate the micro- with the macro-economic aspects of a bioeconomy.[48,49] The specificity of these models is to integrate top-down neoclassical growth model assumptions with bottom-up methodologies of modelling primary energy input factors. Game-theory behaviour features are also included to analyse optimal energy investment policies. Energy prices would be determined endogenously as well as technological changes to include renewables. Main findings show that when the adoption of renewables is not mandatory, free-riding occurs across industries. Another study suggests that hybrid models are important tools for policy makers to implement optimal environmental policies. In addition, these models can also contribute to investigate parameters estimates for substitution between capital and energy on one hand and the adoption of renewable energy technologies on the other.[50]

6.3.2 A Theoretical CGE Model for a Bio-based Economy

This section illustrates a theoretical macroeconomic model of an economy with biorefinery schemes for Europe. It is a static multi-country and multi-sector model of EU countries and an aggregate of Rest of the World (ROW) countries. It also incorporates CO_2, NO_x and SO_2 emissions (arising from production processes), which are responsible for GHGs across Europe.

The model consists of a series of sub-blocks constituting the global bio-based economy: production, consumption, trade, government, environment, market clearing and model main closure rules. The following notations are used to indicate main sets: activities or sectors *a*, regions *r*, regions excluding the globe region *rng* and input factors *f*.

6.3.2.1 The Production Structure

The production function of an economy with biorefineries is illustrated in Figure 6.3.

Total output $QX_{a,r}$ is obtained by employing a Leontief-type technology (with zero substitution elasticity) between non-energy intermediates $QINT_{a,r}$ and a nested constant elasticity of substitution (CES) function for value added $QVA_{a,r}$.

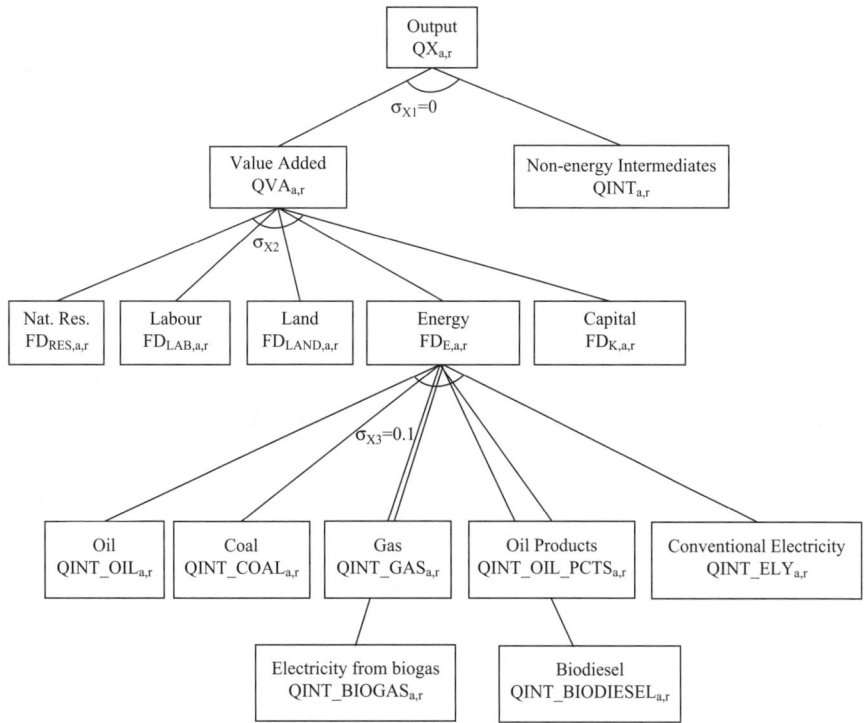

Figure 6.3 Production structure with biorefinery schemes (biogas for bioelectricity production and biodiesel).

QINT$_{a,r}$ is obtained as:

$$QINT_{a,r} = \sum_a comacto_{a,r} \times QX_{a,r} \qquad (6.1)$$

In eqn (6.1) the demand for intermediate non-energy inputs is given by the sum of input-output coefficients comactco$_{a,r}$ among sectors, multiplied by activity output QX$_{a,r}$. The value added (VA$_{a,r}$) is obtained by a nested CES function that allocates FDLAND$_{a,r}$ (land), FDLAB$_{a,r}$ (labour), FDRES$_{a,r}$ (natural resources) and FDKE$_{a,r}$ (a capital and energy bundle). These belong to the set f in total demand factors FD$_{f,a,r}$.

$$QVA_{a,r} = F_{2a,r}\left[\alpha_{x2LAB_{f,a,r}}FD_{LAB_{a,r}}^{\rho_{x2a,r}} + \alpha_{x2RES_{f,a,r}}FD_{RES_{a,r}}^{\rho_{x2a,r}} + \alpha_{x2LAND_{f,a,r}}FD_{LAND_{a,r}}^{\rho_{x2a,r}} \right.$$
$$\left. + \alpha_{x2KE_{f,a,r}}FD_{KE_{a,r}}^{\rho_{x2a,r}}\right]^{\frac{1}{\rho_{x2a,r}}}$$
$$(6.2)$$

In eqn (6.2) all $\alpha_{x2f,a,r}$ represent the share of the respective factor demand in value added and all $\rho_{x2a,r}$ refer to the elasticity of substitution across input factors. The tangency conditions in the value added nest satisfy the condition that the price of factor demands WF$_{f,r}$ equate the value of the marginal product (the First Order Conditions in eqn (6.2)) such that:

$$WF_{f,r} = PV_{a,r}F_{2a,r}\left[\sum_f \alpha_{x2f,a,r}FD_{f,a,r}^{\rho_{x2a,r}}\right] \times \alpha_{x2f,a,r}FD_{f,a,r}^{\rho_{x2a,r}-1} \qquad (6.3)$$

where PV$_{a,r}$ is the price of the value added. The respective factor demand equations will result as follows:

$$FD_{RES_{a,r}}^{\rho_{x2a,r}-1} = \left(\frac{QVA_{a,r}}{F_{2a,r}}\right)WF_{RES_{a,r}}^{-\rho_{x2a,r}}\left[\alpha_{x2RES,a,r}WF_{RES,r}\right]^{\frac{\rho_{x2a,r}}{1-\rho_{x2a,r}}} \qquad (6.4)$$

$$FD_{LAND_{a,r}}^{\rho_{x2a,r}-1} = \left(\frac{QVA_{a,r}}{F_{2a,r}}\right)WF_{LAND_{a,r}}^{-\rho_{x2a,r}}\left[\alpha_{x2LAND,a,r}WF_{LAND,r}\right]^{\frac{\rho_{x2a,r}}{1-\rho_{x2a,r}}} \qquad (6.5)$$

$$FD_{LAB_{a,r}}^{\rho_{x2a,r}-1} = \left(\frac{QVA_{a,r}}{F_{2a,r}}\right)WF_{LAB_{a,r}}^{-\rho_{x2a,r}}\left[\alpha_{x2LAB,a,r}WF_{LAB,r}\right]^{\frac{\rho_{x2a,r}}{1-\rho_{x2a,r}}} \qquad (6.6)$$

$$FD_{K_{a,r}}^{\rho_{x2a,r}-1} = \left(\frac{QVA_{a,r}}{F_{2a,r}}\right)WF_{K,r}^{-\rho_{x2a,r}}\left[\alpha_{x2K,a,r}WF_{K,r}\right]^{\frac{\rho_{x2a,r}}{1-\rho_{x2a,r}}} \qquad (6.7)$$

where WF$_{RES,r}$, WF$_{LAND,r}$, WF$_{LAB,r}$, WF$_{K,r}$ $\in$ WF$_{f,r}$.

Modelling Stakeholders' Interplay and Policy 303

On the third nest of the production function, the energy bundle $FDE_{a,r}$ is obtained by combining energy and bioenergy commodities:

$$\begin{aligned}FDE_{a,r} = F_{3a,r}[&\alpha_{x3OIL,r}QINT\_OIL_{a,r}^{\rho_{x3a,r}} + \alpha_{x3GAS,r}QINT\_GAS_{a,r}^{\rho_{x3a,r}} \\&+ \alpha_{x3COAL,r}QINT\_COAL_{a,r}^{\rho_{x3a,r}} + \alpha_{x3OIL\_PCTS,r}QINT\_OIL\_PCTS_{a,r}^{\rho_{x3a,r}} \\&+ \alpha_{x3ELY,r}QINT\_ELY_{a,r}^{\rho_{x3a,r}} + \alpha_{x3BIOGAS,r}QINT\_BIOGAS_{a,r}^{\rho_{x3a,r}} \\&+ \alpha_{x3BIODIESEL,r}QINT\_BIODIESEL_{a,r}^{\rho_{x3a,r}}]\end{aligned}$$

(6.8)

where $\alpha_{x3OIL,r}$, $\alpha_{x3GAS,r}$, $\alpha_{x3COAL,r}$, $\alpha_{x3OIL\_PCTS,r}$, $\alpha_{x3ELY,r}$, $\alpha_{x3BIOGAS,r}$, $\alpha_{x3BIODIESEL,r} \in \alpha_{x3a,r}$. The tangency condition in the energy composite nest is such that:

$$WF_{f,r} = PQINT_{a,r}F_{3a,r}\left[\sum_f \alpha_{x3a,r}QINT_{a,r}^{\rho_{x3a,r}}\right] \times \alpha_{x3a,r}QINT_{a,r}^{\rho_{x3a,r}-1}$$

(6.9)

where $PQINT_{a,r}$ is the price of intermediate energy and bioenergy commodities. The price system of the production block is given by the composite price of output $PX_{a,rng}$ defined in terms of commodity prices $PXC_{c,rng}$ and by the value added price for activity $PVA_{a,rng}$. The corresponding equations are determined as follows:

$$PX_{a,rng} = PXC_{a,rng}$$

(6.10)

$$\begin{aligned}PVA_{a,rng} = &PX_{a,rng} \\&\times \left[1 - (tx_{a,rng}) - \left(\frac{\sum_f tfu_{f,a,rng}WF_{f,rng}FD_{f,a,rng}}{QX_{a,rng}PX_{a,rng}}\right)\right. \\&\left.- \left(\frac{\sum_{ptype} Emissions_{ptype,rng,a}EmRate_{rng,ptype}FD_{f,a,rng}}{QX_{a,rng}PX_{a,rng}}\right)\right] \\&- \sum_a PQ_{a,rng}comacto_{c,a,rng}\end{aligned}$$

(6.11)

In eqn (6.11), $PVA_{a,rng}$ is the result of the activity price $PX_{a,rng}$ over the amount of total production $QX_{a,rng}PX_{a,rng}$ net of: a) production taxes ($tx_{a,rng}$); b) factor taxes ($tfu_{f,a,rng}$) resulting on the volume of demand factors $WF_{f,rng}FD_{f,a,rng}$; and c) emissions (on production activities) multiplied by their emission factor rates ($Emission_{ptype}EmRate_{rng,ptype}$). From the obtained value, intermediate input prices $PQ_{a,rng}$ (multiplied by Leontief's input-output coefficients $comacto_{a,rng}$) are subtracted.

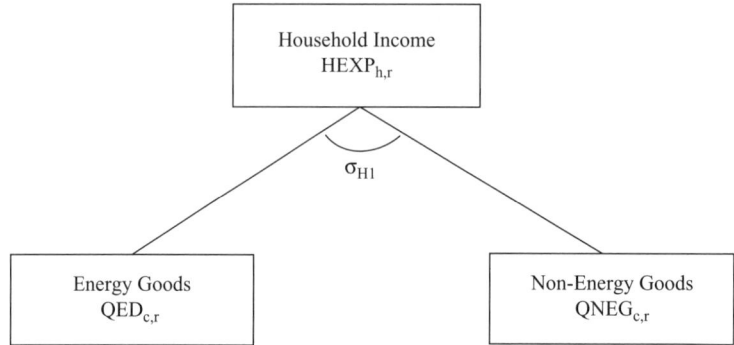

Figure 6.4 Consumer structure.

6.3.2.2 Consumer Structure

Household income YH_{rng} is considered as the difference between the total regional household's income $REGHIN_{rng}$ net of government taxes $YGOV_{rng}$:

$$YH_{rng} = REGHIN_{rng} - YGOV_{rng} \qquad (6.12)$$

$$HEXP_{rng} = YH_{rng}(1 - (SADJ_{rng} kaphsh_{rng})) \qquad (6.13)$$

In eqn (6.13), household expenditures $HEXP_{rng}$ are defined net of savings. Saving rates are computed on household income YH_{rng} by a given saving rate scaling factor, $SADJ_r$ multiplied by $kaphsh_{rng}$, the share of household income saved (after taxes).

Consumer structure is shown in Figure 6.4.

Each commodity (energy and non-energy good) is demanded according to a Stone–Geary utility function:

$$PQ_{a,rng}QCD_{a,rng} = (PQ_{a,rng}qcdcons0_{a,rng}) + \alpha_{h0_{a,r}}\left(HEXP_{rng} - \sum_a PQ_{a,rng}qcdcons0_{a,rng}\right) \qquad (6.14)$$

where $PQ_{a,rng}QCD_{a,rng}$ represents the value of total commodities, $PQ_{a,rng}qcdcons_{a,rng}$ indicates the subsistence commodity value and $\alpha_{h0_{a,k}}$ corresponds to the marginal budget share for commodities.

6.3.2.3 Trade Structure

The trade structure is shown in Figure 6.5. Each region trades final commodities with countries within and outside the EU. Demand for commodities is determined *via* Armington assumption. This assumption allows domestic and

Modelling Stakeholders' Interplay and Policy

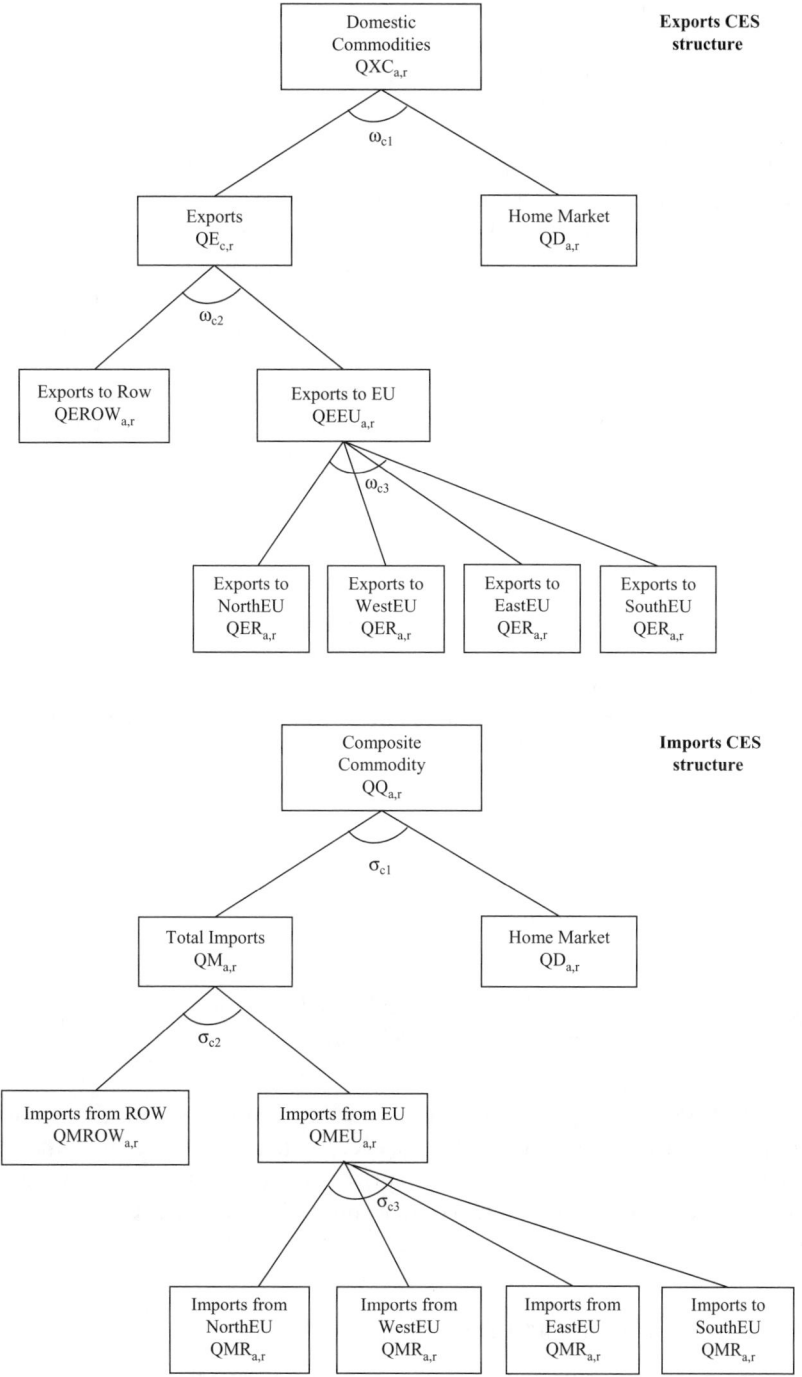

Figure 6.5 Trade structure.

imported goods to be considered as imperfect substitutes. Aggregate production (composite commodity) is then allocated between domestic consumption and imported goods. In turn, composite imported goods are given by imports from ROW and imports from the EU market. Finally, at the third nest, the composite of imported commodities from the EU are allocated among each Member State. On the export side, aggregate consumption is the result, *via* Constant Elasticity of Transformation (CET) functions, of domestic requirements in the home market and composite exported goods to ROW and EU.

The value of traded goods is determined net of import tariffs and inclusive of transport margins. Transport goods are considered homogenous and imported by all countries within a globe region. This region is an artefact to sort out the problem, for a commonly used database, of lack of country of origin and country of destination for the transactions occurring for trade in margins services related to transportation commodities. Without this region, equilibrium in global demand and supply for transport services for traded goods would not occur.

6.3.2.4 Government Module

The government has the sole function to collect various taxes from different sources: activities (production), pollution, factors use, domestic sales and trade. The main assumption is that government income always balances (there is no deficit) and its expenditures are the result of total government consumption.

6.3.2.5 Environment Block

Environmental accounts consider various types of emissions such as NO_x, SO_2 and CO_2. Total emissions, $EMISSIONS_{ptype,r,a}$, are defined in terms of pollution type *ptype*, from activity *a* in country *r*:

$$Emissions_{ptype,rng,a} = \sum_a \left(\frac{comacto_{a,rng} QX_{a,rng} PX_{a,rng}}{PQINT_{a,r}} EmRate_{r,a,ptype} \right) \quad (6.15)$$

where $EMISSIONS_{ptype,rng,a}$ are computed in proportion to the value of commodity activities produced, $QX_{a,rng} PX_{a,rng}$, taking into account the technology matrix $comactco_{a,r}$ and the price of intermediate energy factors $PQINT_{a,r}$. The obtained value is then multiplied by the emission factor $EmRate_{r,a,ptype}$.

6.3.2.6 Commodity Market Block

Final composite consumption is allocated to intermediate production use, households, government, trade (exports net of imported commodities) and investment. All are subject to taxation when present. The solution to commodity markets considers a market-clearing price, which clears all demand and supply as illustrated in previous sections.

6.3.2.7 Model Main Closure Rules

Model main closure rules determine the feature of the economy that is demand-driven. Input factors to production are fully employed. Government tax rates and savings do not change (the model is static) as well as households' propensity to save. Exchange rates are fixed within the EU and flexible when countries trade outside the EU.

6.3.3 Summing Up

This section illustrated the main features of a theoretical CGE model for a biorefinery economy. This includes production, households, government, trade, environment and market clearing blocks. The main discussion is hinged upon the production structure. This describes a composite output obtained *via* a Leontief's technology in value added and non-energy intermediates. The value added is then obtained as the result of various input factors which are mixed *via* a CES function. To account for biorefineries, the energy bundle is further disaggregated into fossil fuels (oil, coal, gas, oil products and conventional electricity) and biofuels (biogas supplying bioelectricity and biodiesel). Finally, the elasticity of substitution ($\sigma_{X3=0.1}$) among fossil and biofuel commodities would indicate a low degree of substitution between old and renewable energy technologies. To promote higher substitution between conventional and biorefinery production processes, enhancement in R&D for second- and third-generation biofuels (biomass) is strongly needed and financial support should be encouraged at all levels of government.

6.4 Conclusions

Over the last decades the scientific community has shown a great interest in the study of biobased economies, technologies and markets, to help governments implementing effective bioenergy policies. Interest in biofuels, more than any other renewable, has grown rapidly due to the success of Brazil, which, since the 1970s, has re-launched its economy. More recently, other countries (in both the developed and developing world) are setting up policies for biofuels or other bioenergy commodities to comply with international environmental agreements (*i.e.* The Kyoto Protocol). The USA, for example, adopted a new round of revisions for its Renewable Fuel Standards policy.[51] The novelty of this policy is the adoption of a strategic approach integrating economic, environmental and social issues to optimise biofuels at all levels of production chain. In the EU (a joint worldwide leader of biofuels with Brazil and the USA), a new set of energy policies would change present and future scenarios of energy supply and consumption within the community. The new regulation

(Directive 2009/28/EC) promotes the use of energy from renewable sources and establishes reference values of energy from renewables for each Member State to comply with the EU '20/20/20' strategy.[52] This strategy is an important voluntary commitment of the EU to reach the Kyoto Protocol's goals and decrease by 20% greenhouse gas (GHG) emissions by 2020 with the use of 20% of renewables. Furthermore, in the transport sector, the 2007 Renewable Energy Road Map establishes the adoption of a minimum 10% biofuels use. Biofuels use in the transport sector contribute already to 14% of EU total market fuels.[53] This percentage may nonetheless increase from both bioethanol production in Sweden and biodiesel production in Germany as well as from any other EU country using other feedstock such as straw, rapeseed and palm oils and/or wood processes.[54,55]

The growing and well-documented interest of scientists, local and national governments and supranational organisations towards biofuel as a primary alternative energy source calls for an extra effort from economists and technology analysts to model policy scenarios and assess the impact of well-crafted policy actions. In this chapter we proposed two different, yet complementary, approaches to policy modelling. First, we developed a protocol for a microeconomic assessment of biorefinery production; subsequently we presented a CGE model for assessing a biorefinery economy in the EU to help in achieving policy objectives of the new EU energy strategy.

Both methodologies should be considered, in the authors' view, as parts of a holistic approach able to reconcile the bottom-up or micro-analysis with the top-down, macro decision making. In fact, macro actions should be guided and informed on well-established behavioural economic interactions.

On the other hand, micro analysts should always be familiar with and at least aware of the overall situation, which is always best represented by keeping the macro picture in mind. Researchers should embrace the view that local systems are not isolated islands but rather part of a macro system that, to some extent, influence their dynamics. At the same time, macro policy scenarios should be fully based on instances coming from the micro dimension. This suggests that a hybrid approach, able to combine the fine-grain insights that can be obtained from a micro-based analysis with the overall picture captured by a macro approach, should provide the best answers to policy makers.

References

1. Biorefinery Euroview Consortium, *Biorefinery Euroview, Current Situation and Potential of the Biorefinery Concept in the EEC*, Paper presented at 16th European Biomass Conference, 5/6/08, Valencia, Spain, 2008.
2. M. Bauer, *Resistance to New Technology*, Cambridge University Press, Cambridge, 1995.
3. D. Elliot and R. Elliot, *The Control of Technology*, Chapman and Hall, London, 1976.

4. J. Wajcman, *Feminism Confronts Technology*, Polity Press, Cambridge, 1996.
5. C. Freeman and L. Soete, *Technical Change and Full Employment*, Basil Blackwell, Oxford, 1987.
6. F. Berkhout, A. Smith and A. Stirling, *SPRU Electronic Working Paper*, 2003, 106.
7. A. Rip and R. Kemp, in *Human Choice and Climate Change*, ed. S. Rayner and E. L. Malone, Battelle Press, Columbus, OH, 1998, vol. 2, p. 327.
8. F. W. Geels, *Res. Pol.*, 2002, **31**, 1257.
9. F. W. Geels and J. Schot, *Res. Pol.*, 2007, **36**, 399.
10. R. Kemp and J. Rotmans, *International Conference Towards Environmental Innovation Systems*, Garmisch-Partenkirchen, September, 2001.
11. R. Kemp, J. W. Schot and R. Hoogma, *Tech. Anal. Strat. Manag.*, 1998, **10**, 175.
12. J. W. Schot, *Hist. Tech.*, 1998, **14**, 173.
13. W. W. M. Van der Laak, R. P. J. M. Raven and G. P. J. Verbong, *Energ. Pol.*, 2007, **35**, 3213.
14. A. Smith, A. Stirling and F. Berkhout, *Res. Pol.*, 2005, **34**, 1491.
15. R. P. J. M. Raven, PhD Thesis, Eindhoven University of Technology, 2005.
16. E. Giuliani and M. Bell, *Res. Pol.*, 2005, **34**, 47.
17. N. Krajnc and J. Domac, *Energ. Pol.*, 2007, **35**, 6010.
18. R. Axelrod, *Structure of Decision, the Cognitive Maps of Political Elites*, Princeton University Press, Princeton, NJ, 1976.
19. O. Coban and G. Secme, *Inform. Sci.*, 2005, **169**, 131.
20. B. Kosko, *Int. J. Man Mach. Stud.*, 1986, **24**, 65.
21. J. J. Heckman, *J. Econometrics*, 2001, **100**, 3.
22. C. Kadushin, *Am. Socio. Rev.*, 1968, **33**, 685.
23. M. Grossetti, *Soc. Network.*, 2005, **27**, 289.
24. L. C. Freeman, *Soc. Network.*, 1979, **1**, 215.
25. P. Morone and R. Taylor, in *Encyclopedia of Information Science and Technology*, ed. Mehdi Khosrow-Pour, 2nd edn, IGI Global, USA, 2009.
26. P. Dicken and A. Malmberg, *Econ. Geogr.*, 2001, **77**, 345.
27. A. Malmberg and P. Maskell, *Environ. Plann.*, 2002, **34**, 429.
28. A. Amin and P. Cohendet, *Architectures of Knowledge. Firms, Capabilities and Communities*, Oxford University Press, Oxford, 2004.
29. M. C. J. Caniels and H. A. Romijn, *Schumpeter Conference 2006*, 21–24 June, Nice, France.
30. U. Ozesmi and S. L. Ozesmi, *Ecol. Model.*, 2004, **176**, 43.
31. E. C. Tolman, *Psychol. Rev.*, 1948, **55**, 189.
32. P. Hage and F. Harary, *Structural Models in Anthropology*, Oxford University Press, New York, 1983.
33. C. J. Puccia, in *Analysis of Ecological Systems: State-of-the-Art in Ecological Modelling*, ed. W. K. Lauenroth, G. V. Skogerboe and M. Flug, Elsevier, Amsterdam, 1983, p. 719.
34. EEA, European Environment Agency, *EEA Technical Report*, 2007, 12.
35. H. Scholes, *Biomass Bioenergy*, 1998, **15**, 333.

36. E. K. Yiridoe, R. Gordon and B. B. Brown, *Energ. Pol.*, 2009, **37**, 1170.
37. F. Harary, R. Z. Norman and D. Cartwright, *Structural Models: An Introduction to the Theory of Directed Graphs*, John Wiley & Sons, New York, 1965.
38. A. Lopolito, M. Prosperi, R. Sisto and P. Pazienza, in *Renewable Raw Materials – New Feedstocks for the Chemical Industry*, ed. R. Ulber, D. Sell and T. Hirt, Wiley-VCH, Weinheim, Germany, 2010.
39. A. Charnes, W. W. Cooper and E. Rhodes, *Eur. J. Oper. Res.*, 1978, **2**, 429.
40. K. J. Arrow and G. Debreu, *Econometrica*, 1954, **22**, 265.
41. S. Kakutani, *Duke Math. J.*, 1941, **8**, 457.
42. M. Banse, H. Van Meijl, A. Tebeau and G. Woltjer, *Impact of EU Biofuels Policies on World Agricultural and Food Markets*, Agricultural Economics Research Institute (LEI), The Hague, 2008.
43. OECD, *Agricultural Market Impact of Future Growth in the Production of Biofuels*, Paris, France, 2006.
44. A. Gohin and G. Moschini, in *Proceedings of Biofuels, Food & Feed Tradeoffs Conference*, Farm Foundation and the USDA, 12–13 April, St. Louis, Missouri, 2007.
45. D. Rajagopal and D. Zilberman, *World Bank Policy Research Working Paper*, 2007, 4341.
46. D. K. Birur, T. W. Hertel and W. E. Tyner, *Global Trade Analysis Project Working Paper*, 2008, 53.
47. S. McDonald, S. Robinson and K. Thierfelder, *Energ. Econ.*, 2006, **28**, 243.
48. V. Bosetti, C. Carraro, M. Galeotti, E. Massetti and M. Tavoni, *Energ. J.*, 2006, **27**, 13.
49. A. Schafer and H. D. Jacoby, *Energ. J.*, 2006, **27**, 171.
50. C. Bataille, M. Jaccard, J. Nyboer and N. Rivers, *Energ. J.*, 2006, **27**, 93.
51. Environmental Protection Agency, *Renewable Fuel Standard Program (RFS2) Regulatory Impact Analysis*, EPA-420-R-10-006, 2010.
52. European Commission, *Official Journal of the European Union*, 2009, **L 140**, 16.
53. http://ec.europa.eu/energy/energy_policy/doc/03_renewable_energy_roadmap_en.pdf. Accessed July 2010.
54. C. De Lucia, *Environmental Policies for Air Pollution and Climate Change in the New Europe*, Routledge-Taylor & Francis, London, 2010.
55. C. De Lucia, in *Handbook of Biofuels Production – Processes and Technologies*, ed. R. Luque, J. Campelo and J. H. Clark, Woodhead Publishing Limited, Cambridge, 2010, p. 13.

Subject Index

References to figures are given in *italic* type. References to tables are given in **bold** type.

abrasive dehullers, 103
acidity of rapeseed oil, 106
acrolein production from
  glycerol, 178, 181
activated carbon, purifying glycerol
  with, 173–4
acylglycerols, 182
adaptations to existing mills,
  151–61
  alcohols as hexane
    alternative, 156–7
  anaerobic digestion, 159–60
  cold pressing, 154
  dehulling, 151–4
  gas-assisted oil pressing, 156
  gum recovery, 160
  integrated scheme, 160–1
  marc hexane retention
    reduction, 154–5
  oleosome isolation, 158–9
  supercritical CO_2
    extraction, 155–6
  transesterification and
    extraction, 157–8
  water extraction, 159
adjacency matrices, 291
advanced oil crops
  biorefineries, 20–3. *see also*
    biorefineries
agent-based model (ABM)
  design, 292–5

alcohols as hexane alternative, 156–7,
  157
alkanes
  production of, 192–3
  in straw waxes, 79
  straw yields, 78
alkyd resin production, 1, 2–3
alkyl benzene sulfonate
  alternatives, 2
almond oil, 36
alperujo (olive waste)
  valorisation, 123–4
Alternaria spp., 58
Amberlite, purifying glycerol
  with, 170
amino acids. *see also* protein and
  amino acid isolation
  chemicals derived from, 139–41,
    140
  extraction process, 135–41, *136–7*
  fractionation of, 139
anaerobic digestion. *see also* biogas
  biogas production with, 81–4
  integrating into existing
    plant, 159–60
  SWOT analysis, **160**
  wet *vs.* dry, 82
annealing algorithm, 206–7
aqueous extraction, 131–2, 137–9
  advantages of, 137
  of sunflower oil, 90–3, 126–7

Argan oil, 39
aspartic acid, 9
Aspen Plus, 205

barley straw, 77
beeswax, 78
Berber's gold (Argan oil), 39
biodiesel manufacturing
  byproducts, utilising, 3
  capital costs for, **212**
  in conventional refineries, 192–3
  cost analysis model, 210–4
  economic optimisation, 223–7, **242**
  environmental
    optimisation, 240–9, **242**
  enzyme catalysis for, 187–8
  Fischer-Tropsch synthesis
    (FTS), 176, 194–6
  flowsheet for, 28, *210*
  glycerol incorporation, 187–97
  green methods, 2
  holistic assessment of process
    options, 263–7
  Hydro Thermal Upgrading
    (HTU), 196
  input and output flowrates, **224**
  materials and utility prices, 212–3
  multi-objective optimisation, **249**
  oils commonly used for, 28
  production process, 28
  purified glycerol co-production
    optimisation, 227–9, 249–55
  pyrolysis for, 193–4
  rapeseed oil for, 29
  recycled greases for, 28
  second-generation
    technologies, 193–7
  succinic acid co-production
    vs. crude glycerol
      production, 266–7
    optimisation of, 213–4, 229–32
  sunflower oil for, 31
biodiesels
  distilling, modelling for cost
    analysis, 212–3
  glycerol in, 188–9
  low-temperature behaviour, 29
  lubricity of, 187, 189, 190
Biofine process, 147–9
biofuels
  glycerol incorporation, 189–92
  legislation promoting, 29
  lipase production, 189–92
  producing *via* microbial
    biotechnology, 193
biogas. *see also* anaerobic digestion
  anaerobic digestion production
    of, 81–4
  applications for, 83
  benefits of, 84
  combustion reaction, 256
  composition of, 216–7
  cost analysis model, 215–8
  definition of, 81
  digestate split, **233**
  economic considerations, 84
  economic optimisation, 232–4, **256**
  emissions, 256–7
  environmental
    optimisation, 255–7
  input and output flowrates, **233**
  multi-objective optimisation, 257,
    258
  parameter ranges for production,
    216
  plants for, 81–2
  process flowsheet for, *215*, *232*,
    256
  purification of, 83
  rapeseed hull production of, 109
  vs. supercritical CO_2
    production, 270–3, **271**
  sustainability of, 83
biolubricants, 2
biomass
  CO_2 extraction of, 71–2
  component extraction, economic
    considerations of, 3
  composition of, 7, 139
  cost considerations, 12, 17
  economic dependence on, 4
  efficient utilisation of, 6

Subject Index 313

as feedstock, 139-41
green, 17
hydrothermal conversion of, *197*
pelletising, 73-5
photosynthetic reaction, 5
preservation methods, 17
product flowchart, 10
production of, 5
products based on, *5*
pyrolysis, 193-4
separation of, 11
similarity to petroleum, 6-7
types of, 5
waste of, 4, 5
biomethane. *see* biogas
BIOPOL, 6
biopol production, 15
biorefineries. *see also* advanced oil crops biorefineries
compared to petroleum-refineries, 6-7
design of, 11
efficiency improvements with, 2
integrated process for, 160-1
biorefinery systems
green biorefineries, 17-9
Lignocellulosic Feedstock (LCF), 11-4, *12*
two-platform concept, 19-20
Whole Crop Biorefinery, 14-7
biosorption, 88-90
broomrape, 58, 65-6
building blocks, 9

cake meal, 93-6
canola oil. *see also* rapeseed oil
area harvested, 35
definition of, 27
glucosinolate levels, 49
origin of term, 49
capital costs for equipment, **212**, 216, 224
carbon sequestration via straw incorporation, 75
carnauba palm, 78-9
castor oil, 37-8

cellulose
hydrolysation of, 8
plant fibre composition of, 86
as precursor, 12
uses for, 3
centrifugal propellers for dehulling, 103
centrifugation systems, 120-1, 123-4
cereal fractionation, 14-5. *see also* Whole Crop Biorefinery
CGE (Computable General Equilibrium) models, 299-307
applying to biofuels, 299-300
for bio-based economy, 300-7
commodity market block, 306-7
consumer structure, 304
environment block, 306
government module, 306
hybrid nature of, 300
model main closure rules, 307
production structure, 301-4
trade structure, 304-6
charcoal rot, 58-9
chromatography, purifying glycerol with, 173-4
citric acid, producing with glycerol, 182-3
CLEARFIELD sunflowers, 63
CO_2
as extraction solvent, 68-70. *see also* supercritical CO_2 extraction
phase diagram, *69*
cognitive maps, 290. *see also* fuzzy cognitive maps (FCMs)
cold pressing
cost analysis model, 220
vs. hexane extraction, 269-70
integrating into existing plant, 154
rapeseed, 106
SWOT analysis for, **154**
combined heat and power (CHP) engines
biogas conversion, 83
modelling for cost analysis, 217
commodity market block of CGE models, 306-7

commodity oils. *see also specific oils*
  cottonseed oil, 31–2
  groundnut oil, 32
  linseed oil, 34
  olive oil, 32–4
  rapeseed oil, 26–8
  sesame oil, 32
  soybean oil, 29–30
  sunflower oil, 30–1
  tall oil, 34–6
Common Agricultural Policy (CAP), 23
communication relations, 287–8
composting sunflower hulls, 114
coniferous trees, producing tall oil from, 34–6
consumer structure of CGE model, 304
corn
  as feedstock, 6
  products yielded from, 16
  stalk composition, 86
  wet-milling, 16–7
cost analysis of biorefineries, 203–77
  biorefinery schemes analyzed in, 209–23
    biodiesel production, 210–4
    biogas production, 215–8
    levulinic acid production, 222–3
    oil extraction, 220–1
    protein extraction, 219
    supercritical CO_2 extraction, 218–9
    thermomoulding, 221
  conclusions of, 276–7
  economic optimisation, 223–39
    of biodiesel and purified glycerol production, 227–9
    of biodiesel and succinic acid production, 229–32
    of biodiesel production, 223–7
    of biogas production, 232–4
    of supercritical CO_2 extraction, 234–9
  environmental and multi-objective optimisation, 240–63
    of biodiesel and purified glycerol production, 249–55
    of biodiesel production, 240–9
    of biogas production, 255–7
    of supercritical CO_2 extraction, 257–63
  holistic assessment of process options, 263–76
    of biodiesel production, 263–7
    economic comparison, 274–5
    environmental impact, 275–6
    of oil extraction, 267–70
    small-scale, 274
    of straw consumption, 270–3
  methodology for, 205–23
    life cycle analysis, 207–8
    multi-objective optimisation, 208–9
    optimisation methods, 205–7
    simulation software, 205
  overview of, 203–5
cottonseed oil
  area harvested, 35
  demand for, 32
  world production of, 31
crambe oil, 38
crop rotation
  of rapeseed, 52–3
  of sunflowers, 61
Crude Sulfate Turpentine (CST) distillation, 35
crude tall oil (CTO), 34–6

Data Envelopment Analysis (DEA), 296–7
degumming, 160
dehulling, 103–15
  economic evaluation of, **107**
  equipment for, 103–4
  hull separation after, 105
  integrating into existing plant, 151–4
  rapeseed, 103–10
  sunflower seeds, 104, 110–4
  SWOT analysis for, 151–4
dehydrations of glycerol, 178

Subject Index 315

densification of straws, 73–5
deterministic optimisation method, 206–7
digestate recycling, modelling for cost analysis, 217–8
dihydroxyacetone (DHA), 181
direct thresh of rapeseed, 21
diseases. *see also* pest and disease control
 of rapeseed, 51–2
 of sunflowers, 58–60
DMC-Biod, 191
downy mildew, 59
dry-milling whole cereal crops, 14. *see also* Whole Crop Biorefinery

Ecodiesel, 191
economic analysis of biorefineries. *see* cost analysis of biorefineries
economic optimisation, 223–39, **275**
 of biodiesel and purified glycerol production, 227–9
 of biodiesel and succinic acid production, 229–32
 of biodiesel production, 223–7, **242**
 of biogas production, 232–4, **256**
 of levulinic acid production, 274–5
 of protein extraction, 274–5
 of supercritical CO_2 extraction, 234–9, **259**
 of thermomoulding, 274–5
electrostatic hull separation, 105
emissions. *see also* environmental and multi-objective optimisation
 calculating, 207–8
 comparison of, **275**
 economic and environmental optimisation of, 241–9
 greenhouse gas, 75
 methanol feed and, *247*, 264–6, *265*
 vs. profits, 275–7
 recycle fraction effect on, 256–7
 from straw processing, 270–3
 from supercritical CO_2 extraction, 259–63
energy costs, 208
environment block of CGE models, 306
environmental and multi-objective optimisation, **208**, 240–63, **275**. *see also* emissions
 of biodiesel and purified glycerol production, 249–55
 of biodiesel production, 240–9, **242**
 of biogas production, 255–7, **258**
 overview of, 275–6
 of supercritical CO_2 extraction, 257–63, **259**, **262**
environmental impact, 207–8. *see also* emissions
enzymatic pre-treatment, 118
 of rapeseed, 125
 of sunflower seeds, 127
error function (EF) in simulations, 206
erucamide production, 26
esterifications of glycerol, 179
ethanol
 glycerol production of, *177*, 180–1
 uses for, 13
 via biomass-nylon-process, 13
etherifications of glycerol, 179
EU Biofuel Directive, 29, 299–300
evening primrose oil, 37
expeller presses, 127–8
extraction, oil. *see* oil recovery
extraction, straw. *see* supercritical CO_2 extraction
extrusion extraction, 91, 126–7, 128–9

falling film evaporators, 169, *171*
fats, historic importance of, 23. *see also* plant oils
fatty acid methyl ester (FAME), 24

fatty acids
  in Argan oil, 39
  effect on biodiesel at low
    temperature, 29
  in jatropha oil, 38
  in lallemantia oil, 37
  in rapeseed oil, 27
  in safflower oil, 37
  in sunflower oil, 31
  in tall oil, 35–6
feed-in-tariff remuneration
  system, 84
feedstocks
  biogas production potential of, 81
  biomass as, 139–41
  carbohydrates as, 5–6
  cost considerations, 12
  flowchart for, *10*
  lignocellulosic. *see* lignocellulosic
    feedstock
fertilisers, 60
Fischer-Tropsch synthesis
  (FTS), 176, 194–6
flaking seeds, 115
flaxseed oil, 34
flowrates
  for biodiesel production, **224**
  for biogas production, **233**
  effect on profit, *231*
  for supercritical CO_2
    extraction, **235**
Fluid Catalytic Cracker (FCC), 24
Foggia case study, 297–8
food-feed-fuel conflicts,
  minimising, 6
foodprocessing residues, 6
fractionation
  of amino acids, 139
  of cereals, 14–5
  of green biomass, 17–8
  of oilseeds, 22
  overview of process, 8
  of sunflower plants, 92–6
free fatty acids (FFA), removing from
  glycerol, 168–9
Friolex process, 159

fructose, 144–5
Fuel Quality Directive, 29
fumaric acid, 9
fungicides
  efficacy of, 52
  for Phomopsis stem canker, 61
2,5-furan dicarboxylic acid, 9
furfural, 13
furfuryl alcohol, preparing levulinic
  acid via, 143–4
fuzzy cognitive maps (FCMs), 290–1

Gas Assisted Mechanical Expression
  (GAME), 129–30, 132
gas-assisted oil pressing, 129–30
  integrating into existing plant, 156
  SWOT analysis for, **156**
genetically modified (GM) crops
  rapeseed, 54–5
  sunflowers, 66–7
Gliperol, 191
glucaric acid, 9
glucose
  isomerisation into fructose, 144–5
  products accessible by, 8
  versatility of, 7
glucosinolates
  in canola oil, 49
  extraction from rape meal, 3
glutamic acid, 9
glycerol
  applications of, 166–87
  aqueous phase reforming
    (APR), 176
  in biodiesel
    manufacturing, 187–97
  biofuels incorporating, 189–92
  biotransformations, 180
  as building block, 9, 174–87
  chemicals derived from, 180–5
  commodity chemicals derived
    from, 176
  composition of, **168**
  continuous concentration of, *171*
  continuous distillation of, *172*
  co-production with biodiesel, 227–9

crude, utilisation of, 186
crude vs. purified economic
  comparison, 264–6
dehydrations, 178
esterifications, 179
ethanol production from, *177*
etherifications, 179
Fischer-Tropsch synthesis
  (FTS), 176
food applications, 175
future market of, 186–7
gel permeation, 174, *175*
generation of, *167*
global production of, 175
as green solvent, 186
halogenations, 177–8
oxidations, 179
price of, **189**
purification technologies, 167–74,
  213
  catalyst removal, *240*
  chromatography and
    regenerative column
    adsorption, 173–4
  conventional processes,
    169–70
  economic optimisation, **249**
  environmental
    optimisation, 240–55
  modelling for cost
    analysis, 212–3
  recent developments in, 170–3
  separation units, *241*
  soap splitting, 168–9
pyrolysis of, 179–80
selective reductions, 177
succinic acid conversion, 213–4
vs. succinic acid, in biodiesel
  production, 266–7
transforming into
  products, 174–87
glycerol tertiary butyl ether
  (GTBE), 182
GM crops
  rapeseed, 54–5
  sunflowers, 66–7

government module of CGE
  models, 306
gravimetric hull separators, 105
green biomass fractionation, 17–8
green biorefineries, 17–9
green certificate remuneration
  system, 84
green chemistry, 1–4. *see also*
  emissions; environmental impact
green juice production, 18
green solvents, 186
greenhouse gas (GHG) emissions, 75
groundnut oil
  area harvested, 35
  uses for, 32
  world production of, 32
gum recovery, 160

halogenations of glycerol, 177–8
harvest index for sunflowers,
  57–8, 65
harvesting olives, 33–4
hazelnut oil, 36
health effects
  of high oleic sunflower oil
    (HOSO), 31
  of olive oil, 32–3
HEAR (High Erucic Acid
  Rape), 26–7
heating seeds as pre-treatment. *see*
  thermal pre-treatment of seeds
hemicellulose/polyoses
  plant fibre composition of, 86
  as precursor, 12
hemp oil, 37
herbicides, 63. *see also* pest and
  disease control
hexane extraction
  vs. cold pressing, 269–70
  cost analysis model, 220–1
  of rapeseed oil, 22
  of sunflower oil, 126
hexoses, producing levulinic acid
  from, 144–8
high oleic sunflower oil (HOSO), 31
HIPLEX system, 156

hot-pressing sunflower cake
  meal, 93–6
hulls, valorising. *see* dehulling
hulls boiler, 112
Hydro Thermal Upgrading
  (HTU), 196
hydrogen, producing with
  glycerol, 180
hydrolysis of proteins, 134–5
3-hydroxy propionic acid, 9
3-hydroxybutyrolactone, 9

innovation niches, 282–4
  assessing development
    status, 286–91
  behavioural rules for, 293–5
  questionnaire for investigating, **288**
insecticides, 61. *see also* pest and
  disease control
integrated scheme biorefinery,
  160–1
integrating biorefinery
  technology, 151–61
  alcohols as hexane
    alternative, 156–7
  anaerobic digestion, 159–60
  cold pressing, 154
  dehulling, 151–4
  gas-assisted oil pressing, 156
  gum recovery, 160
  marc hexane retention
    reduction, 154–5
  oleosome isolation, 158–9
  supercritical CO_2
    extraction, 155–6
  water extraction, 159
ion exclusion chromatography, 169
irradiation as seed pre-
  treatment, 117–8, 125
irrigation, 60, 63–4
itaconic acid, 9

jatropha oil, 38–9
jojoba, 78–9

Kraft pulping, 34–5

lallemantia oil, 37
landscape, in policy scenario
  modelling, 282
LCF (lignocellulosic feedstock). *see*
  lignocellulosic feedstock
LEAR (Low Erucic Acid Rape), 27
levulinic acid
  as building block, 9
  formation of, from fructose, 144–5
  history of development, 142–3
  properties of, 141–2
  uses for, 142
levulinic acid production, 18, *21*,
  141–50, *142*
  Biofine process, 148–9
  cost analysis model, 222–3
  economic optimisation, 274–5
  from hexoses, 144–8
  at high temperature, 147–8
  history of development, 142–3
  from lignocellulosic
    feedstock, 149–50
  at low temperature, 146–7
  material prices, 222
  preparation routes, 143–4
  process flowsheet for, *222*
  from rapeseed, 21
  reaction for, 223
life cycle analysis (LCA), 207–8
lignin
  plant fibre composition of, 86
  as precursor, 12
  in sunflower seeds, 113
  uses for, 3–4
lignocellulosic feedstock, 9
  biofuel production from, *196*
  levulinic acid production
    from, 149–50
  treating for biogas production, 82
Lignocellulosic Feedstock (LCF)
  biorefinery, 11–4, *12*
linoleic sunflower oil. *see* sunflower oil
linseed oil
  area harvested, 35
  uses for, 34
  world production of, 34

lipases, and biofuel
    production, 189–92
Low Erucic Acid Rape (LEAR), 27

macadamia nut oil, 36
macro-economic policy
    modelling, 299–307
    CGE model application,
        299–300
    CGE model for bio-based
        economy, 300–7
    overview of, 307
MacSharry reforms, 24
malic acid, 9
marc hexane retention
    reduction, 154–5
materials and utility prices, 212–3
Mcgyan Process, 2
methanation of biomass, 159–60
methanol recovery, 169
    crude glycerol *vs.* purified glycerol
        analysis, 264–6
    distillation column for, **212**
    emissions effect, 264–6, *265*
    energy requirements for, *243–4*,
        250–1
    modelling for cost analysis,
        225–6, *228*
methodology for cost
    analysis, 205–23
    life cycle analysis, 207–8
    multi-objective
        optimisation, 208–9
    optimisation methods, 205–7
    simulation software, 205
microdiesel biosynthesis, 193
micro-economic policy
    modelling, 281–99
    application of, 297–9
    Data Envelopment Analysis
        (DEA), 296–7
    theoretical framework, 281–5
    three-steps methodology,
        285–97
        agent-based model (ABM)
            design, 292–5

assessing niche development
    status, 286–91
evaluating policy actions,
    295–7
microwave and radio
    pre-treatment, 117–8, 125
microwave pyrolysis, 4
Miscella production, 22
modelling biorefinery schemes for
    cost analysis, 209–23
    biodiesel production, 210–4
    biogas production, 215–8
    levulinic acid production, 222–3
    oil extraction, 220–1
    protein extraction, 219
    supercritical CO_2 extraction, 218–9
    thermomoulding, 221
modelling policy scenarios. *see* policy
    scenario models
mono- (MAG) and di-acylglycerol
    (DAG), 182
multi-level approach (MLA) for
    micro-economic policy
    modelling, 282

network indices, 291
niches, in policy scenario modelling.
    see innovation niches
nitrogen fertiliser, 51
NuSun sunflower oil, 31
nylon production, 13

oil bodies in seeds (oleosomes), 116,
    158–9
oil cake. *see* press cake
oil crops. *see also specific oils*
    area harvested, **35**
    climatic requirements for, 25
    economic potential of, 25–6
    European cultivation of, 24–6
    historic importance of, 23
    world production of, 24
oil of dragonhead (lallemantia
    oil), 37
oil pressing processes. *see* pressing
    processes

oil recovery, 119–33. *see also specific oils*
  cold pressing *vs.* hexane extraction, 269–70
  cost analysis model, 211, 220–1
  flake water content, effect on, **117**
  holistic assessment of process options, 267–70
  from olives, 121–4
  pressing processes, 127–30
  process overview, *119*
  from rapeseed, 124–5
  residual, 132
  simultaneous with protein extraction, 137–9, **158**
  solvent extraction, 130–2
  from sunflower seed, 125–7
oilseed rape. *see* rapeseed; rapeseed oil
oleiferous crops. *see* oil crops
oleochemical industry, 2
oleosomes, 116, 158–9
olive cake, 122–4
olive leaves, 122
olive oil
  area harvested, 35
  centrifugation process, *120–1*
  culinary uses for, 32–3
  extraction process, 121–4
  harvest of, 33–4
  historic importance of, 23
  quality definitions for, 33
  subsidisation of, 34
  uses for, 32–3
  world production of, 32, 122
olives
  anatomical parts of, 121–2
  byproducts of, 122–4
  composition of, 133
  mill wastewater, 123
  oil content of, 121, **122**
  waste valorisation, 123–4
optimisation, economic. *see* economic optimisation
optimisation, environmental. *see* environmental and multi-objective optimisation

optimisation methods in cost analysis, 205–7
  deterministic and stochastic, 206–7
  identifying parameters for, 205
  multi-objective optimisation, 208–9
  objective function, constructing, 205–6
organic solvent extraction, 130
organosolv process, 3–4
Orobanche cumana, 58
oxidations of glycerol, 179

palm oil production, 24
peanut oil (groundnut oil)
  area harvested, 35
  uses for, 32
  world production of, 32
pectins, extracting from sunflower stalks, 88
pelletising straw, 73–5
peptide bonds, *134*
peptide extraction process, 135–41
pest and disease control. *see also* diseases
  birds and game animals, 62
  fungicides, 61
  GM modifications for, 54–5, 66
  insecticides, 61
  prophylactic methods, 65
  with wheat wax extracts, 79
petroleum-refineries, compared to biorefineries, 6–7
phenols, extracting from olive waste, 124
Phoma black stem, 59
Phomopsis stem canker, 59, 61
phosphorus, cooking, 116
pig pancreatic lipase (PPL), biofuel production with, 190
pith, sunflower, 87
plant oils. *see also specific oils*
  European production of, 24–6, **25**
  historic importance of, 23

polymer production with, 2–3
  uses for, 24
plant waxes, 75–80
  in sunflower oil, 110–1
  supercritical CO_2 extraction optimisation, 234–9
platform chemicals, 16. *see also* levulinic acid
policy scenario models, 280–308
  macro-economic approach, 299–307
    CGE model application, 299–300
    CGE model for bio-based economy, 300–7
    overview of, 307
  micro-economic approach, 281–99
    application of, 297–9
    theoretical framework, 281–5
    three-steps methodology, 285–97
poly-3-hydroxybutyricacid production, 15
polycosanols in straw waxes, 79
polymers, green production of, 2–3
polytrimethylenterephthalate (PTT), 13
pomace olive oil, 33. *see also* olive oil
precursors, 7, 9
press cake
  amino acid extraction from, *137*
  from aqueous extraction of sunflower oil, 93–6
  fibre extraction from, 22–3
  production of, 18
  residual oil recovery from, 132
pressing processes, 127–30
  cold pressing. *see* cold pressing
  extrusion, 128–9
  gas-assisted oil pressing, 129–30
  screw pressing, 127–8
pre-treatment processes, 103–18
  dehulling, 103–15
  enzymatic, 118, 125, 127
  microwave and radio frequency, 117–8, 125

pulsed electric field, 118
thermal, 115–8, 124–5
processing oil-bearing plants, 102–61
  biorefinery integration in existing plants, 151–61
    alcohols as hexane alternative, 156–7
    anaerobic digestion, 159–60
    cold pressing, 154
    dehulling, 151–4
    gas-assisted oil pressing, 156
    gum recovery, 160
    integrated scheme, 160–1
    marc hexane retention reduction, 154–5
    oleosome isolation, 158–9
    supercritical CO_2 extraction, 155–6
    transesterification and extraction, 157–8
    water extraction, 159
  levulinic acid production, 141–50
    Biofine process, 148–9
    from hexoses, 144–8
    history of development, 142–3
    LCF conversion process, 149–50
  oil recovery processes, 119–33
    from olives, 121–4
    pressing processes, 127–30
    from rapeseed, 124–5
    residual oil recovery, 132
    solvent extraction, 130–2
    from sunflower seeds, 125–7
  pre-treatment processes, 103–18
    dehulling, 103–15
    enzymatic, 118, 125, 127
    microwave and radio frequency, 117–8
    pulsed electric field, 118
    thermal, 115–7
  protein and amino acid isolation, 133–41

processing oil-bearing plants (*continued*)
    peptide and amino acid extraction, 135–41
    protein hydrolysis, 134–5
  waste stream valorisation, 119–33
product flowchart, *10*
production structure of CGE model, 301–4
profit analysis. *see* cost analysis of biorefineries
propan-1,2,3-triol. *see* glycerol
1,3-propanediol, 13, 81
propylene glycol production, 13, 181
proteases, protein hydrolysis with, 134
protein and amino acid isolation, 133–41, *138*
  after oil recovery, 136–7
  cost analysis model, 219
  economic optimisation, 274–5
  peptide and amino acid extraction, 135–41
  protein hydrolysis, 134–5
  simultaneous with oil extraction, 137–9
pulsed electric field pre-treatment, 118
pumpkin seed oil, 36
punctual indices, 291
purification of glycerol, 167–74, *213*
  with activated carbon, 173–4
  adsorption techniques, 173–4
  catalyst removal, *240*
  chromatography and regenerative column adsorption, 173–4
  conventional processes, 169–70
  vs. crude glycerol, 264–6
  economic optimisation, **249**
  environmental optimisation, 240–9
  modelling for cost analysis, 212–3
  recent developments in, 170–3
  separation units, *241*
  soap splitting, 168–9

pyrolysis
  of biomass, 193–4
  of glycerol, 179–80
  microwave, 4

quantitative trait locus (QTL), 66

radiation interception of rapeseed, 53
radio pre-treatment. *see* microwave and radio pre-treatment
rape straw
  alkane yields from, 78
  composition of, **51**, 215
  yields, 51
rapeseed
  aqueous extraction of, 138
  biogas production with hulls, 109
  climatic hardiness of, 26, 50
  cold pressing, 106
  composition of, 105–6, 133, 210
  cost analysis model for biogas production, 215–8
  crude fibre extraction, 22–3
  cultivars, breeding for yield, 53–4
  cultivars of, 49
  cultivation problems, 50
  dehulling, 103–10
  diseases affecting, 51–2
  enzymatic pre-treatment, 125
  European production of, 50
  flaking, 115
  GM cultivars of, 54–5
  growing conditions for, 49
  harvest of, 20–1
  HEAR (High Erucic Acid Rape), 26–7
  hull utilisation, 108–10
  hybrids, yield from, 54
  LEAR (Low Erucic Acid Rape), 27
  microwave and radio pre-treatment, 125
  oil content of, 121, **122**
  oil recovery, temperature effect on, **116**
  protein extraction, 219

quantity of straw production, 20
rotational breaks for, 52–3
sulfur, effect on, 51
SWOT analysis for
  dehulling, 151–4, **152**
thermal pre-treatment, 124–5
world production of, 49
yields from, 22, 29, 50–6
rapeseed oil
  acidity of, 106
  area harvested, 35
  as biodiesel source, 29
  decentralised production of, 22–3
  extraction process, 124–5
  fatty acid composition, 27
  fractionation of, 22
  levulinic acid production, 21
  predicted consumption of, 29
  uses for, 27–8
  world production of, 24
refining glycerol. see purification of glycerol
Renewable Energy Directive, 29
residual oil recovery, 132
roll dehullers, 103–4

safflower oil, 37
sclerotinia, 65, 66
screw pressing, 127–8
Sequential Quadratic Programming (SQP), 206–7, 226, 230
sesame oil
  area harvested, 35
  growing conditions for, 32
  uses for, 32
  world production of, 32
simulated annealing (SA), 206–7, 226, 230
simulation software for cost analysis, 205
soap splitting, as glycerol pre-treatment, 168–9
social circles, 287
social network analysis (SNA), 286–9

socio-technical regime (ST-regime), 282
solid state fermentation of alperujo, 124
solvent extraction of oils, 130–2
  levulinic acid recovery with, 150
  with organic solvents, 130
  with supercritical fluids, 130–1
  with water (aqueous extraction), 131–2
sorbitol, 9
sorghum composition, 86
soybean oil
  area harvested, 35
  European production of, 30
  uses for, 30
  world production of, 24
  yields from, 30
speciality oils, 36
sterols in straw waxes, 79
stochastic optimisation method, 206–7
straws. see also specific straws
  component extraction, economic considerations of, 3
  defined, 14
  densification of, 73–5
  holistic assessment of process options, 270–3
  levulinic acid production from, 141–50
  nutrient value, **75**
  pelletising, 73–5
  unutilised amount of, 75
  waxes from, 75–80
succinic acid
  as building block, 9
  vs. crude glycerol in biodiesel production, 266–7
  glycerol production of, 180, *214*
  glycerol production of, modelling for cost analysis, 213–4, 229–32
sugar platform interaction with syngas platform, 19–20
sulfur levels, effect on rapeseed yield, 51

SUNFLO model for sunflower
  planting, 64–5
sunflower oil
  aqueous extraction, 90–3, 126–7
  area harvested, 35
  European production of, 30–1
  extraction process, 125–7
  extrusion extraction, 126–7
  fatty acid composition, 31, 57
  hexane extraction, 126
  supercritical CO_2 extraction, 126
  thermomoulding, 221
  uses for, 31
  waxes in, 110–1
  world production of, 24
sunflower seeds
  composition of, 110–1, 113, 133
  crude fibre extraction, 23
  dehulling, 104, 110–4
  enzymatic pre-treatment, 127
  hullability of, 111
  oil content of, 121, **122**
  SWOT analysis for
    dehulling, 151–4
  yield from, **64**
sunflower stalks
  applications for, 88–90
  biosorption with, 88–90
  composition of, 85–7, *88*
  pectin extraction, 88
  pith/straw separation, 87–8
  potential harvest size, 85
  pulp production with, 90
  structure of, 85–6
sunflower straw
  alkane yields from, 78
  separation from pith, 87–8
sunflowers
  breeding for disease resistance, 66
  cake meal composition, 94
  composition of, **85**, **94**
  disease control, 61
  diseases of, 58–60
  European production of, **57**
  farming practices *vs.*
    recommendations, 62

  fertiliser for, 60
  fibre composition, **86**
  GM and, 66–7
  growing conditions for, 30–1, 58
  harvest index, 57–8
  historic cultivation of, 57
  hulls, uses for, 113–4
  irrigation of, 60, 63–4
  physical characteristics of, 56–7
  pith, application for, 88, **89**
  pith composition, **87**
  rotational breaks for, 61
  water availability, adapting
    practices to, 64–5
  weed control, 61, 63
  yield increase, 60–7
supercritical CO_2 extraction
  vs. biogas production, 270–3, **271**
  CO_2 density, *70*
  co-solvents, 70
  cost analysis model, 218–9
  economic considerations, 71–3
  economic optimisation, 234–9, **259**
  energy requirements for, *260*
  environmental
    optimisation, 257–63, **259**
  equipment components, *71*
  extraction time considerations, 69
  extractor arrangements, 70–1
  input and output flowrates, **235**
  integrating into existing
    plant, 155–6
  multi-objective optimisation, **262**
  operating costs, *73*, **236**
  process flowsheet for, *218*, *235*,
    258
  raw material costs, 75
  as solvent, 130–1
  straw densification for, 73–5
  straw extractives from, 75–8
  of sunflower oil, 126
  SWOT analysis for, **155**
surfactant production, 2
surplus reduction, 24
sustainability, and biomass
  increase, 48

Subject Index

SUSTOIL project policy objectives, 281
swath thresh of rapeseed, 21
SWOT analysis
 for alcohols as hexane alternatives, 157
 for anaerobic digestion, **160**
 for cold pressing, **154**
 for dehulling strategies, 151–4
 for gas-assisted oil pressing, **156**
 for marc hexane retention, **155**
 for oleosome isolation, **158**
 for simultaneous extraction and transesterification, **158**
 for supercritical CO_2 extraction, **155**
 for water extraction, **159**
syngas
 production of, 18, *20*, 195–6
 via pyrolysis, 14
syngas platform, interaction with sugar platform, 19–20

tail-end separation, for dehulling, 104
tall oil, 34–6
tall oil fatty acids (TOFA), 35–6
thermal pre-treatment of seeds, 115–8, 124–5
thermomoulding
 cost analysis model, 221
 economic optimisation, 274–5
thermoplastic extrusion, 131–2
thin film distillation, 170
trade structure of CGE models, 304–6
transesterification and extraction
 integrating into existing plant, 157–8
 modelling for cost analysis, 211–2
triglycerides
 as biofuels, 187
 processing in oil-refining plants, 192–3
 transesterification reaction, *167*
trombin, specificity of, *135*

trypsin, 134, *135*
twin-screw extruder, *91*
two-platform concept, 19–20

vacuum flash evaporators, 170
vegetable oils. *see* plant oils
virgin olive oils, 33. *see also* olive oil
viscosity, 188

walnut oil, 36
waste biomass, 4–6
waste stream valorisation, 119–33
wastewater from olive mills, 123
water conservation, 63–4
water extraction, 159
waxes from plants, 75–80
 in sunflower oil, 110–1
 supercritical CO_2 extraction optimisation, 234–9
weed control, in sunflower crops, 61, 63
wet-milling whole cereal crops, 16–7. *see also* Whole Crop Biorefinery
wheat straw
 fibre composition, 86
 wax composition, *77*
 wax extraction
  economic optimisation of, 234–9
  environmental optimisation of, 257–63
  process flowsheet for, *258*
 wax extracts, pest reduction with, 79
 wax fraction, 77
 yields from, *55*
Whole Crop Biorefinery, 14–7

xylitol/arabinitol, 9

yield increase
 average per annum, 55
 for rapeseed, 50–6
 for sunflowers, 57–8, 60–7